KU-546-499

AS GEOGRAPHY
Concepts & Cases

PAUL GUINNESS
GARRETT NAGLE

Hodder & Stoughton
A MEMBER OF THE HODDER HEADLINE GROUP

AS GEOGRAPHY
Concepts & Cases

PAUL GUINNESS
GARRETT NAGLE

Hodder & Stoughton
A MEMBER OF THE HODDER HEADLINE GROUP

Acknowledgements

The publishers would like to thank the following individuals, institutions and companies for permission to reproduce copyright photographs in this book:

1.12 Gil Moti, Still Pictures; 1.31, 1.33, 1.34, 2.1, 2.11, 2.34, 2.36, 3.2, 3.15, Garrett Nagle; 3.28 (top right and bottom right) University of Dundee, NERC Satellite Station; David Moore 5.1; 5.11, 5.14, 5.15, Garrett Nagle; 5.18, 5.19 John Connor Press Asssociates, Lewes; 5.21, 5.22, 5.24, 5.27 David Moore; 5.31, 5.34, John Farmer, Skyscan Photolibrary; 5.35, 5.36, 5.50 Garrett Nagle; 6.1, 6.2 Garrett Nagle; 6.7 Wayne Shakell/Life File; 6.8 Caroline Field/Life File; 6.11 Garrett Nagle; 7.4 Paul Guinness; 7.5 Hidajet Delic/Associated Press AP; 7.13, 7.20 7.23 Paul Guinness; 7.28 Penny Tweedie/Colorific; 7.33, 7.44, 7.45, 7.47, 7.51, Paul Guinness; 7.52 Lina Faria 7.53 Eric Draper/Associated Press AP; 7.59, 7.60, 7.68, Paul Guinness; 7.71 Richard Powers/Life File; 7.74 A. Rake/The Hutchison Library; 8.3, 8.4, 8.8, 8.9, 8.10, 8.19, 8.20, 8.21, 8.23, 8.24, 8.25, 8.27, 8.37, 8.38, 8.41, 8.44, 8.48, 8.49, 8.51, 8.52, 8.57, 8.59, 8.60, 8.61, 8.76, 8.80, 8.87, 8.88, 8.91, 8.92, 8.96, 8.98, 8.101, 8.106, 8.107, 8.108, 8.109, 8.113, 8.114, 8.121, 8.122, 8.129 Paul Guinness; 8.130 Trackair Aerial Surveys; 8.133 Sealand Aerial Photography; 9.1, 9.3, 9.4 Paul Guinness; 9.11 Eric Draper/Associated Press AP; 9.15, 9.23, 9.24, Paul Guinness; 9.40 Museum of Labour History (check exact credit); 9.42, 9.43, 9.45 Paul Guinness; 9.56 Sue Cunningham/SCP; 9.58, 9.59, Paul Guinness; 9.63 Derek Cullen/Bord Fáilte; 9.66, 9.67 Paul Guinness.

The publishers would also like to thank the following for permission to reproduce material in this book:

Atlantic Syndication Partners for the reproduction of 'Living in the Brown Belt', Evening Standard, 11/10/99 and 'Our dying village', Daily Mail, 5/7/99; Maps reproduced from Ordnance Survey mapping with the permission of the Controller of Her Majesty's Stationery Office, © Crown copyright (399450); New Earnings Survey/United Kingdom: Ratio of males to females/Male and Female chart/Labour Force Survey/England and Wales: Children per woman, Office for National Statistics © Crown Copyright 1999; Philip Allan Updates for the extract from Case Study 2, 'Suburbanisation', G. Nagle in *Geography Review Vol. 12:1* (September 1998); Telegraph Syndication for extracts from the Daily Telegraph 1/12/99, 11/12/99 and the Sunday Telegraph 24/8/97.

Every effort has been made to trace and acknowledge ownership of copyright. The publishers will be glad to make suitable arrangements with any copyright holders whom it has not been possible to contact.

Dedication

To Mary, Courtenay and Christopher;
Angela, Rosie, Patrick and Bethany.

Orders: please contact Bookpoint Ltd, 78 Milton Park, Abingdon, Oxon OX14 4TD. Telephone: (44) 01235 827720, Fax: (44) 01235 400454. Lines are open from 9.00 – 6.00, Monday to Saturday, with a 24 hour message answering service. Email address: orders@bookpoint.co.uk

British Library Cataloguing in Publication Data
A catalogue record for this title is available from The British Library

ISBN 0 340 78091 6

First published 2000
Impression number   10 9 8 7 6 5 4 3 2 1
Year   2005 2004 2003 2002 2001 2000

Copyright © 2000 Paul Guinness, Garrett Nagle

Typeset by Fakenham Photosetting Limited, Fakenham, Norfolk.
Printed in Italy for Hodder & Stoughton Educational, a division of Hodder Headline Plc, 338 Euston Road, London NW1 3BH by Printer Trento.

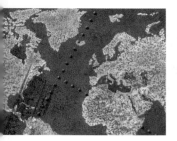

contents

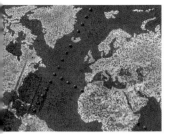

1 hydrology and *rivers*

The hydrological cycle

The **hydrological cycle** is the cycle of water between atmosphere, lithosphere and biosphere (Figure 1.1). At a local scale the cycle has a single input, **precipitation** (PPT), and two major losses (outputs), **evapotranspiration** (EVT) and runoff. A third output, leakage, may also occur from the deeper subsurface to other basins.

Water can be stored at a number of stages or levels within the cycle. These stores include vegetation, surface, soil moisture, groundwater and water channels.

Human modifications are made at every scale. Good examples include large-scale changes of channel flow and storage, irrigation and land drainage, and large-scale abstraction of groundwater and surface water for domestic and industrial use.

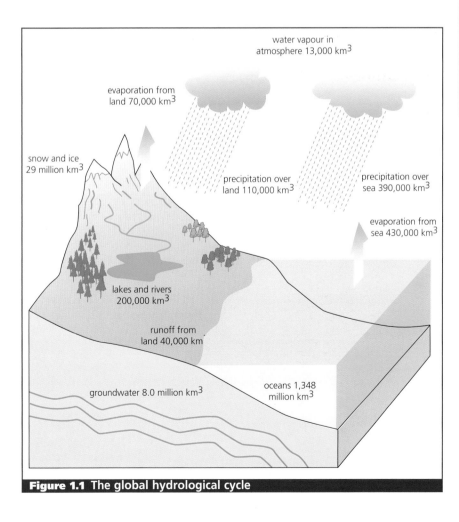

water vapour in
atmosphere 13,000 km³

evaporation from
land 70,000 km³

snow and ice
29 million km³

precipitation over
land 110,000 km³

precipitation over
sea 390,000 km³

evaporation from
sea 430,000 km³

lakes and rivers
200,000 km³

runoff from
land 40,000 km

groundwater 8.0 million km³

oceans 1,348
million km³

Figure 1.1 The global hydrological cycle

Precipitation

Precipitation is considered in detail in Chapter 3. Here it is important to mention the main characteristics that affect local hydrology. These are:

- the total amount of precipitation
- seasonality
- intensity
- type (snow, rain, etc.)
- geographic distribution
- variability.

Interception

Interception refers to water that is caught and stored by vegetation. There are three main components:

- interception loss – water which is retained by plant surfaces and which is later evaporated away or absorbed by the plant
- throughfall – water which either falls through gaps in the vegetation or which drops from leaves, twigs or stems
- stemflow – water which trickles along twigs and branches and finally down the main trunk.

Interception loss varies with different types of vegetation (Figure 1.2). Interception is less from grasses than from deciduous woodland owing to the smaller surface area of the grass shoots. From agricultural crops, and from cereals in particular, interception increases with crop density.

Evaporation

Evaporation is the process by which a liquid or a solid is changed into a gas. It is the conversion of solid and liquid precipitation (snow, ice and water) to water vapour in the atmosphere. It is most important from oceans and seas. Evaporation increases under warm, dry conditions and decreases under cold, calm conditions.

Factors affecting evaporation include meteorological factors such as temperature, humidity and windspeed (Figure 1.3). Of these temperature is the most important. Other factors include the amount of water available, vegetation cover and colour of the surface (albedo or reflectivity of the surface).

Evapotranspiration

Transpiration is 'the process by which water vapour escapes from the living plant, principally the leaves and enters the atmosphere'. The combined effects of evaporation and transpiration are normally referred to as **evapotranspiration** (EVT). EVT represents the most important aspect of water loss, accounting for the loss of nearly 100% of the annual precipitation in arid areas and 75% in humid areas. Only over ice and snow fields, bare rock slopes, desert areas, water surfaces and bare soil will purely evaporative losses occur.

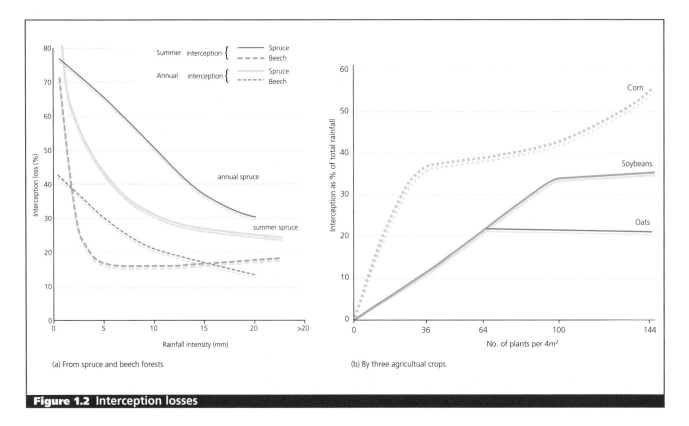

(a) From spruce and beech forests

(b) By three agricultural crops

Figure 1.2 Interception losses

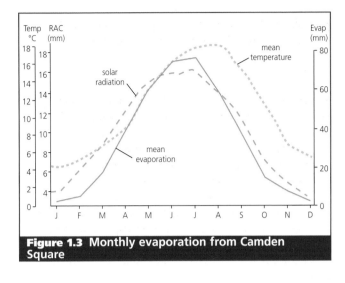

Figure 1.3 Monthly evaporation from Camden Square

Potential evapotranspiration (P.EVT)

The distinction between actual EVT and P.EVT lies in the concept of **moisture availability**. Potential evapotranspiration is the water loss that would occur if there were an unlimited supply of water in the soil for use by the vegetation. For example, the actual evapotranspiration rate in Egypt is less than 250 mm, because there is less than 250 mm or rain annually. However, given the high temperatures experienced in Egypt, if the rainfall were as high as 2000 mm, there would be sufficient heat to evaporate that water. Hence the potential evapotranspiration rate there is 2000 mm. Rates of potential evapotranspiration for Britain are shown in Figure 1.4. The factors affecting evapotranspiration include all those which affect evaporation. In addition, some plants such as cacti have adaptations to help them reduce moisture loss.

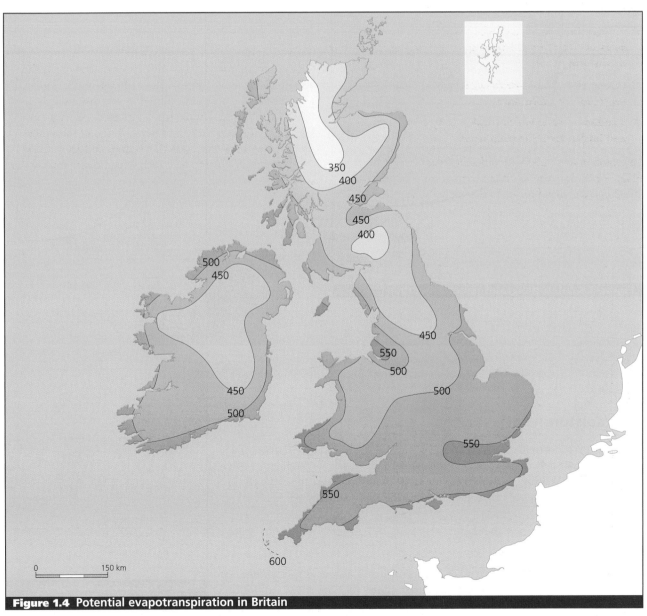

Figure 1.4 Potential evapotranspiration in Britain

Infiltration

Infiltration is the process by which water soaks into or is absorbed by the soil. The **infiltration capacity** is the maximum rate at which rain can be absorbed by a soil in a given condition.

Infiltration capacity decreases with time through a period of rainfall until a more or less constant value is reached (Figure 1.5). Infiltration rates of $0–4\,mm^{-1}\,h^{-1}$ are common on clays whereas $3–12\,mm$ per hour are common on sands. Vegetation also increases infiltration. On bare soils where rainsplash impact occurs, infiltration rates may reach $10\,mm$ per hour. On similar soils covered by vegetation rates of between 50 and $100\,mm$ per hour have been recorded. Infiltrated water is chemically rich as it picks up minerals and organic acids from vegetation and soil.

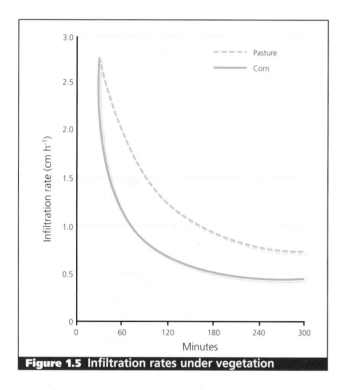

Figure 1.5 Infiltration rates under vegetation

Infiltration is inversely related to overland runoff and is influenced by a variety of factors such as duration of rainfall, antecedent soil moisture (pre-existing levels of soil moisture), soil porosity, vegetation cover, raindrop size and slope angle (Figure 1.6).

Soil moisture

Soil moisture refers to the subsurface water in the soil. **Field capacity** refers to the amount of water held in the soil after excess water drains away, i.e. saturation or near saturation. **Wilting point** refers to the range of moisture content in which permanent wilting of plants occurs. They define the approximate limits to plant growth.

Groundwater

Groundwater refers to subsurface water. The permanently saturated zone within solid rocks and sediments is known as the phreatic zone. The upper layer of this is known as the water table. The water table varies seasonally. It is higher in winter

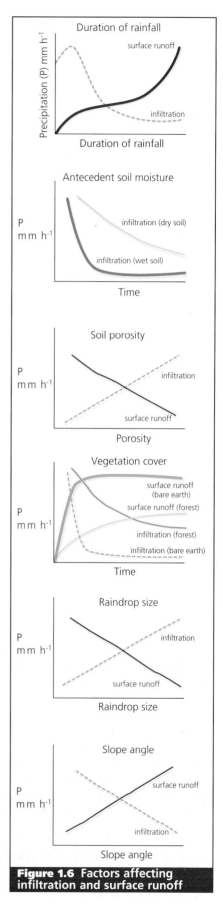

Figure 1.6 Factors affecting infiltration and surface runoff

following increased levels of precipitation. The zone that is seasonally wetted and seasonally dries out is known as the aeration zone. Most groundwater is found within a few hundred metres of the surface but has been found at depths of up to 4 km beneath the surface (Figure 1.7).

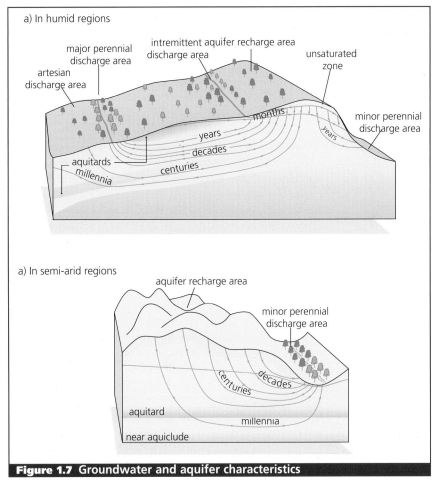

Figure 1.7 Groundwater and aquifer characteristics

1 Define the following hydrological characteristics:
   (a) interception
   (b) evaporation
   (c) infiltration.

2 Figure 1.1 shows the global hydrological cycle. How much fresh water is locked in:
   (a) snow and ice
   (b) groundwater
   (c) lakes and rivers?
   (d) How does the amount of freshwater compare with the amount of sea water?

3 Study Figures 1.3 and 1.4.
   (a) Define the terms: evapotranspiration and potential evapotranspiration.
   (b) Describe the pattern of evaporation over the year.
   (c) How is this pattern explained by mean annual temperatures?
   (d) Describe and suggest reasons for the geographic variations in potential evapotranspiration, as shown in Figure 1.4.

4 Study Figure 1.6 which shows factors influencing infiltration and overland runoff. Write a paragraph on each of the factors, describing and explaining the effect it has on infiltration and overland runoff.

5 Figure 1.13 shows a graph of the sediment yield from a small forested catchment during and after the felling and replanting of trees.
   (a) Explain why sediment yields are high in the period of felling and replanting.
   (b) Explain why sediment yields decline in subsequent years.
   (c) How would you expect the hydrological response of the catchment to have changed during the 25-year period.
   (d) Suggest ways in which forestry practices might be designed to reduce sediment yield during felling and replanting.

Groundwater is very important. It accounts for 96.5% of all freshwater on the earth. However, while some soil moisture may be recycled by evaporation into atmospheric moisture within a matter of days or weeks, groundwater may not be recycled for as long as 20 000 years. Hence, in some places, where **recharge** is not taking place, groundwater is considered a non-renewable resource.

Aquifers (rocks which contain significant quantities of water) provide a great reservoir of water. Aquifers are permeable rocks such as sandstones and limestones. The water in aquifers moves very slowly and acts as a natural regulator in the hydrological cycle by absorbing rainfall which otherwise would reach streams rapidly. In addition, aquifers maintain stream flow during long dry periods.

Groundwater recharge occurs as a result of:

■ **infiltration of part of the total precipitation at the ground surface**

■ **seepage through the banks and bed of surface water bodies such as ditches, rivers, lakes and oceans**

■ **groundwater leakage and inflow from adjacent rocks and aquifers**

■ artificial recharge from irrigation, reservoirs, etc.

Losses of groundwater result from:

■ evapotranspiration, particularly in low-lying areas where the water table is close to the ground surface

■ natural discharge by means of spring flow and seepage into surface water bodies

■ groundwater leakage and outflow through aquicludes and into adjacent aquifers

■ artificial abstraction, such as in the London Basin.

The influence of human activity

The human impact on precipitation

There are a number of ways in which human activity affects precipitation. Cloud seeding has probably been one of the more successful. Rain requires either ice particles or large water droplets on which to form. Seeding introduces silver iodide, solid carbon dioxide (dry ice) or ammonium nitrate to attract water droplets. The results are somewhat unclear. In Australia and the United States seeding has increased precipitation by 10–30% on a small scale and on a short-term basis. On the other hand, the increase in precipitation in one place might decrease precipitation elsewhere. In urban and industrial areas precipitation is often increased by up to as much as 10% due to increased cloud frequency and amount and because of the addition of pollutants, the heat island effect and turbulence.

Evaporation and evapotranspiration

The human impact on evaporation and evapotranspiration is relatively small in relation to the rest of the hydrological cycle but is nevertheless important. There are a number of impacts.

1 **Dams** – the construction of large dams increases evaporation. For example, Lake Nasser behind the Aswan Dam loses up to a third of its water due to evaporation. Water loss can be reduced by using chemical sprays on the water, by building sandfill dams and by covering the dams with some form of plastic cover.

2 **Urbanisation** leads to a huge reduction in evapotranspiration due to the lack of vegetation. There may also be a slight increase in evaporation because of higher temperatures and increased surface storage.

Human impact on interception

Interception is determined by vegetation, density and type. Most vegetation is not natural but represents some disturbance by human activity. In farmland areas, for

example, cereals intercept less than broad leaves. Row crops, such as wheat or corn, leave a lot of soil bare. In the Mississippi Basin, for example, in woodland areas sediment yields are just one tonne per unit area, whereas pasture produces 30 units and areas under corn produces 350 units of sediment. Deforestation leads to:

■ a reduction in evapotranspiration

■ an increase in surface runoff

■ a decline of surface storage

■ a decline in time lag.

Afforestation has the opposite effect, although the evidence does not necessarily support this. For example, in parts of the Severn catchment sediment loads increased four times after afforestation. Why was this? The result is explained by a combination of an increase in overland runoff, little ground vegetation, young trees, access route for tractors and fire breaks and wind breaks. All of these created a lot of bare ground. Yet after only five years the amount of erosion declined.

The human impact on infiltration and soil water

Human activity has a great impact on infiltration and soil water. Land use changes are important. Urbanisation creates an impermeable surface with compacted soil. This reduces

Figure 1.8 The potential hydrological effects of urbanisation	
Urbanising influence	**Potential hydrological response**
Removal of trees and vegetation	Decreased evapotranspiration and interception; increased stream sedimentation
Initial construction of houses, streets and culverts	Decreased infiltration and lowered groundwater table; increased stormflows and decreased baseflows during dry periods
Complete development of residential, commercial and industrial areas	Decreased porosity, reducing time of runoff concentration, thereby increasing peak discharges and compressing the time distribution of the flow; greatly increased volume of runoff and flood damage potential
Construction of storm drains and channel improvements	Local relief from flooding; concentration of flood waters may aggravate flood problems downstream

infiltration and increases over land runoff and flood peaks (Figure 1.9). Infiltration is up to five times greater under forests compared with grassland. This is because the forest channels water down its roots and stems. Deforestation causes reduced interception, increased soil compaction and more overland flow. Land use practices are also important. Grazing leads to a decline in infiltration due to compaction and pounding of the soil. By contrast, ploughing increases infiltration because it loosens soils. Waterlogging and salinisation are common if there is poor drainage. When the water table is close to the surface, evaporation of water leaves salts behind and may form an impermeable crust. Human activity also has an increasing impact on surface storage. Surface storage is growing due to the building of large-scale dams. These dams are being built with an increasing size and volume and number. This leads to a number of effects:

- increased storage of water

- decreased flood peaks (a decline of 71% in the Cheviots)

- low flows in rivers, for example in the River Hodder in Lancashire, the flow declined 10% in winter, but 62% in summer

- decreased sediment yields (clear water erosion)

- decreased losses due to evaporation and seepage leading to changes in temperature and salinity of the water

- increased flooding of the land

- triggering of earthquakes (see page 97)

- salinisation – for example in the Indus Valley of Pakistan, 1.9 million hectares are severely saline and up to 0.4 million hectares are lost per annum to salinity

- large dams can cause local changes in climate.

In other areas there is a decline in the surface storage, for example in urban areas water is channelled away over impermeable surfaces in to drains and gutters very rapidly.

Changing groundwater

Human activity has seriously reduced the long-term viability of irrigated agriculture in the High Plains of Texas. Before irrigation development started in the 1930s, the High Plains groundwater system was stable, i.e. in a state of dynamic equilibrium, with long-term recharge equal to long-term discharge. However, groundwater is now being used at a rapid rate to supply **centre-pivot irrigation schemes**. In less than 50 years, the water level has fallen by 30–50 metres in a large area to the north of Lubbock, Texas. The aquifer has narrowed by more than 50% in large parts of certain counties, and the area irrigated by each well is contracting as well yields fall.

In some industrial areas, by contrast, recent reductions in industrial activity have led to less groundwater being taken out of the ground. As a result, groundwater levels in such areas have begun to rise, adding to the problems caused by leakage from ancient, deteriorating pipe and sewer systems. This is happening in many British cities including London, Liverpool and Birmingham. In London, a 46% reduction in groundwater abstraction has led to the water table in the chalk and Tertiary beds rising by as much as 20 metres (Figure 1.10). Such rises have numerous implications, including:

- increase in spring and river flows

- re-emergence of flow from 'dry springs'

- surface water flooding

- pollution of surface waters and the spread of underground pollution

- flooding of basements

- increased leakage into tunnels

- reduction in stability of slopes and retaining walls

- reduction in bearing capacity of foundations and piles

- swelling of clays as they absorb water

- chemical attack on building foundations.

There are various methods of recharging groundwater resources, providing that sufficient surface water is available. Where the materials containing the aquifer are permeable (as in some alluvial fans, coastal sand dunes or glacial deposits) water-spreading (a form of infiltration and seepage) is used. By contrast, in sediments with impermeable layers such water-spreading techniques are not effective, and the appropriate method may then be to pump water into deep pits or into wells. This method is used extensively on the heavily settled coastal plain of Israel, both to replenish the groundwater reservoirs when surplus irrigation water is available, and to attempt to diminish the problems associated with salt-water intrusions from the Mediterranean.

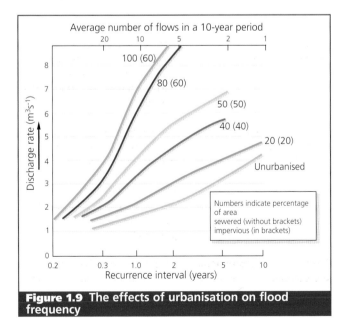

Figure 1.9 The effects of urbanisation on flood frequency

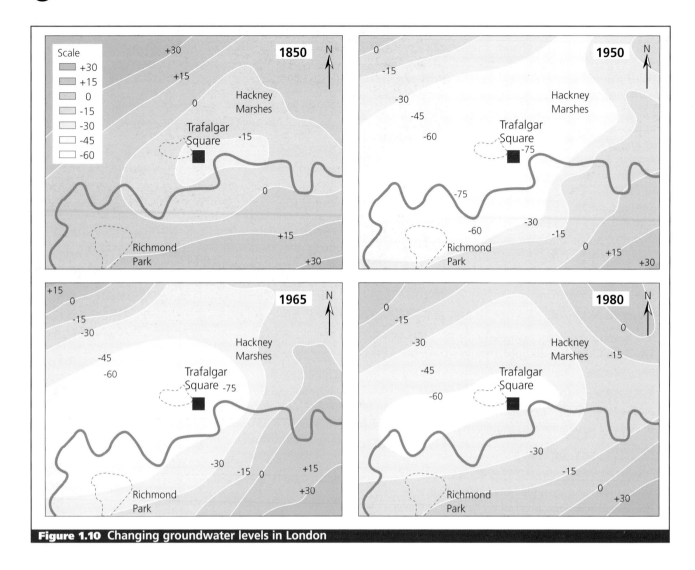

Figure 1.10 Changing groundwater levels in London

Changing hydrology of the Aral Sea

The Aral Sea began shrinking in the 1960s when Soviet irrigation schemes took water from the Syr Darya and the Amu Darya. This greatly reduced the amount of water reaching the Aral Sea. By 1994, the shorelines had fallen by 16 m (nearly 50 feet), the surface area had declined by 50% and the volume had been reduced by 75% (Figure 1.11). By contrast, salinity levels had increased by 300%.

Increased salinity levels killed off the fishing industry. Moreover, ports such as Muynak are now tens of kilometres from the shore (Figure 1.12). Salt from the dry seabed has reduced soil fertility and frequent dust storms are ruining the region's cotton production. Drinking water has been polluted by pesticides and fertilisers and the air has been affected by dust and salt. There has been a noticeable rise in respiratory and stomach disorders and the region has one of the highest infant mortality rates in the former Soviet Union.

Flood hydrographs

A flood hydrograph shows how the discharge of a river varies over a short time (Figure 1.14). Normally it refers to an individual storm or group of storms of not more than a few days in length. Before the storm starts the main supply of water to the stream is through groundwater flow or **baseflow**. This is the main supplier of water to rivers. During the storm some water infiltrates into the soil while some flows over the surface as overland flow or runoff. This reaches the river quickly as **quickflow.** This causes a rapid rise in the level of the river.

The effect of urban development on hydrographs is to increase peak flow and decrease time lag (Figure 1.15). This is due to an increase in the proportion of impermeable ground in a drainage basin as well as an increase in the drainage density. Storm hydrographs also vary with a number of other factors such as basin shape (Figure 1.16), drainage density, and gradient.

Figure 1.11 The decline of the Aral Sea

Figure 1.12 Boats stranded by the former edge of the Aral Sea

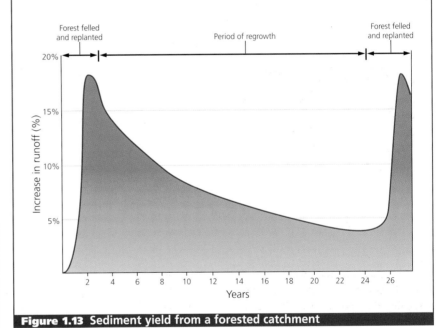

Figure 1.13 Sediment yield from a forested catchment

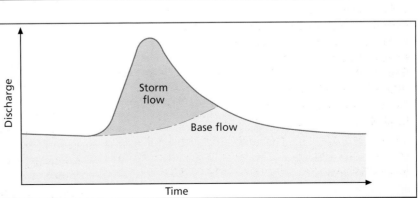

Figure 1.14 Typical form of flood hydrograph

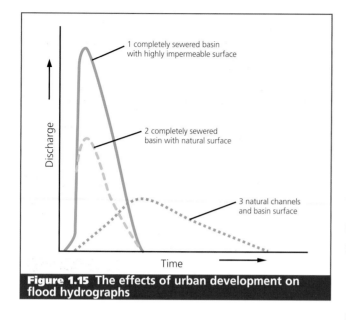

Figure 1.15 The effects of urban development on flood hydrographs

Labels on figure:
- 1 completely sewered basin with highly impermeable surface
- 2 completely sewered basin with natural surface
- 3 natural channels and basin surface

Axes: Discharge (vertical), Time (horizontal)

River regimes

The character or **regime** of the resulting stream or river is influenced by several variable factors:

- the amount and nature of precipitation

- the local rocks, especially porosity and permeability

- the shape or morphology of the drainage basin, its area and slope

- the amount and type of vegetation cover

- the amount and type of soil cover.

On an annual basis the most important factor determining stream regime is climate (Figure 1.18). Figure 1.19a shows a simple regime, based upon a single river with one major peak flow. By contrast, Figure 1.19b shows a complex regime for the River Rhine. It has a number of large tributaries which flow in a variety of environments, including alpine, Mediterranean and temperate. By the time the Rhine has travelled downstream it is influenced by many, at times contrasting, regimes.

Stream flow

Stream flow occurs as a result of overland runoff, groundwater springs, from lakes and from meltwater in mountainous or Arctic environments. Stream flow and associated features of erosion are complex.
The velocity and energy of a stream are controlled by:

- the gradient of channel bed

- the volume of water within channel, which is controlled largely by the precipitation in the drainage basin (e.g. 'bankfull' gives rapid flow whereas low levels give lower flows)

- the shape of the channel

- channel roughness, including friction.

Turbulent flow occurs in streams with high velocities and complex shapes, such as a meandering channel with alternating pools and riffles. Turbulence causes marked variations in pressure within the water. As the turbulent water swirls (eddies) against the bed or bank of the river, air is trapped in pores, cracks and crevices and put momentarily under great pressure. As the eddy swirls away, pressure is released and the air expands suddenly, creating a small explosion which weakens the bed or bank material. Thus turbulence is associated with **hydraulic action**.

Vertical turbulence creates hollows in the channel bed. Hollows may trap pebbles which are then swirled by eddying, grinding at the bed. This is a form of vertical corrasion or **abrasion** and given time may create pot-holes. Cavitation and vertical abrasion may help to deepen the channel, allowing the river to downcut its valley. If the downcutting is dominant over the other forms of erosion (i.e. vertical erosion exceeds lateral erosion) then a gulley or gorge will develop.

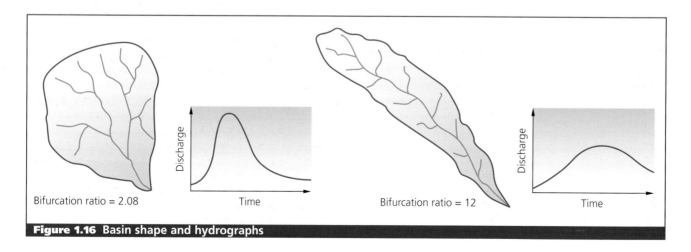

Bifurcation ratio = 2.08

Bifurcation ratio = 12

Figure 1.16 Basin shape and hydrographs

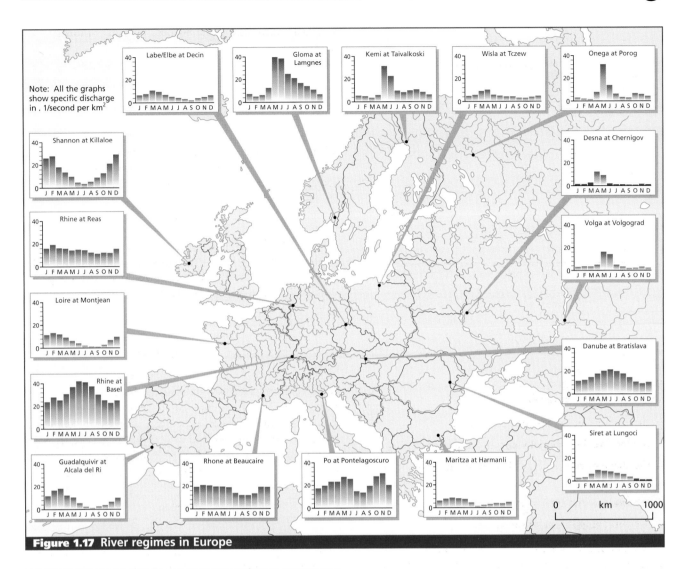

Figure 1.17 River regimes in Europe

Figure 1.18 Precipitation and runoff for Delaware river

Date	Time	Duration of rainfall	Total (cm)
29 Sept.	6 a.m.	12 hours	0.1
29 Sept.	6 p.m.	12 hours	0.9
30 Sept.	6 p.m.	24 hours	3.7
30 Sept.	12 p.m.	6 hours	0.1
			Total 4.8

Date	Stream runoff (m³ s⁻¹)
28 Sept.	28.3 (baseflow)
29 Sept.	28.3 (baseflow)
30 Sept.	339.2
1 Oct.	2094.2
2 Oct.	1330.1
3 Oct.	584.3
4 Oct.	367.9
5 Oct.	254.2
6 Oct.	198.1
7 Oct.	176.0
8 Oct.	170.0
9 Oct.	165.2 (baseflow)

Questions

1 Study Figure 1.9 which shows the effect of urban development on flooding. Describe and explain the changes in flood frequency and flood magnitude that occur as urbanisation increases.

2 **(a)** Why do you think the former Soviet Union embarked on such a programme of large-scale irrigation? Use an atlas to produce detailed information.

   (b) Why have salinity levels increased so much?

   (c) What problems does the shrinking of the Aral Sea cause for towns such as Aralsk and Muynak?

   (d) What is the likely effect of the irrigation scheme on the two rivers in terms of velocity, erosion, sediment transport and deposition?

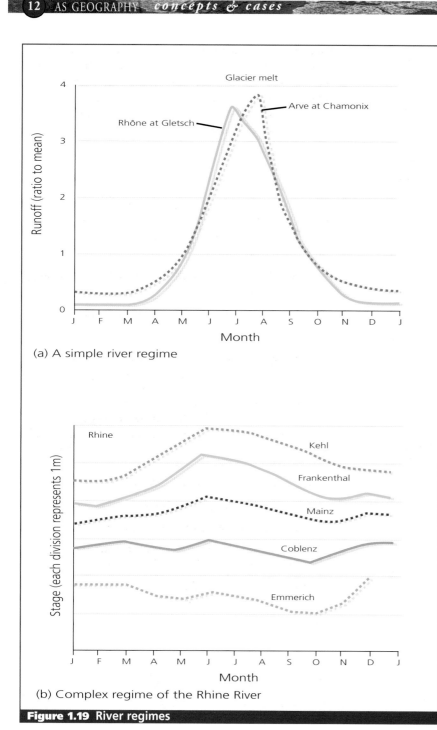

(a) A simple river regime

(b) Complex regime of the Rhine River

Figure 1.19 River regimes

Factors affecting erosion

- **Load** – the heavier and sharper the load the greater the potential for erosion.

- **Velocity** – the greater the velocity the greater the potential for erosion (Figure 1.20).

- **Gradient** – increased gradient increases the rate of erosion.

- **Geology** – soft, unconsolidated rocks such as sand and gravel are easily eroded.

Key Definitions ②

River regime The annual variation in the flow of a river.

Flood or **storm hydrograph** Shows how a river changes over a short period, such as a day or a couple of days. Usually it is drawn to show how a river reacts to an individual storm. Each storm hydrograph has a series of parts.

Rising limb How quickly the flood waters begin to rise.

Peak flow The maximum discharge of the river as a result of the storm.

Time lag The time between the height of the storm (not the start or the end) and the maximum flow in the river.

Recessional limb The speed with which the water level in the river declines after the peak.

Baseflow The normal level of the river, which is fed by groundwater.

Quickflow or **stormflow** The water which gets into the river as a result of overland runoff.

Questions

1 The data in Figure 1.18 show precipitation and runoff data for a storm on the Delaware River, New York. Make a copy of Figure 1.18 and plot the storm hydrograph for this storm.

2 Study Figure 1.8 and Figure 1.9 which show the impact of urbanisation on floods and flood hydrographs. Describe and explain the differences in the relationship between discharge and time following urbanisation.

Key Definitions ③

Abrasion The wearing away of the bed and bank by the load carried by a river.

Attrition The wearing away of the load carried by a river. It creates smaller, rounder particles.

Capacity The total load that a stream can carry.

Competence The size of the largest particle that a stream can carry.

Hydraulic action The force of air and water on the sides of rivers and in cracks.

Solution The removal of chemical ions, especially calcium.

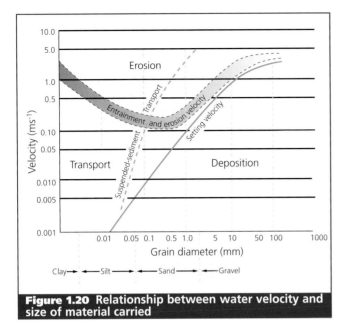

Figure 1.20 Relationship between water velocity and size of material carried

and parts of Australia to in excess of 10 000 tonnes per km² per year in parts of Taiwan, New Zealand and China. In the two former cases steep slopes, high rainfall and tectonic instability are major influences while in the latter, the deep loess deposits and the almost complete lack of natural vegetation cover are important.

Features of erosion

Plunge flow occurs where the river spills over a sudden change in gradient, undercutting rocks by hydraulic impact and abrasion, thereby creating a waterfall (Figure 1.22). There are many reasons for this sudden change in gradient along the river:

- **a band of resistant strata, such as the resistant limestones at Niagara Falls**
- **a plateau edge, such as at Victoria Falls (Figure 1.23)**
- **a fault scarp, such as at Gordale**
- **a hanging valley, such as at Glencoyne, Cumbria**
- **coastal cliffs, such as at Kimmeridge Bay, Dorset.**

The undercutting at the base of the waterfall creates a precarious overhang which will ultimately collapse. Thus a waterfall may appear to migrate upstream, leaving a gorge of recession downstream. The Niagara Gorge is 11 km long due to the retreat of the Niagara Falls.

Gorges may also be formed as a result of:

- **antecedent drainage (Rhine Gorge)**
- **glacial overflow channelling (Newtondale)**
- **collapse of underground caverns in carboniferous limestone areas (River Axe at Wookey Hole)**
- **periglacial run-off (Cheddar).**

- **pH – rates of solution are increased when the water is more acidic.**
- **Human impact – deforestation, dams and bridges interfere with the natural flow of a river and frequently end up increasing the rate of erosion.**

Global rates of erosion

The global average rate of erosion is estimated at about 250 tonnes per km² per year. However, there is a great deal of variation in sediment yields (Figure 1.21). These range from 10 tonnes per km² per year in such areas as northern Europe

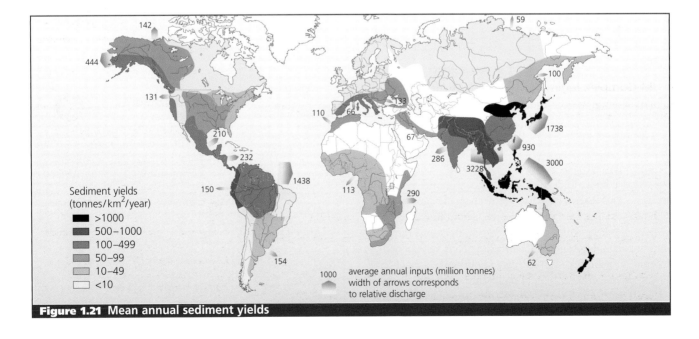

Sediment yields (tonnes/km²/year)
- >1000
- 500–1000
- 100–499
- 50–99
- 10–49
- <10

1000 average annual inputs (million tonnes) width of arrows corresponds to relative discharge

Figure 1.21 Mean annual sediment yields

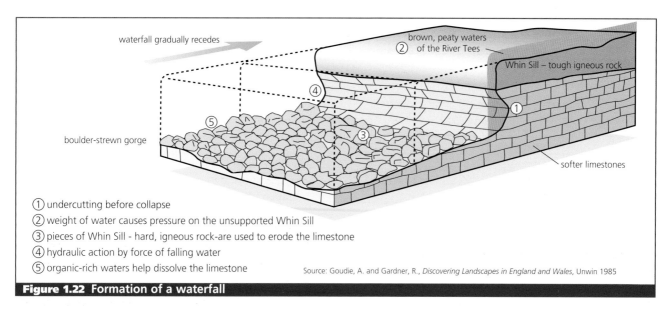

waterfall gradually recedes

brown, peaty waters
② of the River Tees

Whin Sill – tough igneous rock

④

①

⑤

boulder-strewn gorge

③

softer limestones

① undercutting before collapse
② weight of water causes pressure on the unsupported Whin Sill
③ pieces of Whin Sill - hard, igneous rock-are used to erode the limestone
④ hydraulic action by force of falling water
⑤ organic-rich waters help dissolve the limestone

Source: Goudie, A. and Gardner, R., *Discovering Landscapes in England and Wales*, Unwin 1985

Figure 1.22 Formation of a waterfall

Figure 1.23 Victoria Falls

Meanders

Water flow in a river often takes the form of **helicoidal flow**, a 'corkscrewing' motion. This is associated with the presence of alternating pools and riffles in the channel bed, and where the river is carrying large amounts of material. The erosion and deposition by helicoidal flow creates meanders (Figure 1.24).

Meanders occur as the top current comes in contact with the channel bank, undercutting it and causing the bank to collapse. This leaves river cliffs at the channel edge. As the water heaps up against this bank area, a bottom or subsidiary current develops, which carries energy and material to the opposite bank but slightly downstream. Thus a corkscrew motion is established. Immediately opposite the area of bank caving there will be an area of slow water where deposition occurs, forming a slip-off slope. Hence, as one bank area is being eroded, the opposite bank area is being built out – thus

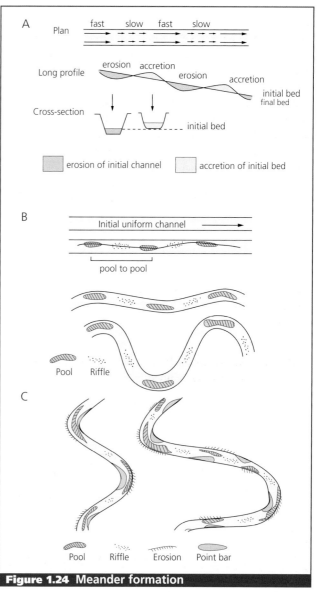

Figure 1.24 Meander formation

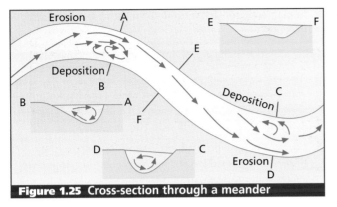

Figure 1.25 Cross-section through a meander

a meander is allowed to develop, with a characteristic asymmetric cross section (Figure 1.25).

Once established, the meandering pattern is self-perpetuating. As a result, the meanders become accentuated i.e. the amplitude of the bends are increased. Sometimes this happens to such an extent that the meander swings round almost to meet itself, leaving only a 'swan neck' of land between. The flow round this very accentuated meander becomes increasingly tortuous and only a slight increase in volume or velocity will cause the water to flood the swan's neck, eroding a new, straighter channel, and abandoning the meander as an ox-bow lake (cutoff). Eventually, this ox-bow lake will silt up and be colonised by vegetation.

As well as the change in amplitude, there is a change over time. Meanders migrate downstream and as they do they trim intervening spurs, creating a flood plain marked at its edges by a series of bluffs (Figure 1.26). Eventually the river will meander freely over an extensive plain.

Transport

Erosion by the river will provide loose material. This eroded material (plus other weathered material that has moved downslope from the upper valley sides) is carried by the river as its load. The load is transported downstream in a number of ways (Figure 1.27):

- the smallest particles (silts and clays) are carried in suspension as the **suspended load**

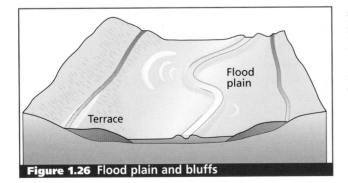

Figure 1.26 Flood plain and bluffs

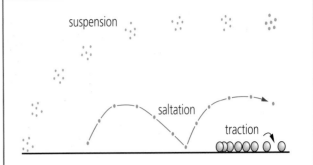

Figure 1.27 Transport in a river

- larger particles (sands, gravels, very small stones) are transported in a series of 'hops' as the **saltated load**
- pebbles are shunted along the bed as the **bed or tracted load**
- in areas of calcareous rock, material is carried in solution as the **dissolved load**.

Deposition

There are a number of causes of deposition, such as:

- a shallowing of gradient which decreases velocity and energy
- a decrease in the volume of water in the channel
- an increase in the friction between water and channel.

Many types of deposition are found along the course of a river. **Riffles** are small, ridges of material deposited where the river velocity is reduced midstream, in between pools (the deep parts of a meander). If many such ridges are deposited, the river is said to be 'braided'. On the inner bends of meanders, water flow is slack and **slip off slopes** are deposited.

Levees and flood plain deposits are formed when a river bursts its banks over a long period of time. Water quickly looses velocity, leading to the rapid deposition of coarse material (heavy and difficult to move a great distance) near the channel edge. These coarse deposits build up to form embankments called **levees**. The finer material is carried further away to be dropped on the **flood plain**, sometimes creating **backswamps**.

In estuaries, there is a constant mixing of fresh river water and saline sea water. When the tide is flowing in, the river flow is slowed down, losing energy. The mixing of the waters results in a chemical reaction that causes the salts and clays (the suspended load) to clot together (a process known as **flocculation**) and become too heavy for further transport. Thus the flocculated material is deposited as **mudflats** or **mudbanks**, uncovered at low tide and dissected by many creeks. If the mudflats are built up, they may become colonised by marshland plants to create salt marsh. The non-clay load is carried a little further seaward to the outer estuary and then dropped off as sandbanks.

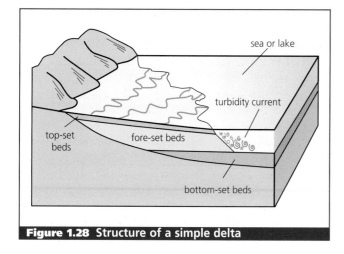

Deltas

Deltas are river sediments deposited when a river enters a standing body of water such as a lake, a lagoon, a sea or an ocean (Figure 1.28). They result from the interaction of fluvial and marine processes. For a delta to form there must be a heavily laden river, such as the Nile or the Mississippi, and a standing body of water with negligible currents, such as the Mediterranean or the Gulf of Mexico. Deposition is enhanced if the water is saline, since salty water causes small clay particles to flocculate or adhere together. Other factors include the type of sediment, local geology, sea level changes, plant growth and human impact.

The material deposited as a delta can be divided into three types.

1 **Bottomset beds** – the lower parts of the delta are built outwards along the sea floor by turbidity currents (currents of water loaded with material). These beds are composed of very fine material.

2 **Foreset beds** – over the bottom set beds, inclined/sloping layers of coarse material are deposited. Each bed is deposited above and in front of the previous one, the material moving by rolling and saltation. Thus the delta is built seaward.

3 **Topset beds** – composed of coarse material, they are really part of the continuation of the river's flood plain. These topset beds are extended and built up by the work of numerous distributaries (the main river has split into several smaller channels).

The character of any delta is influenced by the complex interaction of several variables:

■ the rate of river deposition

■ the rate of stabilisation by vegetation growth

■ tidal currents

■ the presence or not of longshore drift

■ human activity (deltas often form prime farmland when drained).

There are many delta types, but the three 'classics' are:

■ **Arcuate delta** – fan-shaped. These are found in the areas where regular longshore drift of other currents keep the seaward edge of the delta trimmed and relatively smooth in shape, such as the Nile and Rhone deltas.

■ **Cuspate delta** – pointed like a tooth of cusp, e.g. the Ebro and Tiber deltas, shaped by regular but opposing, gentle water movement.

■ **Bird's foot delta** – where the river brings down enormous amounts of fine silt, deposition can occur in a still sea area, along the edges of the distributaries for a very long distance offshore such as the Mississippi delta.

Deltas can also be formed inland. When a rivers enters a lake it will deposit some or all of its load, so forming a **lacustrine delta**. As the delta builds up and out, it may ultimately fill the lake basin. The largest lacustrine deltas are those which are being built out into the Caspian Sea by the Volga, Ural, Kura and other rivers.

The river course – long and cross profiles

The long profile of a river is the gradient of the channel bed from source to mouth (Figure 1.29). From an intensive study of long profiles in rivers, W. M. Davis proposed that the long profile is attained after a long period of time, during which the river's activities change (in time and place), creating a slope/profile and an associated velocity that allows erosion and deposition to be exactly balanced. In other words, **equilibrium** is reached and the profile is **graded** to the lowest level (base level), which is usually the sea level. Near the source of the river (the upper course) erosion is limited because the volume of water in the channel and the load are small. Erosion is also limited near the mouth (the lower course) because the river is heavily laden and much energy is expanded on transport. In a relative sense, therefore, erosion is at a maximum in the middle course. Thus, the graded profile of a river is typically concave.

In recent years, considerable criticism has been made of this concept of 'grade'. Such a graded profile may never be achieved by some rivers for a number of reasons. These include:

■ the presence of local base levels

■ changes in the general activity of the river by rejuvenation.

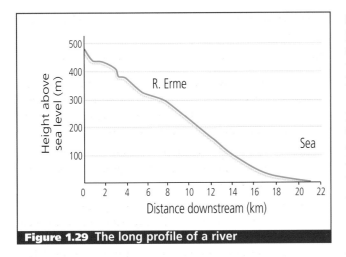

Figure 1.29 The long profile of a river

Waterfalls and lake basins form important local base levels. However, many of these are only temporary interruptions in the long profile. For example, waterfalls may retreat and lakes may fill. Sediment is deposited in the lake as a lacustrine delta which gradually enlarges to fill the lake. Meanwhile, the out-flowing water will accelerate down the steeper slope downstream of the basin, causing headward erosion at the outlet. Both processes will cause the lake and local base level to disappear.

Rejuvenation can seriously interrupt the attainment of a graded profile by causing the river to increase its downcutting activity. Rejuvenation is caused by:

- tectonic activity (folding, faulting, uplift, subsidence, tilting,) causing dynamic rejuvenation

- an increase in the volume of water in the drainage system, due to
  (i) an increase in precipitation in the catchment area because of climatic change, or
  (ii) the result of river capture

- changes in base level (either sea level falls or the land rises) due to
  (i) eustatic fall in sea-level (a worldwide fall in sea level), as in the Pleistocene Ice Age when water was 'locked up' as ice on the land, or
  (ii) an isostatic change in the level of the land (the local uplift of the land, in late and post glacial times as a result of the ice melting, relieving pressure on the land).

Rejuvenation has affected the vast majority of rivers throughout the world. As the last major period of rejuvenation occurred in recent geological time, its effects can still be clearly seen in fluvial landscapes. Indeed, some areas that were heavily glaciated are still experiencing rejuvenation and landscape change.

Features associated with rejuvenation include knickpoints, river terraces, and incised meanders. With the fall in base level, the seaward end of the river will fall over a sharp change in gradient, creating a waterfall. The sudden change in gradient called a knickpoint. As with ordinary waterfalls, the plunging river will cause the river to retreat upstream. If stability continues for a long time, the knickpoint may retreat to the river source, creating a profile 'graded' to the new lower base level.

Rejuvenation causes the river to cut down, leaving portions of the former floodplain untouched and upstanding as terraces. These paired terraces will converge at the associated knickpoint. In one period of rejuvenation, one knickpoint and one pair of terraces will be created. However, there are often cycles of rejuvenation, creating a series of knickpoints and terraces. This is known as polycyclic relief (i.e. many cycles of rejuvenation).

Incised meanders are formed where the meander pattern of the river is maintained as the river increases its downcutting. There are two types of incised meander, ingrown and entrenched meanders. In an ingrown meander downcutting is slow. As the incision of the river continues, there is time for the ordinary processes of meander development and migration to operate. Hence, greatly enlarged river cliffs (bluffs) and slipoff slopes are created, and the meander has an asymmetric cross-section as with normal meanders. By contrast, entrenched meanders occur where downcutting is so rapid that meander migration is not allowed. Thus the river cuts more of a winding gorge with a symmetric cross-section. Incised meanders are very well developed along the course of the River Wye.

As the long profile changes, so does the cross profile. The general characteristics of this cross-profile are controlled by:

- the type and rate of river activity

- the type and rate of weathering and downslope transport of this weathered material

- local geology.

Davis' *Cycle of Erosion*

In his *Cycle of Erosion*, W. M. Davis stated that a river's activity would change over time. In the early stages the river would be 'youthful', engaged primarily in cutting narrow, V-shaped valleys into the general land level. In the middle part of the time cycle, the river begins to 'mature', eroding and depositing material, while the action of weathering has begun to open up the valley sides, generally subduing the relief. In the last stages of the cycle, the river becomes more sluggish ('senile'), flowing over the land that has become degraded by other agents of weathering and erosion. These three stages are, according to Davis, reflected in the long and cross profiles of many rivers: the upper reaches reflect the 'youthful' stage, the middle reaches the 'mature' stage, while the lower reaches reflect the 'senile' stage.

Drainage patterns in drainage basins

Each river/drainage system is contained within a **drainage basin** or catchment area. Neighbouring systems are separated by a drainage divide or **watershed**. Some watersheds are major such as the Western Cordillera in North America and its continuation into South America as the Andes, while others are much smaller in scale. These smaller divides are often modified by the erosional activity of the headwaters of the streams within the catchment areas.

River capture

Since both systems on either side of a divide are constantly at work modifying the divide by headward erosion, the divide may migrate towards one or the other, eventually leading to breaching. The possibility of river capture will depend upon a number of factors:

- the relative power of erosion of the two systems
- the relative amount of precipitation in each of the basins (more precipitation leads to more energy)
- the geological structure of each basin and the resistance of the local rocks
- human activity such as deforestation in one catchment.

Some form of river capture is almost certain to occur anywhere, since it is highly unlikely that two adjacent systems will experience identical conditions. For one reason or another, one system will dominate over the other.

Classification of drainage patterns

There are various ways of classifying drainage patterns. The most common ones are by shape of river networks, accordancy with geological and topographic structures, and with morphometry (order and plan) (Figure 1.30).

Floods as hazards

Floods are one of the most common of all environmental hazards. This is because so many people live in fertile river valleys and in low lying coastal areas. However, the nature and scale of flooding varies greatly. For example, less than 2% of the population of England and Wales and in Australia live in areas exposed to flooding, compared to 10% of the population of the USA. The worst problems occur in Asia where floods damage about 4 million ha of land of each year and affect the lives of over 17 million people. Worst of all is China where over 5 million people have been killed in floods since 1860.

Some environments are more at risk than others. The most vulnerable include the following:

- **Low lying parts of active flood plains and river estuaries. For example in Bangladesh 110 million people live relatively unprotected on the flood plain of the Ganges and Brahmaputra rivers. Floods caused by the monsoon**

Dendritic
Irregular branching of channels ("treelike") in many directions. Common in massive rock and in flat-lying strata. In such situations, differences in rock resistance are so slight that their control of the directions in which valleys grow headward is negligible.

Parallel
Parallel or subparallel channels that have formed on sloping surfaces underlain by homogenous rocks. Parellel rills, gullies, or channels are often seen on freshly exposed highway cuts or excavations having gentle slopes.

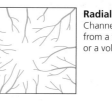

Radial
Channels radiate out, like spokes of a wheel, from a topologically high area, such as a dome or a volcanic cone.

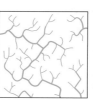

Rectangular
Channel system marked by right-angle bends. Generally results from the presence of joints and fractures in massive rocks or foliation in metamorphic rocks. Such structures, with their cross-cutting patterns, have guided the directions of valleys.

Trellised
Rectangular arrangement of channels in which principal tributary streams are parallel and very long, like vines trained on a trellis. This pattern is common in areas where the outcropping edges of folded sedimentary rocks, both weak and resistant, form long, nearly parallel belts.

Annular
Streams follow nearly circular or concentric paths along belts of weak rock that ring a dissected dome or basin where erosion has exposed successive belts of rock of varying degrees of erodibility.

Centripetal
Streams converge toward a central depression, such as a volcanic crater or caldera, a structural basin, a breached dome, or a basin created by dissolution of carbonate rock.

Deranged
Streams show complete lack of adjustment to underlying structural or lithologic control. Characteristic of recently deglaciated terrain whose preglacial features have been remodeled by glacial processes.

Source: Skinner, J. & Porter, S., *The dynamic earth*, Wiley 1989

Figure 1.30 Drainage patterns

regularly cover 20–30% of the flat delta. In very high floods up to half of the country may be flooded. In 1988, 46% of the land was flooded and over 1500 people were killed.

■ Small basins subject to flash floods. These are especially common in arid and semi-arid areas. In tropical areas some 90% of lives lost through drowning are the result of intense rainfall on steep slopes.

■ Areas below unsafe dams. In the USA there are about 30 000 large dams and 2000 communities are at risk from dams.

■ Low-lying inland shorelines such as those along the Great Lakes and the Great Salt Lake in the USA.

In most developed countries the number of deaths from floods is declining. By contrast the economic cost of flood damage has been increasing. In developing countries, on the other hand, the death rate due to flooding is much greater although the economic cost is not as great. It is likely that the hazard in developing countries will increase over time as more people migrate and settle in low lying areas and river basins. Often newer migrants are forced into the more hazardous zones.

Since the Second World War there has been a change in the understanding of the flood hazard, in the attitude towards floods and policy towards reducing the flood hazard. The response to hazards has moved away from physical control (engineering structures) towards reducing vulnerability through non-structural approaches.

Causes of floods

A flood is a high flow of water which overtops the bank of a river (Figure 1.31). The main cause of floods are climatic forces whereas the conditions that intensify floods tend to be drainage basin specific. Most floods in Britain, for example, are associated with deep depressions (low pressure systems) which are both long lasting and cover a wide area. In India, by contrast, up to 70% of the annual rainfall occurs in three months during the summer monsoon. In Alpine and Arctic areas, melting snow is also responsible for widespread flooding.

Flood intensifying conditions cover a range of factors which alter the drainage basin response to a given storm. The factors which influence the storm hydrograph (pages 8–10) determine the response of the basin to the storm. These factors include topography, vegetation, soil type, rock type, characteristics of the drainage basin and so on.

Questions

1 **(a)** Outline the processes by which a river may erode its channel.
   (b) Describe the methods by which a river may transport material.

2 Explain the meaning of the following terms:
   (a) the competence of a river
   (b) the capacity of a river.
   Discuss how competence and capacity are related to the flow of a river.

3 Figure 1.20 shows the relationship between water velocity and the size of material carried. Use this diagram to describe the sequence of processes and their likely consequences in a stream channel with a wide range of sediment sizes, as discharge changes during a flood event from a low flow velocity of 0.05 m sec^{-1}, to 0.5 m sec^{-1}, then to peak velocity of 1.5 m sec^{-1}.

4 What are deltas? Under what conditions are they formed? Account for variations in the size and shape of deltas. Use Figure 1.28 to account for differences in shape.

5 What are meant by the terms:
   (a) dynamic equilibrium
   (b) rejuvenation and
   (c) river capture?
   Under what conditions is trellised drainage likely to occur?

Figure 1.31 River Thames in flood

Figure 1.32 Increased risk of flooding due to urbanisation

Key Definitions

Bankfull stage A condition in which a river's channel fills completely, so that any further increase in discharge results in water overflowing the banks.

Channel The passageway in which a river flows.

Channelisation Modifications to river channels, consisting of some combination of straightening, deepening, widening, clearing, or lining of the natural channel.

Discharge The quantity of water that passes a given point on the bank of a river within a given interval of time.

Drainage basin The total area that contributes water to a river.

Flash flood A flood in which the lag time is exceptionally short – hours or minutes.

Flood A discharge great enough to cause a body of water to overflow its channel and submerge surrounding land.

Flood-frequency curve Flood magnitudes that are plotted with respect to the recurrence interval calculated for a flood of that magnitude at a given location.

Flood plain The part of any stream valley that is inundated during floods.

Hazard assessment The process of determining when and where hazards have occurred in the past, the severity of the physical effects of past events of a given magnitude, the frequency of events that are strong enough to generate physical effects, and what a particular event would be like if it were to occur now; and portraying all this information in a form that can be used by planners and decision makers.

Load The particles of sediment and dissolved matter that are carried along by a river.

Natural hazards The wide range of natural circumstances, materials, processes, and events that are hazardous to humans, such as locust infestations, wildfires, or tornadoes, in addition to strictly geologic hazards.

Risk assessment The process of establishing the probability that a hazardous event of a particular magnitude will occur within a given period and estimating its impact, taking into account the locations of buildings, facilities and emergency systems in the community, the potential exposure to the physical effects of the hazardous situation or event, and the community's vulnerability when subjected to those physical effects.

The potential for damage by flood waters increases exponentially with velocity. The physical stresses on buildings are increased even more when rough rapidly flowing water contains debris such as rock, sediment and trees.

Other conditions which intensify floods include changes in land use (Figure 1.32). Urbanisation, for example, increases the magnitude and frequency of floods in at least three ways:

■ **creation of highly impermeable surfaces, such as roads, roofs and pavements**

■ **smooth surfaces served by a dense network of drains, gutters and underground sewer increasing drainage density**

■ **natural river channels are often constricted by bridge supports or riverside facilities reducing their carrying capacity.**

Deforestation is also a cause of increased flood runoff and a decrease in channel capacity. This occurs due to an increase in deposition within the channel. However, the evidence is not always conclusive. In the Himalayas, for example, changes in flooding and increased deposition of silt in parts of the lower Ganges–Brahmaputra are due to the combination of high monsoon rains, steep slopes, and the seismically unstable terrain. This ensure that runoff is rapid and sedimentation is high irrespective of the vegetation cover.

Coping with floods

Economic growth and population movements throughout the 20th century have caused many flood plains to be built on. However, in order for people to live on flood plains there needs to be flood protection. This can take many forms such as loss-sharing adjustments or event-modifications.

Loss sharing adjustments include disaster aid and insurance. **Disaster aid** refers to any aid, such as money, equipment, staff and technical assistance that is given to a community following a disaster. In developed countries **insurance** is an important loss-sharing strategy. Even so, not all flood-prone households have insurance and many of those that are insured may be underinsured. In the floods of central England in 1998 (Figure 1.33) many of the affected households were not insured against losses from flooding because the residents did not believe that they lived in an area that was likely to flood. Hence they had very limited flood insurance.

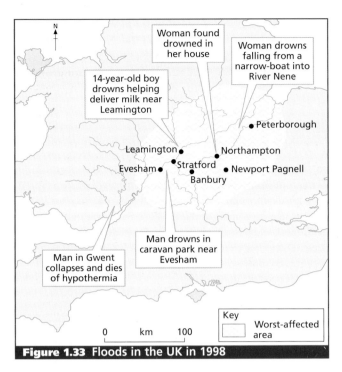

Figure 1.33 Floods in the UK in 1998

Labels on map:
- Woman found drowned in her house
- Woman drowns falling from a narrow-boat into River Nene
- 14-year-old boy drowns helping deliver milk near Leamington
- Man drowns in caravan park near Evesham
- Man in Gwent collapses and dies of hypothermia
- Peterborough
- Leamington
- Northampton
- Evesham
- Stratford
- Newport Pagnell
- Banbury

Key: Worst-affected area

0   km   100

Event modification adjustments include environmental control and hazard resistant design. Physical control of floods depend on two measures: flood abatement and flood diversion. **Flood abatement** involves decreasing the amount of runoff, thereby reducing the flood peak in a drainage basin. There are a number of ways of reducing flood peaks. These include:

- reforestation
- reseeding of sparsely vegetated areas to increase evaporative losses
- treatment of slopes such as contour ploughing or terracing to reduce the runoff co-efficient
- comprehensive protection of vegetation from wild fires, overgrazing, and clear cutting of forests
- clearance of sediment and other debris from headwater streams
- construction of small water and sediment holding areas
- preservation of natural water storage zones, such as lakes.

Flood diversion measures, by contrast, include the construction of levees, reservoirs and the modification of river channels (Figure 1.34). **Levees** are the most common form of river engineering. They can also be used to divert and restrict water to low-value land on the flood plain. Over 4500 km of the Mississippi River has levees. Channel improvements, such as channel enlargement, can increase the carrying capacity of the river. **Reservoirs** store excess rainwater in the upper drainage basin, but may only be appropriate in small drainage networks. It has been estimated that some 66 billion m³ of storage is needed to make any significant impact on major floods in Bangladesh!

Hazard resistant design

Flood proofing includes any adjustments to buildings and their contents, which help reduce losses. Some are temporary, such as:

- blocking up entrances
- sealing doors and windows
- removal of damageable goods to higher levels
- use of sandbags.

By contrast, long-term measures include moving the living spaces above the likely level of the flood plain. This normally means building above the flood level, but could also include building homes on stilts.

Forecasting and warning

During the 1970s and 1980s flood forecasting and warning became more accurate and is now one of the most widely used measures to reduce the problems caused by flooding. Despite advances in weather satellites and the use of radar for forecasting, over 50% of all of unprotected dwellings in England and Wales have less than six hours of flood warning time. In most developing countries there is much less effective flood forecasting. An exception is Bangladesh. Most floods in Bangladesh originate in the Himalayas, so authorities have about 72 hours warning.

Land use planning

Most land use zoning and land use planning has developed since the Second World War. In the USA, land use management has been effective in protecting new housing developments from 1 in 100 year floods (i.e. the size of flood that we would expect to occur once every century). In Britain, thanks to lower population growth there has not been

Figure 1.34 Flood relief channels

the same encroachment on to the flood plain as there has been in the United States. For example, between 1952 and 1982 the population of England and Wales grew by 12% compared with 50% in the USA, 73% in Canada and 78% in Australia. The greater requirement for homes in the USA has led to more encroachment on to flood plains.

One example where partial urban relocation has occurred is at Soldier's Grove on the Kickapoo river in south-western Wisconsin, USA. The town experienced a series of floods in the 1970s, and the Army Corps of Engineers proposed to build two levees and to move part of the urban area. Following further floods in 1978 they decided that relocation of the entire business district was preferable to simple flood damage reduction. Although the levees would have protected the village from most floods, they would not have done so in all cases. Relocation allowed energy conservation and an increase in commercial activity in the area (Figure 1.35).

Figure 1.35 Kickapoo River in SW Wisconsin

Flooding and flood management in Bangladesh

- A **drainage basin** – the area drained by a river.

- A **watershed** – the dividing line between one drainage basin and another.

- An **estuary** – the mouth of a large river which is affected by tides.

- A **tributary** – a smaller river which joins up with a larger one.

Much of Bangladesh has been formed by deposition from three main rivers – the Brahmaputra, the Ganges and the Meghna. The sediment from these and over 50 other rivers forms one of the largest deltas in the world, and up to 80% of the country is located on the delta (Figure 1.36). As a result, much of the country is just a few metres above sea level and is under threat from flooding and rising sea levels (see also Chapter 5, pages 88–9). To make matters worse, Bangladesh is a very densely populated country (over 900 people per km sq) and is experiencing rapid population growth (nearly 2.7% per annum). It contains nearly twice as many people as the UK, but is only about half its size. Hence, the flat, low lying delta is vital as a place to live as well as a place to grow food.

Almost all of Bangladesh's rivers have their source outside the country. For example, the drainage basin of the Ganges and Brahmaputra covers 1.75 million km² and includes the Himalayas, the Tibetan Plateau and much of northern India. Total rainfall within the Brahmaputra–Ganges–Meghna catchment is very high and very seasonal – 75% of annual rainfall occurs in the monsoon between June and September. Cherrapungi, high in the Himalayas, has an average annual rainfall of over 10 000 mm, and this can rise as high as 20 000 mm in a 'wet' year. Moreover, the Ganges and Brahmaputra carry snowmelt waters from the Himalayas. This normally reaches the delta in June and July. Peak discharges of the rivers are immense – up to l00 000 m³ per second (i.e. 100 000 cumecs) in the Brahmaputra, for example. In addition to water, the rivers carry vast quantities of sediment. This is deposited annually to form temporary islands and sand banks.

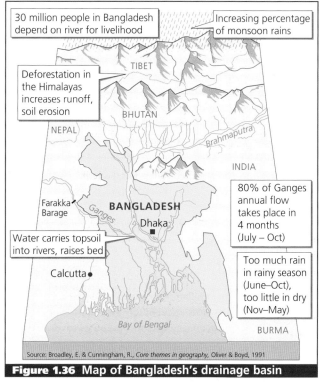

Source: Broadley, E. & Cunningham, R., *Core themes in geography*, Oliver & Boyd, 1991

Figure 1.36 Map of Bangladesh's drainage basin

The advantages of flooding

During the monsoon between 30% and 50% of the entire country is flooded. The floodwaters:

- replenish groundwater reserves
- provide nutrient-rich sediment for agriculture in the dry season
- provide fish (fish supply 75% of dietary protein and over 10% of annual export earnings)
- reduces the need for artificial fertilisers
- flush pollutants and pathogens away from domestic areas.

The causes

There are five main types of flooding in Bangladesh: river floods, overland runoff, flash-floods, 'back-flooding' and storm surges (Figure 1.37). Snow melt in the Himalayas, combined with heavy monsoonal rain, causes peak discharges in all the major rivers during June and July. This leads to flooding and destruction of agricultural land. Outside the monsoon season, heavy rainfall causes extensive flooding

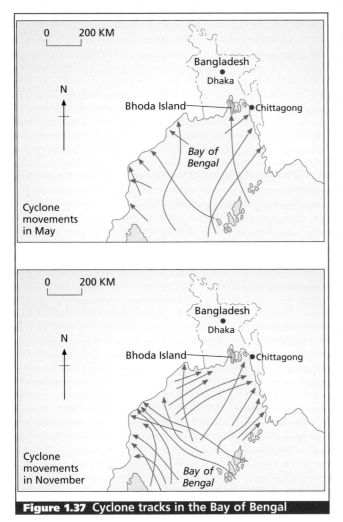

Figure 1.37 Cyclone tracks in the Bay of Bengal

which may be advantageous to agricultural production, since it is a source of new nutrients. In addition, the effects of flash-floods, caused by heavy rainfall in northern India, have been intensified by the destruction of forest which reduces interception, decreases water retention and increases the rate of surface runoff.

Human activity in Bangladesh has also increased the problem. Attempts to reduce flooding by building embankments and dykes have prevented the back flow of flood water into the river. This leads to a ponding of water (also known as 'drainage congestion') and back flooding. In this way, embankments have sometimes led to an increase in deposition in drainage channels, and this can cause large-scale deep flooding. Bangladesh is also subject to coastal flooding (see pages 88–9). Storm surges caused by intense low pressure systems are funnelled up the Bay of Bengal.

The Flood Action Plan

It is impossible to prevent flooding in Bangladesh. The Flood Action Plan attempts to minimise the damage of flooding, and maximise the benefits. The Plan relies upon huge embankments which run along the length of the main rivers. They can withstand the most severe floods, as for example in 1987 and 1988, but at least provide some control over the flooding. The embankments contain sluices that can be opened to reduce river flow and to control the damage caused by flooding.

The embankments are set back from the rivers. This protects them from the erosive power of the river, and has the added advantage of making them cheap both to install and maintain. In addition, the area between the river and embankment can be used for cereal production.

Nevertheless, the scheme has a number of negative impacts. These include:

- increased time of flooding, since the embankments prevent back flow into the river
- not enough sluices have been built to control the levels of the floodwaters in the rivers – this means that there may be increased damage by flooding if the embankments are breached, since the rapid nature of the breach is more harmful than gradual flooding
- sudden breaches of the embankments may also deposit deep layers of infertile sand thereby reducing soil fertility
- by preventing back flow to the river, areas of stagnant water are created which may increase the likelihood of diseases such as cholera and malaria
- embankments may cause some wetlands to dry out, leading to a loss of biodiversity
- decreased flooding reduces the input of fish, which is a major source of protein, especially among the poor.

The rivers of Bangladesh are, in part, controlled by factors beyond the country. There is a delicate balance between the disadvantages that the rivers create, such as death and destruction, and advantages that the rivers bestow, such as a basis for agriculture and export earnings. To date there has been little agreement as to how to control the peak discharges of the rivers. The Flood Action Plan uses embankments to control over the distribution and speed of flooding, even though the embankments have, in turn, led to some serious social, economic and environmental problems.

Management of drainage basins

Human activities and rivers

River landforms may be changed quite considerably by human activity. Such activity can increase or decrease flow, build dams and reservoirs, modify channels, build bridges and upset the natural dynamic equilibrium of rivers. Many of the changes caused are accelerations of natural processes.

Large features such as dams can have many unforeseen consequences. The Hoover Dam, for example, built in 1935, created a number of mild earthquakes (see page 97). The Three Gorges Dam in China (Figure 1.40) illustrates many of the advantages (Figure 1.38) and disadvantages (Figure 1.39) of large-scale engineering projects.

The Three Gorges Dam

The Three Gorges Dam will be over 2 km long and 100 m high and the lake behind it will be over 600 km long. It is important for China. The Yangtze Basin provides 66% of China's rice and contains 400 million people, while the river drains 1.8 million km² and discharges 700 cubic kilometres of water annually.

Figure 1.38 The advantages of the Three Gorges Dam

- It will generate up to 18 000 megawatts, 50% more than the world's largest existing HEP dam.

- It will enable China to reduce its dependency on coal.

- It will supply Shanghai (population over 13 million), one of the world's largest cities, and Chongqing (population 3 million), an area earmarked for economic development.

- It will protect 10 million people from flooding (since 1860 over 5 million people in China have died as a result of flooding).

- It will allow shipping above the Three Gorges; the dams will raise water levels by 90 m, and turn the rapids in the gorges into a lake.

Figure 1.39 The disadvantages of the Three Gorges Dam

- Over 1 million people will have to be removed to make way for the dam and the lake.

- It will take between 15 and 20 years to build and could cost as much as $70 billion.

- Most floods in recent years have come from rivers which join the Yangtze below the Three Gorges Dam.

- The region is seismically active and landslides are frequent.

- The port at the head of the lake may become silted up as a result of increased deposition and the development of a delta at the head of the lake.

- Up to 1.2 million people will have to be moved to make way for the dam.

- Much of the land available for resettlement is over 800 m above sea level, and is colder with infertile thin soils and on relatively steep slopes.

- Dozens of towns, such as Wanxian and Fuling with 140 000 and 80 000 people respectively, will be flooded.

- Up to 530 million tonnes of silt are carried through the gorges annually; the first dam on the river lost its capacity within seven years and one on the Yellow River filled with silt within four years.

- To reduce the silt load afforestation is needed but resettlement of people will cause greater pressure on the slopes above the dam.

- The dam will interfere with aquatic life – the Siberian Crane and the White Flag Dolphin are threatened with extinction.

- Archaeological treasures will be drowned, including the Zhang Fei temple.

The flooding begins
Costing £15bn, the dam is due to be completed in 2009, flooding more than 140 towns and resulting in the forced eviction of 1.2 million people.

Blocking the Yangtze
Trucks dump rocks into water channel between bank and island, constructing two parallel dams.

Channel allows ships to continue to use river during building of dam.

Base of concrete dam wall

Shipping
Canal with five locks will enable cargo and tourist ships to pass the dam.

Construction
Water drained from between the dams – concrete wall of main dam built on dry river bed.

Sandouping
Rock barrage (dam) built across the Yangtze before dam construction

Present route of Yangtze

Wanxian   Badong
Yunyang   Zigui   Yichang
Changsho   Zhong Xian   Shashi
Fengdu
Chongqing   Lake Chang

Jialing   Qu   Fu   Yangtze   Qing

0   200 km

Flooded area
The inland sea will be more than 1,000 km², four times bigger than Hong Kong Island

Lake
Water level will rise 177m above Yangtze river bed.

Dam
Walls 185m high

HEP
26 generators, maximum 18,200MW output

Source: Nagle, G., *The Three Gorges Dam Project*, Stanley Thornes, 1998

Figure 1.40 The Three Gorges Dam

Managing the changing Mississippi Delta

Deltas are difficult areas to manage. The example of the Mississippi (Figure 1.41) illustrates this very clearly. The delta covers an area of 26 000 km², drains an area of 3 220 000 km² and carries 520 km³ of water and 450 million tonnes of sediment into the Gulf of Mexico every year. Much of the sediment is carried in suspension, up to 210 million tonnes. About 40% of the load is silt and 50% clay. The silt is deposited to form a bar up to 7 m high whereas the clay is transported further out the delta. Deposition takes place during floods, forming levees and flood plains, and is enhanced by biological activities, notably colonisation by plants and the trapping of sediment by vegetation. Sediment-charged waters break through the natural levees (crevasse splaying) to find shorter, steeper routes, a process known as avulsion, the channel splits (bifurcates), sediment is deposited and the delta grows (for example, distributaries in West Bay grew 16 km between 1839 and 1875).

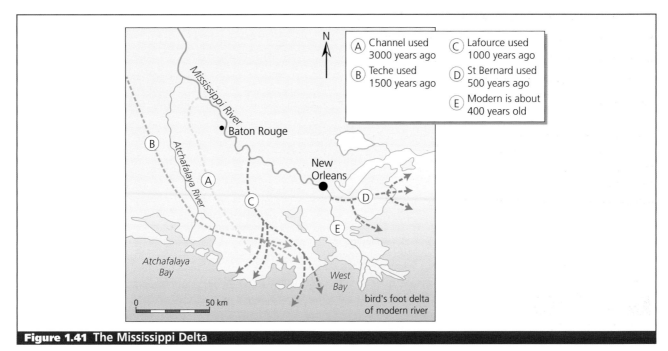

Figure 1.41 The Mississippi Delta

Deltas follow a cycle of development. This may take anything between 100 and 1000 years. First, new channels are created as a result of crevasse splays. Sediment is deposited and new land is created. As a result of changes in gradient, channels are abandoned, deposition declines and a new area is developed. However, this pattern is disrupted by human activities. According to some geographers, if the Mississippi were left to its own devices a new channel would have been created by the mid-1970s, so that the ports at New Orleans and Baton Rouge would be defunct. However, river protection schemes have prevented this. For example, at New Orleans 7 m levees flank the river and 3 m levees abut Lake Pontchartrain. New Orleans is 1.5 m below the average river level and 5.5 m below flood level!

The Mississippi Delta is retreating at rates of up to 25 m per annum. It accounts for 40% of US wetlands and over 100 km² of wetlands are being lost each year. The cause of the delta's decline is a mixture of natural and man-made conditions: rising sea levels, subsidence (due to the weight of the delta on the earth's crust), groundwater abstraction, tropical storms and changes in the location of deposition are all causing the delta to change. Increasingly it appears that the delta is trying to abandon its current course and to develop a new course along the Atchafalaya Channel.

Flood relief measures include:

■ Bonnet Carre Waterway from New Orleans to Lake Pontchartrain and then to the Gulf of Mexico

■ Atchafalaya River, which carries up to 25% of the Mississippi flow, 135 million tonnes of sediment (but this is a fertile area and famed for its bayou landscape and culture)

■ dredging of the Atchafalaya and Mississippi Rivers

■ Morganza floodway between Mississippi and Atchafalaya.

Protecting the Mississippi

It is not just the delta that receives attention from engineers, but the whole of the Mississippi. For over a century the Mississippi has been mapped, protected and regulated. The river flows through ten states and drains one-third of the USA. This contains some of the USA's most important agricultural regions. A number of methods have been used to control flooding in the river, and its effects, including:

Questions

1 Define the following terms: avulsion, crevasse splaying and distributaries.

2 Why does flocculation lead to increased deposition of clay?

3 Why are deltas
   (a) attractive places to live and
   (b) hazardous places to live?
   Illustrate your answer with examples.

4 Using an atlas, explain why the Nile and the Mississippi are heavily laden with sediments.

5 Explain briefly the life cycle of a delta under natural conditions. What was the average annual growth rate for the distributaries in the West Bay part of the Mississippi Delta? Why is silt deposited first rather than the clay?

6 Use examples to explain why deltas, such as the Mississippi, are difficult environments to manage?

7 Using information provided, and/or other examples that you have studied, examine the aim and impact of three of the following human interventions: levees, building new channels, channel straightening, dredging, removal of vegetation and land use management.

8 Why have the authorities protected the Mississippi River? Evaluate the effectiveness of the flood protection methods in the Mississippi Delta.

- stone and earthen levees to raise the banks of the river
- holding dams to hold back water in times of flood
- lateral dykes to divert water away from the river
- straightening of the channel to remove water speedily.

Altogether over US$10 billion has been spent on controlling the Mississippi, and annual maintenance costs are nearly $200 million. But it was not enough. In 1993, following heavy rain between April and July, many of the levees collapsed allowing the river to flood its flood plain. The damage was estimated to be over $12 billion yet only only 43 people died. Over 25 000 sq km of land were flooded. The river was only performing its natural function – people and engineers had modified the channel so much that its normal function did not occur often, and so people thought they were safe from the effects of flooding.

Drainage basin management: the Thames Basin

The national scene

In the UK there is a surplus or over-supply of water in the north and west of the country and a shortage of water in the south and east, especially in the Thames Region which serves almost a fifth of the nation's population. This pattern is likely to become intensified over the next 20–30 years as a result of climatic change. It will become increasingly important to maintain a balance between the needs of the water environment, the demands for water and the supply and development of new water resources.

Water resources in the Thames Region

Diversion of the Thames

The River Thames rises near Cirencester and flows eastwards towards Oxford, then south to Reading and east to London. The Thames Basin consists of two main geological sections – an upper basin consisting of oolitic limestone and a lower basin consisting of mainly Tertiary rocks, sands and gravels. These two sections join at the Goring Gap, a gap through the Chiltern Hills, where the outcrop of chalk has been breached (Figure 1.43).

The Thames hydrological cycle

Being part of southern Britain the Thames Basin has amongst the lowest rainfall in Britain but some of the highest temperatures. This means that evapotranspiration rates are high and therefore there is a potential deficiency of moisture in summer (Figure 1.44). Rainfall varies from about 900 mm in the western part of the basin to below 600 mm near London. By contrast, evaporation figures are higher in the east (over 600 mm per annum) compared with the west (about 500 mm per annum).

Water balance

The water balance of a drainage basin is the result of the inputs (precipitation) and the outputs (evapotranspiration). There may also be short-term and long-term changes in the storage capacity of the basin. The water balance can be shown by a simple line graph.

Figure 1.45 shows annual variations in soil moisture status. In (A) there is a positive

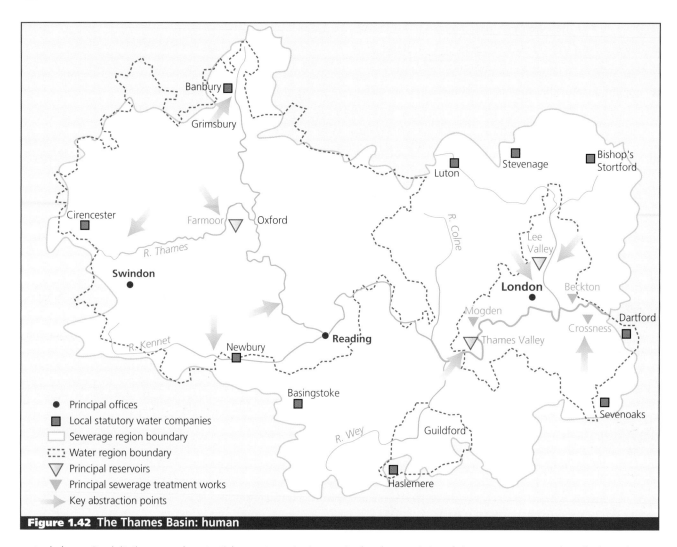

Figure 1.42 The Thames Basin: human

water balance. Precipitation exceeds potential evapotranspiration and there is a plentiful store of water supplying ground water recharge, runoff, and plant use. During period (B) potential evapotranspiration exceeds precipitation. The store of water in the soil is being taken up by plants or else being evaporated. This stage is known as soil moisture utilisation. At (C) all of the soil moisture has been used up. Any rainfall is likely, therefore, to be absorbed by the soil, although in some cases dry soils may generate flash floods. During Stage (D) there is still a deficiency of soil moisture. Farm crops will need to be irrigated whereas wild plants will need mechanisms to survive dry conditions. Towards the end of the year (E) precipitation exceeds potential evapotranspiration and there is a net excess of water again. Consequently, the soil is recharged with moisture. Once the soil store is full (field capacity) water will percolate down to the water table (F).

Runoff from a basin also varies throughout the year. During winter runoff as a percentage of rainfall is very high, whereas during the summer months it is low. Figure 1.46 shows the annual variation in runoff and rainfall for the River Thames near Abingdon.

There are important variations between the hydrographs for the Thames Basin (Figure 1.47). This reflects important differences in the characteristics of the streams measured, and where they are measured. For example,

■ the River Ock, a small tributary of the Thames, mainly flows over impermeable clays through the Vale of the White Horse – runoff is influenced by groundwater abstractions and return of effluents

■ the River Kennet drains an area four times larger than the Ock. Much of it is on Tertiary rocks, sands and gravels – although there is some abstraction of groundwater and use of the water for agriculture and industry, the net impact of these is limited

■ the River Lambourne flows mostly over chalk and consequently has a much lower discharge – the river is maintained by groundwater, as rainwater soaks into the porous chalk rather than flowing over its surface.

Changing hydrology of the Thames

The Thames Basin is one of the most intensively studied basins in the world. This is partly a result of the large number of people who live in the Basin, and partly a result of the importance of the river.

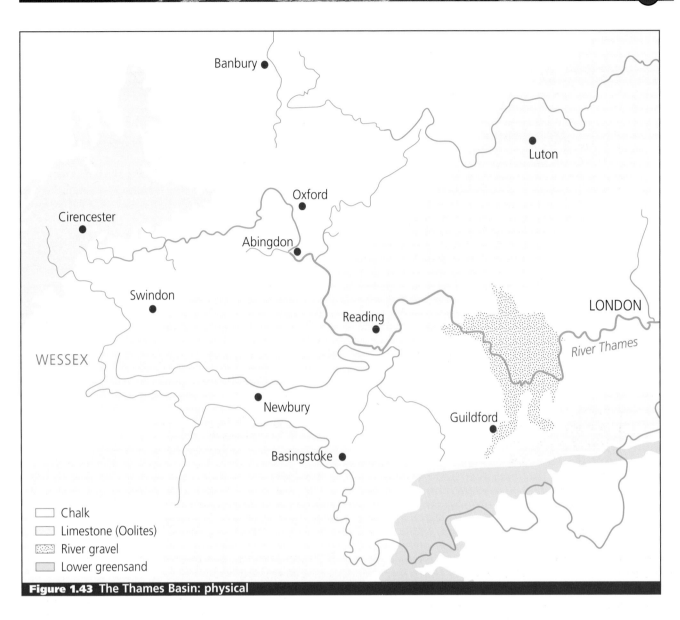

Figure 1.43 The Thames Basin: physical

Legend:
- ☐ Chalk
- ☐ Limestone (Oolites)
- ▦ River gravel
- ▨ Lower greensand

Figure 1.44 Rainfall and evaporation in the Thames Basin		
Month	Average rainfall (mm)	Potential evaporation (mm)
January	69.4	5.5
February	48.4	12.7
March	48.3	33.3
April	48.7	55.4
May	59.3	83.9
June	55.2	100.6
July	57.2	97.8
August	71.9	80.0
September	64.6	48.9
October	67.2	22.2
November	78.5	6.5
December	71.6	3.2

Source: Thames Water

Changing groundwater levels

The aquifer below London is one of the most important in the country (Figure 1.48). About 25% of the water falling on the permeable chalk of the North and South Downs makes its way to the aquifer below London. This water supplies most of the domestic, industrial and commercial demand. However, demand is changing. Increased demand for water during the 19th and early 20th centuries caused the water table to drop by as much as 0.7 m per annum. In some places the aquifer declined by as much as 90 m causing subsidence. In addition, the suburban growth of London led to an increase in the area of impermeable surfaces and a decline in the infiltration of water into the soil. Hence soil moisture replenishment declined. However, since the 1960s there has been a reduced demand for water. Manufacturing industries have declined or relocated causing groundwater levels to rise. In places the levels have risen by as much as 35 m, and London now has to pump water out to keep tunnels dry.

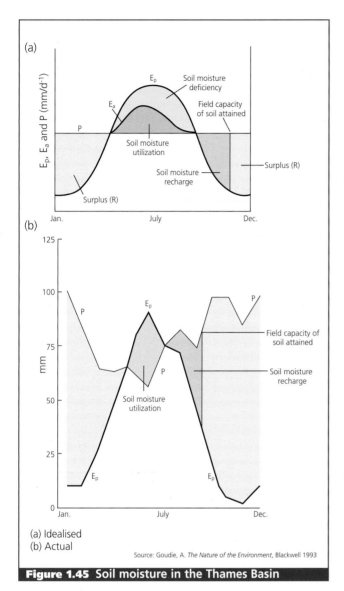

(a) Idealised
(b) Actual

Source: Goudie, A. *The Nature of the Environment*, Blackwell 1993

Figure 1.45 Soil moisture in the Thames Basin

Figure 1.46 Rainfall and runoff at Abingdon

	Rainfall (mm) (1938–94 average)	Runoff (mm) (1938–94 average)
January	67	44
February	47	40
March	53	35
April	47	23
May	58	16
June	55	11
July	54	7
August	65	6
September	61	7
October	65	12
November	70	23
December	72	35
Year	714	259

Source: Institute of Hydrology, 1996

The effects of rising groundwater include:

- flooding of basements

- since 1994 London Underground has spent £18 million on tunnel repairs

- increased chemical weathering of steel and concrete.

In addition, the south-east of Britain is sinking at a rate of about 3 mm per annum, or 30 cm per century. Another threat is related to the greenhouse effect. With increased amounts of atmospheric energy the possibility of storm surges increases. Hence the Thames valley is threatened with the prospect of an increase in tidal flooding. The Thames Barrier was built to protect London from tidal flooding. Between 1982 and 1996 it was closed 26 times in order to protect London from tidal flooding.

There are considerable development pressures in the region and these are likely to increase in the future. Of 25 areas identified as likely to boom before the end of the century nine are in the Thames Region, whereas only one out of 20 areas identified as likely to decline is in the region. Geologically, the area contains much chalk, limestone, sand and gravel, which creates demand for mineral extraction. Growth in housing and infrastructure places additional pressure on the water environment and water resources.

Groundwater resources

Approximately two-thirds of the catchment is permeable and thus subject to direct recharge from rainfall. Polluting discharges may also infiltrate into the ground in these areas. Rainfall varies from 850 mm per annum in western parts of the catchment to less than 650 mm per annum in eastern parts. Rates of recharge to groundwater vary considerably from 524 mm per annum in the north-west to 124 mm per annum in the east.

In much of the catchment a situation has been reached where there is no remaining capacity for abstraction because of the need to protect streamflows and the valley environment. In some areas over-abstraction has led to reduced flows and the drying up of some groundwater-fed rivers, particularly on the chalk aquifer. Abstraction in proximity to the Thames estuary has resulted in the ingress of saline waters several kilometres inland. A notable exception to the above trend is the chalk aquifer in the London Basin. The considerable reduction in abstractions since 1970 has resulted in rising groundwater levels.

Particular groundwater problems have resulted.

- Flows in several rivers have been depleted as a result of large groundwater abstractions close to the headwaters or along the river valleys. Worst affected are the rivers Misbourne, Ver, Wey, Pang and the Letcombe Brook.

- Groundwater has been affected by saline intrusions along the River Thames.

- Rising nitrate concentrations are evident in other parts of the catchment. In July 1996, the EU criticised the UK for allowing the level of pesticides in tap water to exceed 0.1 mg per litre. Certain parts of the country,

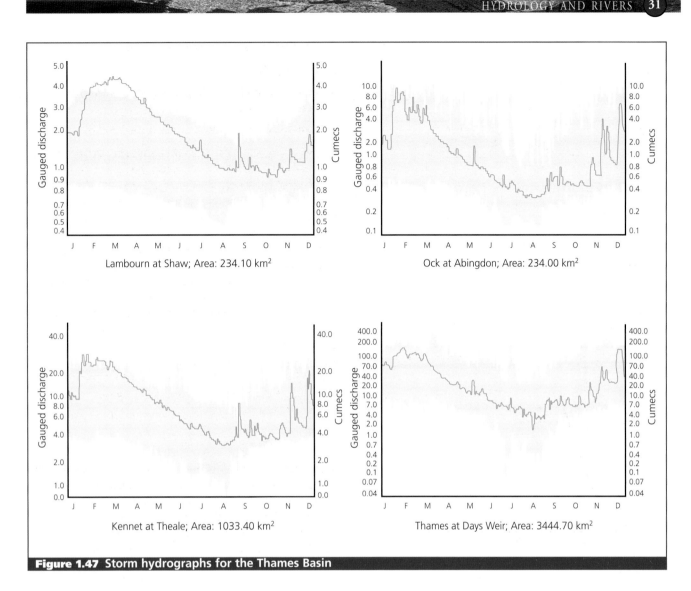

Figure 1.47 Storm hydrographs for the Thames Basin

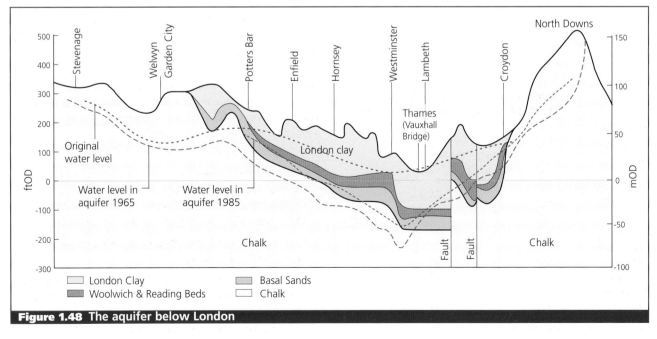

Figure 1.48 The aquifer below London

especially London and the Southeast, were affected.

■ Other chemicals, such as pesticides, are in widespread usage across the catchment and the frequency of detection in groundwater has risen.

■ Groundwater in some urban areas has been contaminated by leakage from sewers and through widespread usage of chemicals such as solvents.

Trends in water use

The vast majority of water abstracted in the region is for drinking water supply. Almost 60% of the water for public supplies comes from surface water supplies, mainly from the Thames and the Lee in association with the major surface storage reservoirs around London. An average of 150 litres are used per person every per day in the home, most of which is used to flush toilets, take baths and showers and use in the washing machine.

Water use in the home accounts for 45% of the total public water supply demand. A further 27% of the public water supply is used by industry and commerce. The remainder, 28% across the region, is lost through leakage from distribution and trunk mains systems, and supply pipes on customer premises.

Over the last 20 years demand for public water supplies has increased by approximately 1.7% each year. Key factors which have influenced demand are:

■ the use of water in the home and garden

■ losses through leakage from distribution systems and consumers' plumbing

■ population growth and household size

■ development pressure and economic activity.

A national strategy for water resources needs must include the following issues:

■ there should be no long-term systematic deterioration of the water environment owing to water resources development or water use (sustainable development)

■ where significant environmental change may occur, but understanding of the issues is incomplete, decisions made or measures implemented should err on the side of caution (precautionary principle)

■ demand on water resources can be managed by measures to minimise losses, for example through distribution systems and by improved efficiency in water use (managing demand).

■ redistribution of water, such as the transfer of Severn/Anglian water to the Thames Region (new schemes).

The Environment Agency (EA) has to manage surface water management, mineral extractions, sewage capacity, water efficiency and demand, waste disposal, habitat enhancement and restoration, and research and development. The EA is concerned with the overall policies on water resources, water quality and surface water management, which includes flood defence issues. The EA has a duty to conserve and enhance the water environment when carrying out any of its functions, and a further duty to promote conservation and enhancement more widely. Increasingly the way to achieve this is through catchment management plans (CMPs).

The role of catchment management plans

The EA had a nationwide programme for the completion of the first round of Catchment Management Plans (CMPs) by the end of 1998. CMPs allow integrated planning of the water environment. The CMP is a locally based document developed from the discussions of national, regional and local organisations. CMPs are closely connected with local sustainable development plans, *Agenda 21s.*

Agenda 21 gives a high priority to fresh water, reflecting the management crisis facing the world's freshwater resources. By the year 2000 all states should have national action programmes for water management, based on catchments, and efficient water-use programmes.

For the EA *Agenda 21* offers:

■ the opportunity to place greater weight on environmental considerations when assessing planning applications

■ the opportunity to increase community involvement in water issues

■ help to implement CMPs

■ the opportunity for integration of land-use and water related issues, leading to full-scale integrated catchment management planning

■ additional opportunities to protect the water environment

■ increased public awareness of the water environment

■ help to identify appropriate environmental indicators

Managing future demands

Managing the growth in demand will require a combination of methods such as leakage control, selective metering and improvements in water efficiency. Many water efficient appliances are now available such as low water use washing machines, low flush toilet cisterns and water-wise gardening products. There is also likely to be a change in the demand for water. Demand for manufacturing industry is likely to decline since the patterns of manufacturing are changing and companies are becoming more efficient at using water. Future agricultural demands depend mainly on changes in agricultural policy. The growth in tourism and recreation will increase the demand for water. For example, the restoration of disused canals may become a pressure on water resources. There are a number of restoration projects currently being considered in the region.

Recent experience of the promotion of major new water resource schemes indicates that it can take up to 15 to 20 years from starting feasibility studies to commissioning for a new scheme. The planning of schemes required by the year 2017 began in 1997.

Water resource development options in the Thames Region

A number of options have been considered but rejected, at least for the present, on financial and/or environmental costs. These include:

- use of gravel workings for storage
- redevelopment of existing resources
- freshwater storage in the tidal Thames Estuary

- inter-regional transfers from Wales via the River Wye, Northumbria (Kielder Water) and Scotland
- desalinisation of water.

Options which are carried forward for further evaluation include:

- London basin groundwater including artificial recharge
- (riverside) groundwater development opportunities
- re-allocation of under-utilised resources
- re-use of effluents presently discharged into the tidal Thames Estuary
- reservoir storage in south-west Oxfordshire.
- Inter-regional transfer from the River Severn and the Anglian Region.

River restoration – the River Cole

River restoration schemes are becoming more and more common as the advantages of natural rivers and their flood plains are realised.

The aims of the River Restoration Project (RRP) are:

- to recreate natural conditions in damaged river corridors
- to improve understanding of the effects of restoration work on nature conservation value, water quality, recreation and public opinion
- to encourage other groups to restore streams and rivers.

The River Cole is one of three river restoration sites in Europe. The others are the River Skerne in Darlington and the River Brede in Denmark. The aim of the RRP for the River Cole near Swindon is to change the water course, improve water quality and manage the bankside vegetation (Figure 1.49).

Why restore rivers?

Over the last 50 years rivers have been seriously affected by urban and agricultural flood defences, land drainage and flood plain urbanisation. The result has been:

- extensive straightening and deepening of river channels, which has damaged wildlife habitats, reduced the value of fisheries and reduced much of the natural appeal of river landscapes
- a major loss of flood plains and wetlands to intensive agriculture and urbanisation, which has destroyed flood

plain habitats and reduced the ability of floodplains to function as areas of flood control

- rivers are used intensively as transport routes, carriers of waste disposal, for industrial purposes, water abstraction, recreation etc.

There are two main ways in restoring rivers, natural and artificial. Natural ways can take hundreds of years, consequently artificial restoration needs to take place. The benefits are greatest when natural river shapes, flows, and loads are copied.

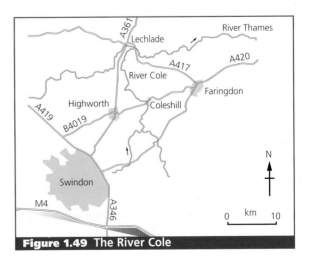

Figure 1.49 The River Cole

Evaluation of the scheme

Benefits of restoration

- greater nature conservation of wetland wildlife in the river and on the flood plain

- increased diversity and numbers of fish

- improved water quality due to increased interception of pollutants by vegetation and natural settling of sediments on flood plain and river bed

- increased flood defence – additional flood storage provided by the enlarged flood plain

- more opportunities for recreation – there is a strong public perception in favour of natural landscapes.

Useful websites include:

River Severn river link at
http://mail.bris.ac.uk/0/6/7Extss/river/.html (check)
US flood insurance data at
http://www.insure.com/home/flood/stats.html

Questions

1 Study the map extract (Figure 1.50) which shows part of the middle course of the Evenlode River.

(a) What river feature is found at 406156?

(b) Describe the pattern of the river (as it appears in plan view) between 386155 (Ashford Bridge) and 410147.

(c) Describe what happens to the stream between 385162 and 386155 (just below Ashford Bridge).

(d) Draw a cross-section from 390150 to 410150 to show variations in relief and drainage.

(e) Draw a sketch map of the map extract and label the following: steep slopes; interlocking spurs; wide flat floored valley; misfit streams; dry valleys (valleys without a stream).

(f) Describe the relief and drainage of the map extract.

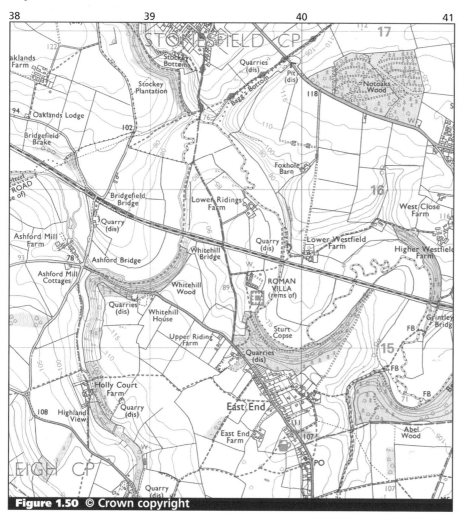

Figure 1.50 © Crown copyright

2 ecosystems *and soils*

The nature of ecosystems: structure and functioning

The structure of an ecosystem

Some ecosystems have distinct boundaries such as a pond, or a tree trunk, whereas others do not such as a desert. Large ecosystems, such as the tropical rainforest or temperate coniferous woodlands (which cover thousands of square kilometres) are known as **biomes**. Where one ecosystem merges into another an **ecotone** is formed. All ecosystems are open systems due to the flow of energy (sunlight) and matter (organisms, abiotic elements) across the ecosystem boundary (Figure 2.1). In addition, there are a number of basic components in all ecosystems. These include:

- **abiotic** elements (the non-living environment) such as water, nutrients and the atmosphere

- **biotic** elements (the living environment) such as:

- **producers** or **autotrophs** – green plants that use energy, water and carbon dioxide to produce carbohydrates through photosynthesis

- **consumers** or **heterotrophs** that obtain their food by eating other plants or animals

- **decomposers** such as bacteria and fungi.

Figure 2.2 shows the biotic and abiotic factors for a deciduous woodland in southern Britain.

The **trophic** structure of an ecosystem refers to the organisation and pattern of feeding in an ecosystem. The **food chain** is the sequence of consumer levels. Normally they are depicted as linear sequences although in reality they are quite complex (Figure 2.3).

In reality some animals, such as otters and foxes, will eat both animals and plants, live as well as decayed matter, and therefore cannot be put into one category. Therefore food webs are found rather than food chains. **Food webs** are very complex. Organisms which feed at the same level on the food chain are said to be at the same **trophic level**.

Energy flow and biomass

Sunlight energy fixed by green plants is passed through the ecosystem from one trophic level to the next (Figure 2.4). As energy is passed through the system it is stored at various layers. The storage of energy – the amount of the living matter present – is referred to as **biomass** or standing crop

Key Definitions 1

Abiotic   Non-living elements in an ecosystem.

Biogeography   The geographical distribution of ecosystems, where and why they are found.

Biome   A large-scale ecosystem.

Biotic   Living matter in an ecosystem.

Dynamic equilibrium   Change in one factor alters the balance of the whole system.

Ecology   The study of organisms in relation to the environment.

Ecosystem   The interrelationship between plants, animals, their biotic and abiotic environment.

Food chain   A linear sequence of feeding from plant to herbivore to carnivore.

Trophic layer   A feeding layer within an ecosystem.

Figure 2.1 Wytham Woods

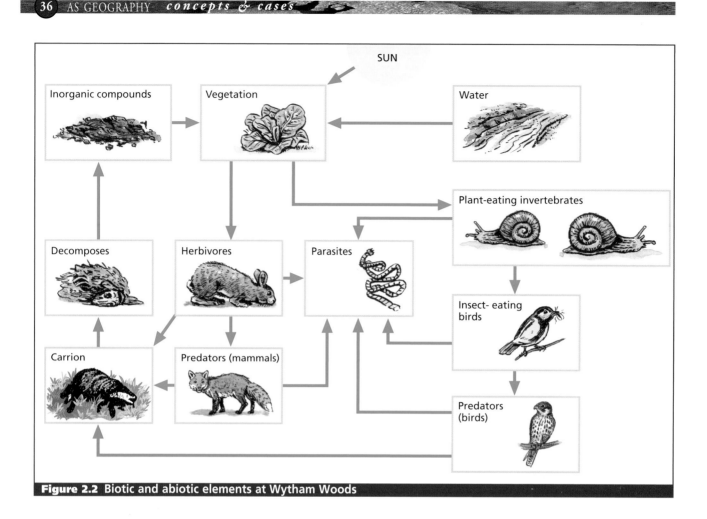

Figure 2.2 Biotic and abiotic elements at Wytham Woods

and is measured in terms of dry weight/calorific value per unit area. Biomass decreases at each trophic layer, producing a trophic pyramid.

The flow of energy decreases with each successive trophic level. Because there is less energy available for use at the later steps on the food chain, less standing crop will be supported at the later trophic levels. The characteristic pyramidal shape of the trophic system (Figure 2.5) is due to:

- the large losses of energy between trophic levels
- large losses within each trophic layer, such as losses due to respiration and mobility.

Productivity

Productivity refers to the rate of production of organic matter. There are two main types:

- **primary productivity** at the autotroph level
- **secondary productivity** at the heterotroph level.

In addition, productivity can be divided into:

- **gross productivity** – the total amount of organic matter produced
- **net productivity which is the amount left after respiration.**

Primary and gross productivity depend on light intensity and duration, and the efficiency of photosynthesis. The potential for primary productivity is greatest at

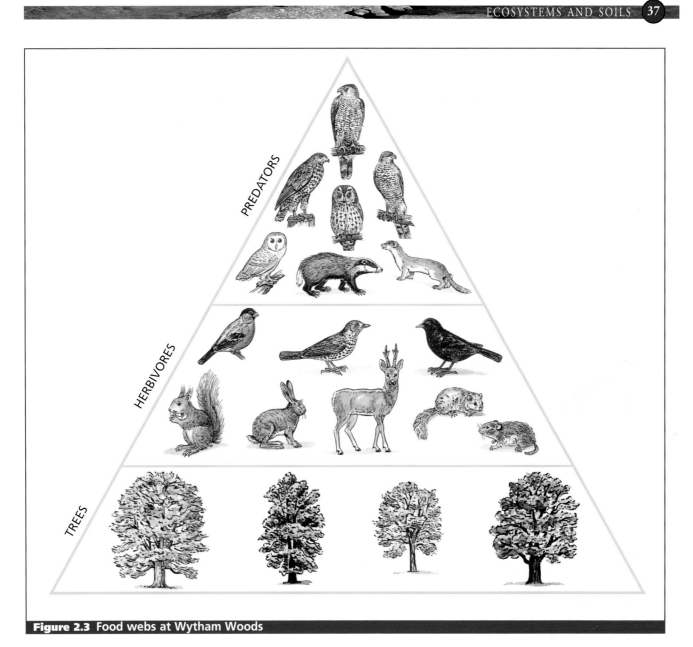

Figure 2.3 Food webs at Wytham Woods

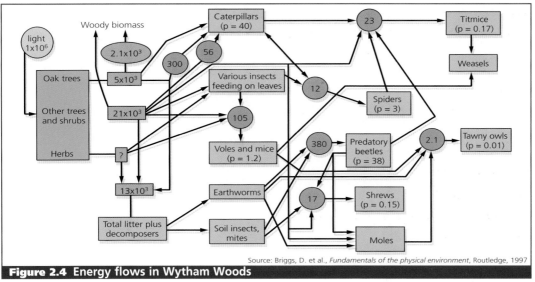

Source: Briggs, D. et al., *Fundamentals of the physical environment*, Routledge, 1997

Figure 2.4 Energy flows in Wytham Woods

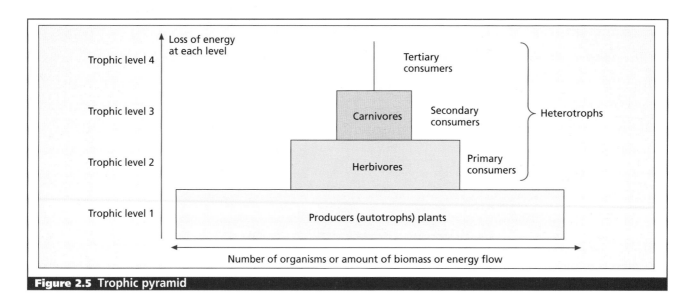

Figure 2.5 Trophic pyramid

the equator (Figure 2.6). By contrast, the efficiency of photosynthesis depends on many factors such as temperature, age of plants and the availability of nutrients.

Of all light energy reaching the vegetation surface, only 1–5% is trapped as food energy (enough to maintain life). Ecological efficiency is the efficiency of transfer of energy from one trophic level to another. For example, the transfer of energy from autotrophs to heterotrophs is only around 10% efficient. Natural systems are more productive and more efficient than artificial ones.

Nutrient cycles

The flow of energy is one way (light energy to heat energy, stored and then lost) whereas nutrients are circulated and reused frequently. The efficiency of nutrient cycling varies between ecosystems (Figure 2.7). Oxygen, carbon, hydrogen and nitrogen are needed in large quantities (**macronutrients**). All others are **trace elements** or **micronutrients** needed only in small doses such as magnesium, sulphur, or phosphorus. Nutrients are taken in by plants and built into new organic matter. When animals eat the plants they take up the nutrients, and eventually return them to the soil when they die and the carcass is broken down by the decomposers.

Phosphorus cycle

The phosphorus cycle is an easily disrupted sedimentary cycle (Figure 2.8). Phosphorus is relatively rare but essential for growth. The source of the phosphorus is the weathering of phosphate rich rocks. This releases phosphorus into the soil, from where it is absorbed by plants and passed up the food chain. It is returned by decomposition (either in the soil or in runoff to the sea). If it goes to the sea it is incorporated into

Ecosystem type	Mean	Mean
Tropical rainforest	2200	45
Temperate evergreen forest	1300	35
Temperate deciduous forest	1200	30
Savanna	900	4
Temperate grassland	600	1.6
Tundra and alpine	1400	0.6
Desert and semi-desert scrub	90	0.7

Figure 2.6 Net productivity of main ecosystem types

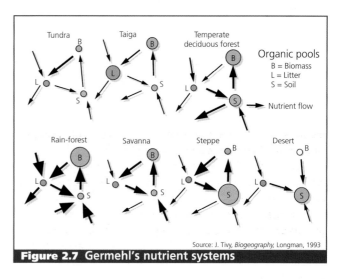

Source: J. Tivy, *Biogeography*, Longman, 1993

Figure 2.7 Germehl's nutrient systems

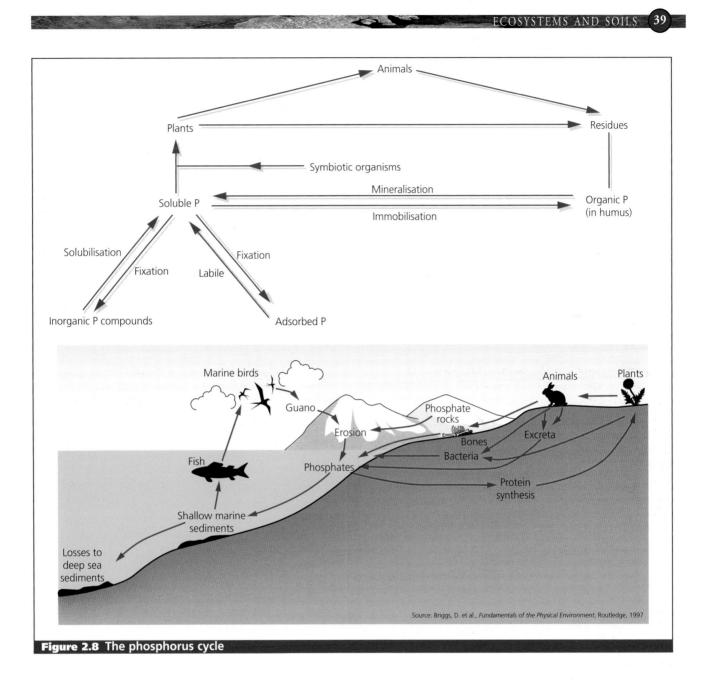

Figure 2.8 The phosphorus cycle

marine sediments and lost from the exchange pool – except where upwelling currents may allow phosphorus to be returned to the land in the form of guano. This, however, is very localised, such as in Peru where it accounts for about 3% of that which is lost from the land.

Phosphate rocks are easily depleted by the use of phosphate for agricultural fertilisers. Phosphate is taken from the rocks for the production of fertiliser, but is rapidly lost because it is easily leached from the soil. This could lead to serious deficiencies in the future.

Nitrogen cycle

The nitrogen cycle is a gaseous cycle, and is also easily disrupted (Figure 2.9). The atmosphere is the source of the gas. Nitrates are absorbed by plants and pass through the food chain. Ultimately, it is released as ammonia when organic matter is decomposed.

Atmospheric nitrogen cannot be used directly by most plants. Thus it needs to be in the form of chemical nitrates first. The conversion of gaseous nitrogen to nitrates occurs by:

■ **electrical action during thunderstorms, or**

■ **conversion by electrical-fixing organisms (bacteria, algae and fungi) operating with leguminous plants. These are extremely important for maintaining soil fertility.**

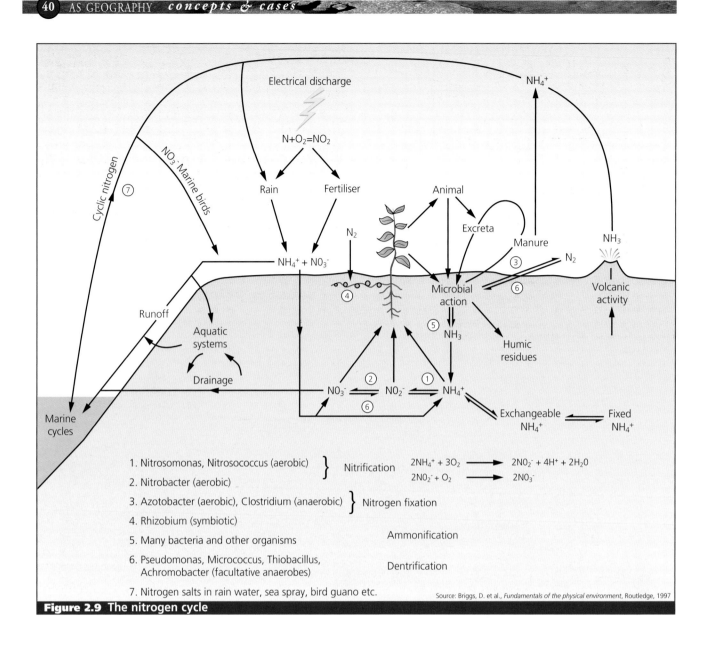

Figure 2.9 The nitrogen cycle

Within the figure:

Electrical discharge

$N + O_2 = NO_2$

Cyclic nitrogen

NO_3^- Marine birds

⑦

Rain   Fertiliser

N_2

Animal

Excreta

NH_4^+

Manure N_2

③

NH_3

⑥

Volcanic activity

$NH_4^+ + NO_3^-$

Runoff

④

Microbial action

⑤ NH_3

Humic residues

Aquatic systems

Drainage

② ① NH_4^+

NO_3^- NO_2^- NH_4^+

Marine cycles

⑥

Exchangeable NH_4^+ Fixed NH_4^+

1. Nitrosomonas, Nitrosococcus (aerobic) } Nitrification $2NH_4^+ + 3O_2 \longrightarrow 2NO_2^- + 4H^+ + 2H_2O$
2. Nitrobacter (aerobic) $2NO_2^- + O_2 \longrightarrow 2NO_3^-$
3. Azotobacter (aerobic), Clostridium (anaerobic) } Nitrogen fixation
4. Rhizobium (symbiotic)
5. Many bacteria and other organisms   Ammonification
6. Pseudomonas, Micrococcus, Thiobacillus, Achromobacter (facultative anaerobes)   Dentrification
7. Nitrogen salts in rain water, sea spray, bird guano etc.

Source: Briggs, D. et al., *Fundamentals of the physical environment*, Routledge, 1997

Succession

Succession refers to the spatial and temporal changes in plant communities as they move towards a seral climax (Figure 2.10). Each **sere** or stage is an association or group of species, which alters the micro-environment and allows another group of species to dominate (Figure 2.11). The **climax community** is the group of species that are at a dynamic equilibrium with the prevailing environmental conditions. In the UK, under natural conditions, this would be oak woodland. On a global scale, climate is the most important factor in determining large ecosystems or biomes such as tropical rainforest or temperate woodland (Figure 2.12). In some areas, however, vegetation distribution may be determined by soils rather than by climate. This is known as **edaphic** control. For example, in savanna areas forests are found on clay soils, whereas grasslands occupy sandy soils.

A **plagioclimax** refers to a plant community permanently influenced by human activity. It is prevented from reaching climatic climax by burning, grazing and so

Key Definitions

Succession Spatial and temporal changes in plant communities as they move towards being the dominant species in an environment.

Climatic climax vegetation The group of species that are at a dynamic equilibrium with the prevailing climatic conditions.

Sere A group of plant species that represent a stage in the development towards succession.

Plagioclimax A succession that has been influenced by human activity.

Arresting factor Any factor, natural or human, that prevents normal succession from occurring.

Secondary succession Plant succession that occurs after the impact of an arresting factor.

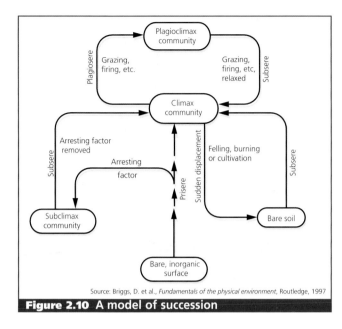

Source: Briggs, D. et al., *Fundamentals of the physical environment*, Routledge, 1997

Figure 2.10 A model of succession

Figure 2.11 Succession

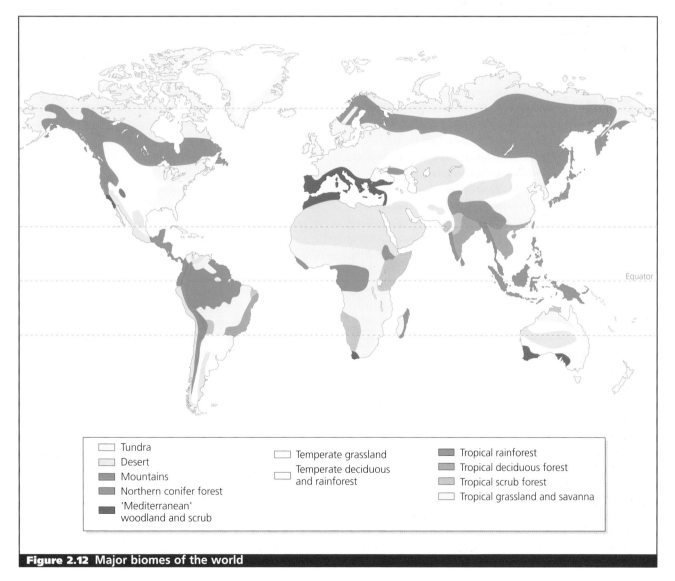

Tundra
Desert
Mountains
Northern conifer forest
'Mediterranean' woodland and scrub

Temperate grassland
Temperate deciduous and rainforest

Tropical rainforest
Tropical deciduous forest
Tropical scrub forest
Tropical grassland and savanna

Equator

Figure 2.12 Major biomes of the world

on. Britain's heathlands are a good example: deforestation, burning and grazing have replaced the original oak woodland. Vegetation succession is clearly evident in salt marshes and sand dunes, as shown on pages 133–136.

Soil characteristics and processes

Soil water

The amount of water in a soil depends on soil texture, organic matter content and density (Figure 2.13). During winter, soils reach **field capacity**. With free drainage, any excess water drains through the soil. Impeded drainage results in waterlogging. During spring, a **soil moisture deficit** (see pages 2–5) develops, as water is used up by plants and some is evaporated.

Soil texture and structure

Soil texture refers to the size of soil particle (Figure 2.14) while soil structure refers to the shape of the particles (Figure 2.15). Silt soils are especially prone to compaction if ploughed when wet. In years of drought the shrinkage of some clays may cause structural damage. Increasingly dry years may result in greater damage. Clay soils are found predominantly in the south and east of England.

Soil fertility

Soft calcareous (lime-rich) rocks are generally quite fertile, as the rate of weathering is sufficient to make up for the loss of nutrients to plants or leaching. These rocks, such as chalk, are often referred to as **base-rich** rocks.

Clay and humus are the main areas of chemical exchange. Clay has a vast surface area in relation to its weight. Nutrients are recycled between the clay and the plants in a process known as **base exchange**. Bases are positively charged ions, such as calcium, magnesium and potassium. By contrast plants release hydrogen ions, which tends to make the soil more acidic over time. Increasing acidity in soils is thought to be a contributory factor in forest decline in central Europe, although there is little clear evidence for this in the UK. Nutrients may be returned to the soil through litter decay, release through weathering or else added as fertilisers.

Soil colour

In some cases soil colour may reflect soil formation or parent material. For example, black soils indicate the presence of humus, and red soils indicate iron. A white crust in a semi-arid area indicates a saline crust, whereas in a humid environment, a white layer indicates heavy leaching. Grey-blue mottling (speckling) points to reduced iron compounds, poor drainage and therefore **gleying**. Finally, the C horizon, the lowest one, often takes on the colour of the parent rock, as it is largely weathered bedrock.

Soil horizons

Soil horizons are layers within a soil, and they vary in terms of texture, structure, colour, pH, and mineral content. The top layer of vegetation is referred to as the Organic (O) horizon and consists of litter in various stages of decay. The mixed mineral-organic layer (A horizon) is generally a dark colour owing to the presence of organic matter. An Ap horizon is one that has been mixed by ploughing, while an Ag horizon is one that is waterlogged.

Questions

1 Figure 2.3 shows a food web. Identify
   (a) the primary producers
   (b) the herbivores
   (c) the carnivores on the diagram.
2 Describe the energy flow of a deciduous woodland, as shown in Figure 2.4.
3 Figure 2.6 shows global variations in NPP.
   (a) Define NPP.
   (b) Identify the ecosystem with (i) the highest and (ii) the lowest NPP/unit area.
   (c) Give reasons to explain why these two ecosystems should differ so much in terms of NPP/unit area.
   (d) Which ecosystem has the lowest mean biomass? Explain two contrasting reasons why the biomass is so low in this ecosystem.

Key Definitions ③

Azonal soils Soils which have not had enough time to properly mature.

Gleyed soil A waterlogged soil.

Horizon A layer in a soil which differs from other layers in terms of structure, texture, pH, colour.

Humus Partially decomposed organic matter derived from the decay of dead plants and animals in soils.

Intrazonal soils A classification stating that within a climatic zone soils vary with rock types.

Parent material The rock and mineral regolith from which soil develops.

Peat An unconsolidated deposit of plant remains with a carbon content of about 60%.

Regolith The irregular cover of loose rock debris that covers the earth.

Rendzina A soil formed on chalk or limestone.

Soil fertility The ability of a soil to provide the nutrients needed for plant growth.

Soil horizons The distinguishable layers within a soil.

Soil profile The succession of soil horizons between the surface and the underlying parent material.

Zonal soils A classification which states that on a global scale soils are determined by climate.

Figure 2.13 Storage capacity of soils
(cm water/30 cm soil depth)

Soil texture	Field capacity	Wilting point	Available water
Sandy loan	5.6	2.8	2.8
Loam	8.4	4.3	4.1
Clay loam	9.9	5.3	4.6
Heavy clay	11.9	6.3	5.6

Source: Briggs, D. et al., *Fundamentals of the physical environment*, Routledge, 1997

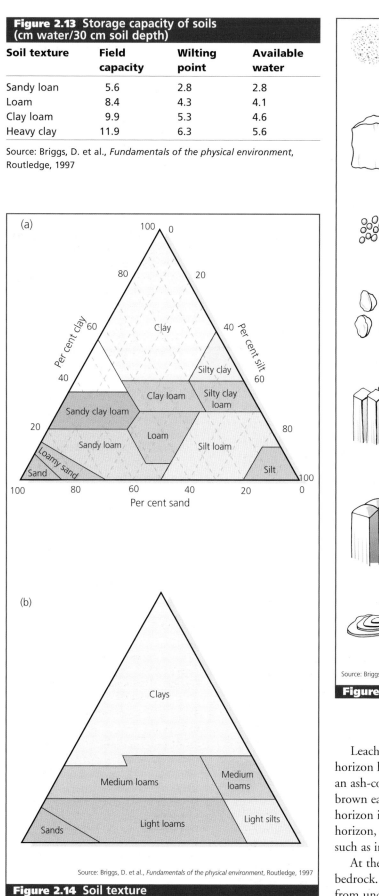

Figure 2.14 Soil texture

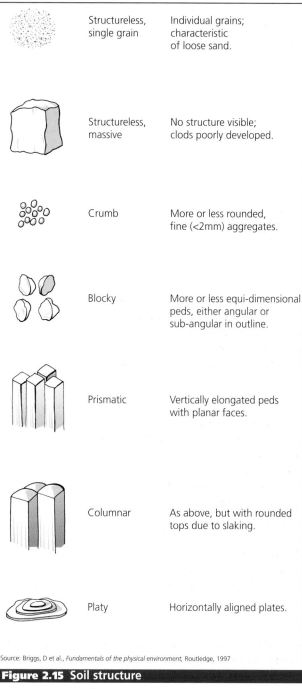

	Structureless, single grain	Individual grains; characteristic of loose sand.
	Structureless, massive	No structure visible; clods poorly developed.
	Crumb	More or less rounded, fine (<2mm) aggregates.
	Blocky	More or less equi-dimensional peds, either angular or sub-angular in outline.
	Prismatic	Vertically elongated peds with planar faces.
	Columnar	As above, but with rounded tops due to slaking.
	Platy	Horizontally aligned plates.

Source: Briggs, D et al., *Fundamentals of the physical environment*, Routledge, 1997

Figure 2.15 Soil structure

Leaching removes material from a horizon and makes the horizon lighter in colour. In a podzol, leaching is intense, and an ash-coloured Ea horizon is formed. By contrast, in a brown earth, leaching is less intense, and a light brown Eb horizon is found. The B horizon is the deposited or illuvial horizon, and contains material removed from the E horizon, such as iron (fe) humus (h) and clay (t).

At the base of the horizon is the parent material or bedrock. Sometimes labels are given to distinguish rock (r) from unconsolidated loose deposits (u).

Figure 2.16 Soil horizons

O	Organic horizon
l	Undecomposed litter
f	partly decomposed (fermenting) litter
h	well decomposed humus

A	Mixed mineral-organic horizon
h	humus
p	ploughed, as in a field or a garden
g	gleyed or waterlogged

E	Eluvial or leached horizon
a	strongly leached, ash coloured horizon, as in a podzol
b	weakly leached, light brown horizon, as in a brown earth

B	Illuvial or deposited horizon
fe	iron deposited
t	clay deposited
h	humus deposited

C	Bedrock or parent material

Soil formation and development

Factors affecting soil formation

There are many factors that affect soil development. These have been divided into:

■ **passive** factors such as soil material, topography and time, and

■ **active** factors such as climate, and biological factors.

Soils and climate

Climate influences soils in two main ways:

■ temperature affects the rate of chemical and biological reactions

■ precipitation effectiveness determines whether water moves up or down through the soil.

The rate of chemical weathering increases two to three times for every increase in temperature of 10°C. In cool climates bacterial action is relatively slow and a thick layer of decomposing vegetation often covers the ground. By contrast, in the humid tropics bacteria thrive and dead organic matter is broken down rapidly.

Precipitation effectiveness is the balance between precipitation and potential evapotranspiration. If precipitation exceeds potential evapotranspiration, leaching occurs. Any soil in which there is a net downward movement through the soil is known as a **pedalfer**. By contrast, if precipitation is less than potential evapotranspiration, there is a soil water deficit. Water is drawn to the surface, bringing with it calcium carbonate. Such soils are known as **pedocals** and are typical of arid and semi-arid environments. In the USA east of the Rockies, pedalfers dominate the east, pedocals the west.

Key Definitions (4)

Calcification A concentration of calcium in the soil as a result of ineffective leaching, in areas of low rainfall, causing the accumulation of calcium in the soil.

Cheluviation The removal of the iron and aluminium sesquioxides under the influence of chelating agents.

Eluviation The removal of material down a soil through solution and suspension

Humification, degradation and mineralisation The process whereby organic matter is broken down and the nutrients are returned to the soil. The breakdown releases organic acids, chelating agents, which break down clay to silica, soluble iron and aluminium.

Illuviation The redeposition of material in the lower horizons.

Leaching The removal of soluble material in solution.

Lessivage The removal in suspension of fine particles of clay.

Podzolisation An intense form of leaching involving the removal of sesquioxides under acidic conditions.

Salinisation The upward movement of soluble salts by capillary action, and their deposition in the surface horizons, forming a toxic crust.

Figure 2.17 The relationship between soil type and climate regions in the British Isles

Climatic region	Mean annual temperature (°C)	Mean annual rainfall (mm)	Soils
Warm/dry	Greater than 8.3	Less than 1000	Leached brown soils
Cold/dry	4.0–8.3	Less than 1000	Semi-podzols, podzols
Warm/wet	Over 8.3	Over 1000	Acid brown soils
Cold/wet	4.0–8.3	Over 1000	Peaty podzols, blanket peat
Very cold/wet	Less than 4.0	Over 1000	Alpine humus soils

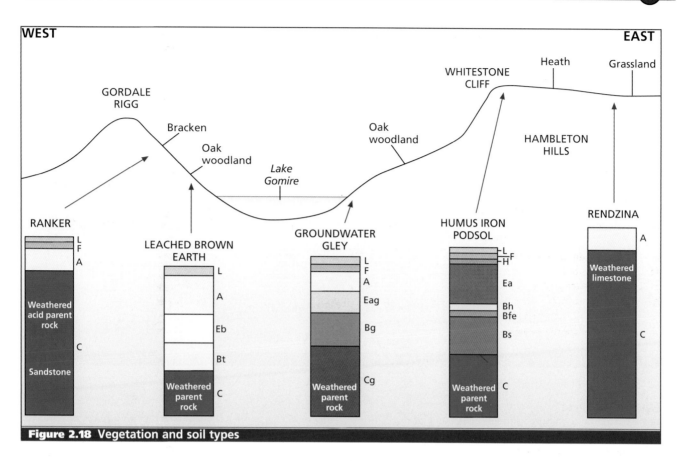

Figure 2.18 Vegetation and soil types

Organic matter

Organic matter is a basic component of soil, although the influences of biotic factors range from microscopic creatures to man. Some influences may be visible, such as interception of precipitation by vegetation. Others are less visible, such as the release of humic acids by decaying vegetation. There is also a relationship between the type of vegetation and the type of soil (Figure 2.18). This is partly because certain types of vegetation require specific nutrients, such as grass in areas rich in calcium and magnesium.

Animals too have an effect on soils. In the top 30 cm of 1 ha of soil there are on average 25 tonnes of soil organisms, that is 10 tonnes of bacteria, 10 tonnes of fungi, 4 tonnes of earthworms and 1 tonne of other soil organisms such as spring tails, mites, isopods, spiders, snails, mice, etc. Earthworms alone can represent 50–70% of the total weight of animals in arable soils. Earthworms can ingest 18–40 tonnes of soil each day in 1 ha and, passed on to the surface, this represents a layer up to 5 mm deep. Man as an animal has obvious effects, ranging from liming, fertiliser application and mulching to mining, deforestation, agricultural practices and extraction for gardening purposes.

Topography

Soils can vary in their nature over quite small distances as any local soil survey will show. This reflects the range of climatic regions. On a very small scale, however, such as within a field, soils are largely affected by drainage and gradient, and position on a slope.

Of these, slope angle is of great importance. Susceptibility to soil erosion increases with gradient. Steeper slopes are associated with thinner soils. Soils on hillsides tend to be better drained whereas those in valley bottoms are at risk of gleying. Aspect, the direction in which a slope faces, has an important bearing on soil formation as it affects the local climate or micro-climate.

The term 'soil catena' refers to the sequence of different soils that varies with relief and drainage, though derived from the same bedrock (Figure 2.19). Such a sequence can be found when following a transect from a mountain or hill top to the valley bottom, reflecting changes in micro-climate, drainage and the position of the water table.

Time

Time is not a causative factor. It does not cause soils to change but allows processes to operate to a greater extent, therefore allowing soils to evolve. The amount of time required for soil formation varies from soil to soil. Coarse sandstones develop soils more quickly than granites or basalts and on glacial outwash a few hundred years may be enough for a soil to evolve. Thin soils are not necessarily young soils,

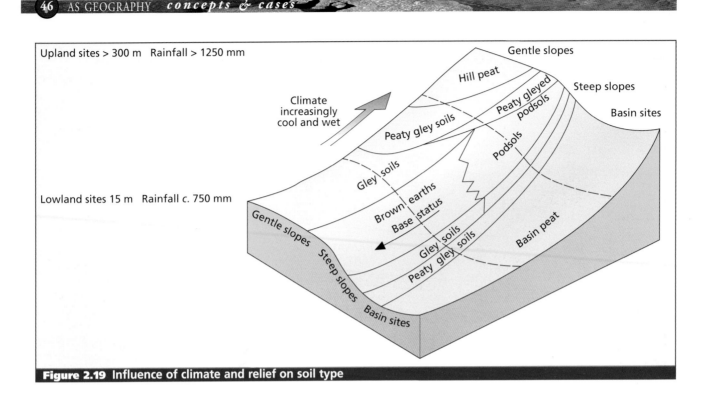

Figure 2.19 Influence of climate and relief on soil type

nor are deep ones 'mature'. Phases of erosion and deposition keep some soils always in a state of evolution. Most mid-latitude soils are referred to as polycyclic, that is they undergo frequent changes as the major soil forming processes change in relation to changing inputs.

Soil classifications

The **zonal** classification states that soils are determined by climatic factor. On a global scale a map of world climates and a map of world soils bears a strong resemblance (Figure 2.20). Indeed, on a general scale this is the case, with brown earths in temperate climates, podzols in cool temperate climates and chernozems in continental climates.

The **intrazonal** classification states that within any climate belt soils vary with respect to local factors such as geology. An excellent example is the case of Purbeck where the geological sequence of limestone, clay, chalk and outwash sands and gravels corresponds to rendzinas, brown earths or gleys, rendzinas and podzols respectively.

The **azonal** classification states that many soils are too young or immature and that there has not been sufficient time for them to develop the characteristics that would relate them to either bedrock or climate.

Soil forming processes

Soils must be considered as open systems in a state of dynamic equilibrium, varying constantly as the factors and processes that influence them alter. All soils have processes in common but they vary in terms of the rates and types of processes. The weathering of bedrock gives the soil its C horizon, as well as its initial bases and nutrients (fertility), structure and texture (drainage). Water movement with soluble chemicals and the mixing of organic matter redistribute materials around the soil.

The range of **soil forming processes** is very large and includes:

- **chemical weathering** such as oxidation and reduction

- **physical weathering** such as hydration and freeze/thaw

- **leaching**, i.e. the downward movement of soluble chemical salts by water

- the **concentration of soluble salts** in the upper layers of the soil; this is the result of water being drawn through the soil by evaporation at the surface and the dissolved minerals in it being deposited

- the incorporation of **decomposing organic matter** from plants and animals

- the downward movement (**translocation**) of solid particles carried by water, especially clay

- **gleying**, i.e. the reduction of red or brown ferric compounds to blue or grey ferrous compounds and waterlogged anaerobic conditions

- **biological disturbance** of the soil by root penetration, the movement of soil animals, especially earthworms, and human cultivation.

The differences between soils are the results of differences in

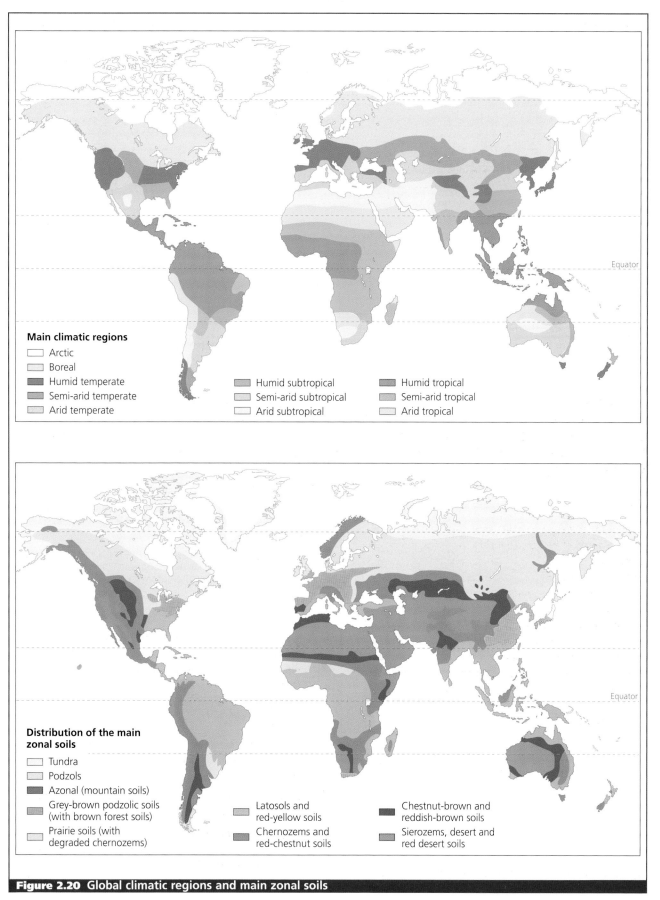

Main climatic regions

- Arctic
- Boreal
- Humid temperate
- Semi-arid temperate
- Arid temperate
- Humid subtropical
- Semi-arid subtropical
- Arid subtropical
- Humid tropical
- Semi-arid tropical
- Arid tropical

Distribution of the main zonal soils

- Tundra
- Podzols
- Azonal (mountain soils)
- Grey-brown podzolic soils (with brown forest soils)
- Prairie soils (with degraded chernozems)
- Latosols and red-yellow soils
- Chernozems and red-chestnut soils
- Chestnut-brown and reddish-brown soils
- Sierozems, desert and red desert soils

Figure 2.20 Global climatic regions and main zonal soils

the relative importance of:

- the movement of water up or down through the soil
- the movement of soluble chemicals, e.g. iron or aluminium up or down through the soil
- the mixing of organic matter in the soil.

Podzolisation

A podzol is a leached soil, with very defined horizons, formed under acidic conditions. Podzolisation is a process widespread on acidic soils. The upper horizons of the soil become rich in silica, taking a characteristic ash-grey colour whereas lower horizons become rich in iron and aluminium. There may even be an iron pan, a thin tough horizon of iron oxides. The cause of this movement and concentration of iron and aluminium is the leaching of certain humic acids, called chelating agents. These are richest in heath plants and coniferous vegetation, and least frequent among grasses and deciduous vegetation. Podzols are thus generally associated with coniferous or heathland vegetation.

Chelation is the process whereby the metallic cations (positively charged irons of iron, calcium, aluminium and sodium for example) are taken into solution by acidic soil water. The water percolates downwards, leaching these minerals and depositing them further down in the soil profile (Figure 2.21). Thus, the typical profile of a podzolised soil has a thick dark layer of organic matter, the pale grey leached or eluviated horizon in which the minerals have been removed, and layers, often coloured, where the minerals have been deposited in an illuviated horizon.

Latosolisation

Latosolisation is the tropical equivalent of podzolisation. However it produces very different results because the soil water is not acidic. Under sustained warm, wet conditions bacterial activity decomposes vegetation so quickly that chelation is rare. For example, in the tropical forests of Costa Rica dead leaves decay in about a month. In the relative absence of these acids, iron remain insoluble and so accumulates in the soil as red clays (hence they are sometimes known as tropical red soils) while silica is washed down through leaching.

Calcification and salinisation

In arid and semi-arid environments potential evapotranspiration exceeds precipitation so the movement of soil solution is upwards through the soil (Figure 2.22). Calcium carbonates and other solutes remain in the soil as

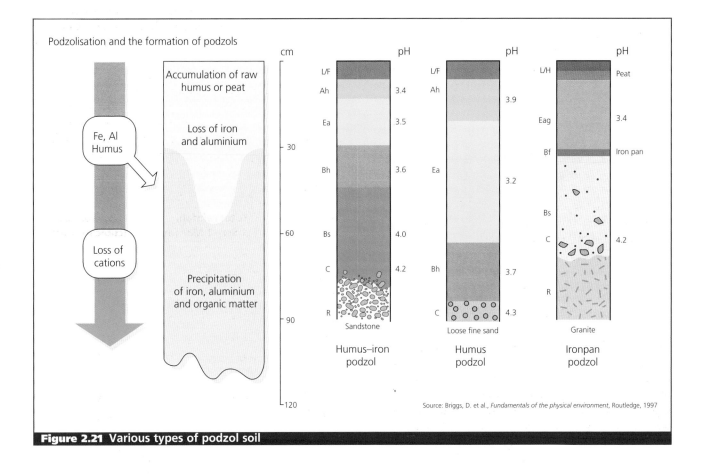

Source: Briggs, D. et al., *Fundamentals of the physical environment*, Routledge, 1997

Figure 2.21 Various types of podzol soil

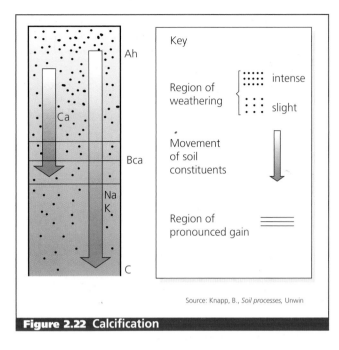

Source: Knapp, B., *Soil processes*, Unwin

Figure 2.22 Calcification

leaching is ineffective. This process is known as calcification. In grasslands, calcification is enhanced because grasses require calcium, draw it up from the lower layers and return it to the upper layers when they die down.

In extreme cases where potential evapotranspiration is intense, sodium or calcium may form a crust on the surface. This may be toxic to plant growth. Excessive sodium concentrations may occur due to the capillary rise of water from a water table that is saline and close to the surface. Such a process is known as salinisation or alkalisation.

Organic changes

Plant litter is decomposed (humified) into a dark shapeless mass. It is also degraded gradually by fungi, algae, small insects, bacteria and worms. Under very wet conditions humification forms peat. Over a long timescale humus decomposes due to mineralisation, which releases nitrogenous compounds and makes them available to plants again.

Gleying

Gleying means waterlogged. Wet or waterlogged areas produce anaerobic (oxygen deficient) conditions, which favour the growth of specialised bacteria. These bacteria reduce ferric iron to the soluble ferrous iron (the process is known simply as reduction). Gley soils are characterised by a thick compact layer of sticky, structureless clay. Blue-grey blotches indicate the presence of reduced iron (red or brown suggests oxidised iron and is associated with periodic drying out of the soil). Gley soils are generally found within the water table or in areas of poor drainage.

Human activities and soils

The main threats to soils include:

- soil erosion
- nitrate and phosphates in water resources
- accumulation of pollutants
- organic matter loss and deteriorating soil structure and associated problems
- organic contaminants in water resources
- acidification
- increasing urban areas, motorways and road building, and industrial development.

Soil erosion

About a third of arable land in England and Wales (20 500 sq km) has been identified as being at risk from wind or water erosion, yet a major difficulty has been measuring the rate at which it is occurring.

Research into the nature of erosion in one field at Albourne Farm, 10 km north-west of Brighton, concluded that soil erosion was largely the result of farm management rather than purely physical factors. The field is over 330 metres long, with a gradient of about 3° and a small stream at the base.

In 1979 the field was planted with strawberries and within a year extensive erosion had occurred. Loss of soil occurred over the whole field, and the furrows between ridges of strawberries had been filled in. Channels had developed at the head of the field, and at the bottom of the field the crop has been buried by deposits up to 10 centimetres deep.

Figure 2.23 Rainfall and soil erosion on a monitored site in the eastern South Downs, England, 1982-91

Year	Total rainfall, 1 Sep–1 Mar (mm)	Total soil loss (cubic metres)
1982–83	724	1816
1983–4	560	27
1984–5	580	182
1985–6	453	541
1986–7	503	211
1987–8	739	13 529
1988–9	324	2
1989–90	621	940
1990–91	469	1527
1991–92	298	112

A number of factors increased the risk of erosion:

- **field size** – prior to 1962 this area had consisted of five fields, but by 1978 it was only one; the length of unbroken slope increased, therefore, from 90 metres to 335 metres
- the **type of crop** meant that large areas and channels of soil were left exposed throughout the winter
- the land was worked **downslope** rather than around the contours
- the soils are highly **susceptible to erosion**, having a low clay and silt content (2–3%), and with a tendency to form a crust or cap
- the farm manager suggested that a prime cause was the large number of **autumn storms** in 1979 – however, although August and December were particularly wet, the frequency of intense rainfall events (days with more than 7.7 millimetres) was below average.

The rates of erosion recorded at Albourne Farm were among the highest recorded in the United Kingdom. Over 200m^{-3} of soil were removed from just 166 ha between July 1979 and March 1980. The surface lowering was about 12 mm with an annual removal rate of 241 tonnes per ha. To put this into perspective, the formation of 2.5 cm of soil may take anything up to 2500 years.

Nitrates in soil

One of the most important impacts of human activity on soils is the increased use of nitrate-rich fertilisers to increase plant productivity. There is considerable concern about the loss of nitrate from soils leading to increased concentrations of nitrates in drainage water. Nitrate increases the productivity of plants in the water systems, especially so when combined with phosphate derived from soils (by erosion) or from domestic sources (in sewage). This could lead to algal blooms whereby algae cover a water surface blocking off light to the lower parts of the water and leading to a loss of plant life at depth.

High levels of nitrate can cause problems to human health, in particular blue baby syndrome and stomach cancer. Nitrate is a greater problem than phosphate because it is more mobile in soil water. Up to a third of nitrates in fertilisers may be lost from the soil by leaching.

Managing soil erosion

Methods of combating soil erosion include simple terraces, brushwood holding fences and gabions (stone filled wire cages), and, probably the most effective, good land use management. This includes non-cultivation of steeper and more vulnerable slopes.

Questions

1. Identify the following horizons: Eb; Bt; Cca; Bg.
2. What abbreviations would you give for
   (a) a deposited horizon containing humus
   (b) a strongly leached horizon
   (c) a waterlogged organic horizon?
3. Study Figure 2.20. Describe how soil types vary with climate.
4. Explain **two** ways in which geology influences soil formation.
5. Define the following terms: leaching; podzolisation; calcification.
6. Study Figure 2.21. Describe the process of podzolisation. How and why does it differ from (i) latosolisation and (ii) calcification?
7. Figure 2.23 shows rainfall and soil erosion for part of the South Downs.
   (a) Using semi-logarithmic graph paper, draw a graph to show the relationship between rainfall and soil loss. (NB Rainfall is the independent variable and soil loss the dependent variable.)
   (b) Describe the results that the graphs show.
   (c) State one statistical technique that could be used to see if there is a relationship between the two variables.
   (d) Using the technique identified in Question **(c)** work out the statistical relationship between the two variables.
   (e) Suggest reasons for the relationship that you have shown.

Dynamic characteristics of ecosystems: deciduous woodlands and tropical rainforests
Case study: Wytham Wood

Wytham Woods, near Oxford, is one of the most intensively studied areas of woodland in the world. The woods consist of 400 ha of protected woodland located in a meander in the River Thames. It covers two low hills, formed of limestone, which rise about 100 m above the river (Figure 2.24).

The majority of the woodland is oak/ash forest with open areas of bracken. Nevertheless, in some areas large stands of sycamore are dominant, and there a few areas where conifers dominate. The University of Oxford Forestry Department extended the areas of woodland cover by planting up several areas of scrubland, mostly in the late 1940s and 1950s.

The main aim of current management is to create a closed canopy forest of native hardwoods, such as oak and beech, planting these in favour of the alien sycamore. At the same time, areas of grassland, small areas of conifer, and areas where succession is occurring are being maintained to provide a wide range of habitats for teaching and research.

Soils and climate

The typical soil of the woodlands is brown earth (Figure 2.25). This occurs in warm temperate climate zones, such as the British Isles, where rainfall exceeds potential evapotranspiration, thereby allowing the downward movement of particles through the soil. Around Wytham precipitation is about 650 mm, and potential evapotranspiration around 550 mm. Summer temperatures reach about 18°C, and winter temperatures 5–7°C. The main type of natural vegetation is temperate broad-leafed deciduous forest.

At the surface the annual shedding of leaves conserves nutrients and the humus that develops is a mildly acidic one called mull, pH 5.5–6.5. The upper horizons are dark brown. The B horizon is light brown owing to the removal of clay, humus, iron and aluminium by leaching. This is not as marked as in the podzol and so the horizon is not as leached. Instead it is light brown. Soil fauna flourish, consequently horizons are mixed. The soil is relatively fertile and may be up to 1m thick.

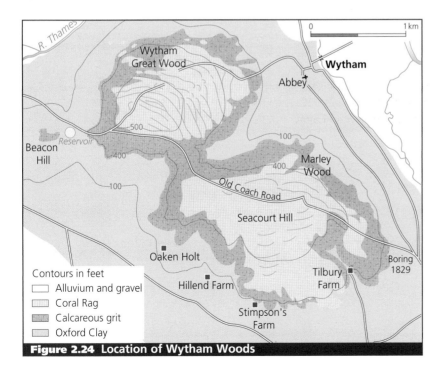

Figure 2.24 Location of Wytham Woods

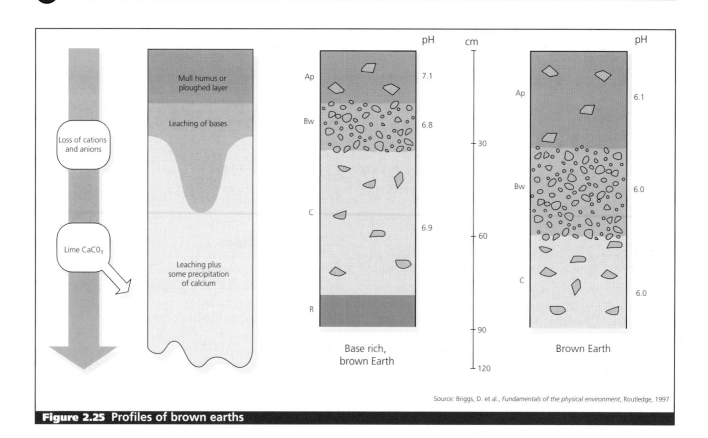

Source: Briggs, D. et al., *Fundamentals of the physical environment*, Routledge, 1997

Figure 2.25 Profiles of brown earths

The flora and fauna

Over 3800 species of animals have been recorded at Wytham. Over 30 species of butterfly and over 950 species of moth have been recorded. In addition, the flora is extensive and over 600 species have been identified. On account of its very diverse ecology, Wytham has been designated a Site of Special Scientific Interest by the Nature Conservancy Council.

Predators and prey: studies of ecosystem dynamics

There have been a number of long-term ecological studies in Wytham Wood. Many of these have examined the energy flow of the ecosystem (Figure 2.4 on page 37) as well as productivity at successive trophic levels (Figure 2.5 on page 38). One investigated the relationship between bank voles, wood mice and tawny owls. In years of high density of mice and voles, the owls breed well, most laying three (or even four) eggs. By contrast, when the rodents were scarce, the owls laid fewer eggs. In 1958, for example, when both mice and voles fell to their lowest densities for 20 years, not a single one of the 31 pairs of owls in the wood attempted to breed.

Another study investigated the relationship between great tits, blue tits and caterpillars. It was thought that by eating many caterpillars the tits might reduce leaf damage and hence increase the growth rate of the trees. If so, increasing the density of tits, by putting up nesting boxes, might be beneficial to foresters. In years of high caterpillar numbers, a large individual oak tree might have half a million caterpillars. However, each pair of tits, which has several oaks trees in its territory, eat about 10 000 caterpillars during the whole of the nesting period, an insignificant proportion of the whole. Thus breeding success is markedly affected by the abundance of their food supply. By contrast, neither the tits nor the owls had much effect on the numbers of their prey.

Another study of a predator and its prey carried out in Wytham was on the effect of sparrowhawks on the tit populations. The sparrowhawks hatch their young at about the time when the young tits are leaving the nest. The hawks feed heavily upon them. However, only those tits which have recently left the nest are easily caught by the hawks. Hawks which hatch their young later do not seem to be able to catch enough tits, nor can they find a good alternative prey.

Sparrowhawks became very rare in the late 1950s as a result of the widespread use of chemicals such as dieldrin, and they ceased to breed in Wytham by 1960. However, after restrictions on the use of dieldrin, the sparrowhawks returned to the woods and some 6 to 8 pairs have been breeding in the wood since the mid 1970s.

Large mammals

Wytham has few large mammals, but there are fallow deer, fox and badger. This is largely on account of the relatively small size of the area compared with the area required by large mammals. The average territory size of the badgers is about 90 ha, and the largest ones usually extend into farmland and their edges. For foxes, the territories are quite small, about 20 ha. Both species rely for a large part of their diet on earthworms. A full-grown badger takes about 200 worms per night. Yet, during a single year, the badgers only take about 5% of the total biomass of worms. Thus once again, the predator does not seem to be having a great effect on the numbers of its prey.

Change over time – natural and man-made

During this century marked changes have occurred in Wytham resulting wholly from things that have happened outside the wood, such as the introduction of new species, new diseases and global warming. In the early part of the century the introduced grey squirrel reached Wytham. These squirrels have a long-term effect on the wood as they strip the bark from the young beech trees. Few, if any, of the beeches planted in the 1950s survived their attacks. Along with the rest of southern Britain, Wytham lost all its elm trees during the 1970s from Dutch Elm Disease. These were an important part of the surrounding hedgerows and field edges, and supported a wide range of insect species (Figure 2.26).

Another introduced mammal, the muntjak deer, appeared in Wytham in the 1970s and has since become quite common. The fallow deer has also become more abundant in recent years and, in places, the two species are having a noticeable effect on the vegetation. In some years, such as 1998, the woods are closed to the public so that the herds of deer can be culled.

An introduced disease, *myxamatosis*, has also had a marked effect on Wytham as it has on many other areas. In the late 1950s it virtually wiped out the rabbit population, and it took over 30 years before the rabbit population re-established itself. When the rabbits disappeared so too did the stoats, which are specialist rabbit predators. In addition, much of the chalk grassland was colonised by invading birches that had formerly been kept in check by the rabbits. Much less expected was the predation of tit nests by weasels. This increased dramatically because either the weasels increased in the absence of stoats, or the weasels need baby rabbits while they themselves are breeding and, in the absence of rabbits, were forced to switch to hunting other prey, such as blue tits.

Global warming is also having an effect on the biogeography of Wytham Woods. For example, badgers have become fewer and the number of blue tits and butterflies had fallen by the mid-1990s compared with earlier years. Scientists at Oxford University predict that by 2050 grasslands will become infested with weeds, common species of birds will become rarer and wetlands will dry up. Badgers have been badly hit. Cubs are unable to dig through the parched earth for worms and have been getting thinner than usual. Thin badgers are much less likely to survive in to adulthood.

The number of butterflies has fallen for several successive years. Species such as the Speckled White normally feed off ground vegetation but the dry soil has meant fewer plants. The dry weather has also hit birds. Caterpillars, the staple food of chicks, have arrived too early in recent years and there has been a food shortage. The result is fewer blue tits and great tits. Streams and wetlands have also been drying up affecting the numbers of frogs, toads and water insects.

Figure 2.26 Insect species associated with common trees and shrubs in Britain

Tree or shrub	Number of insect species
Oak	284
Willow	266
Birch	229
Hawthorn	149
Blackthorn	109
Poplar	97
Crab Apple	93
Scotch Pine	91
Alder	90
Elm	82
Hazel	73
Beech	64
Ash	41
Spruce	*37
Lime	31
Hornbeam	28
Rowan	28
Maple	26
Juniper	20
Larch	*17
Fir	*16
Sycamore	*15
Holly	7
Sweet Chestnut	*5
Horse Chestnut	*4
Yew	4
Walnut	*4
Holme Oak	*2
Plane	*1

* = introduced species

Spatial focus: soils and vegetation of the tropical rainforest

Equatorial soils

The rainforest **biome** was largely unaffected by the Pleistocene glaciations. Hence, not only are the soils well developed but they have been weathered for a long time and therefore lacking in nutrients. Thus they are inherently infertile. More than 80% of the soils have severe limitations of acidity, low nutrient status, shallowness or poor drainage. This is unusual given the richness of the vegetation that they support. The soils of these areas are usually heavily leached and **ferralitic**, with accumulations of insoluble minerals containing iron, aluminium and manganese. The hot humid environment speeds up chemical weathering and decay of organic matter. The nutrients are mainly stored in the biomass due to the rapid leaching of nutrients from the A horizon. There is only a small store of nutrients in the litter or the soil itself (Figure 2.27).

The rate of litter fall is high. The rapid rates of decomposition and the rapid leaching of nutrients from the rooting area has led to an unusual adaption in this ecosystem. Nutrients are passed directly from the litter to the trees by fungi (living on the tree roots). This by-passes the soil storage stage when there is a strong chance that the nutrients will be lost from the nutrient cycle completely.

The rapid decay of litter gives a good plentiful supply of bases. Clay minerals break down rapidly and the silica element is carried into the lower layers. Iron and aluminium sesquioxides which are relatively insoluble remain in the upper layers, as they require acidic water to mobilise them. These leached red or red brown soils are termed ferralitic soils. Deep weathering is a feature of these areas and the depth of the regolith may be up to 150 m deep.

These soils are not easy to manage. If they are ploughed severe soil erosion and a loss of soil fertility occur. Most of the plant nutrients are stored in the vegetation. The leaves and stems falling to the soil surface break down rapidly and nutrients are released during the processes of decomposition. These are almost immediately taken up by the plants. By contrast, the supply of nutrients from the underlying mineral soil is a small component. If the forest cover is removed, the bulk of the systems nutrient store is removed also. This leaves a well weathered, heavily leached soil.

Even when the forest is burnt the nutrients held in the plant biomass store are often lost. During burning there may be gaseous losses and afterwards rainfall may leach nutrients from the ash on the surface. Unless a plant cover is rapidly established, most of the nutrients released from the plant biomass during burning will be lost within a short time. Thus shifting cultivation can only take place for a few years before the overall fertility of the soil is reduced to such an extent that it is not worthwhile

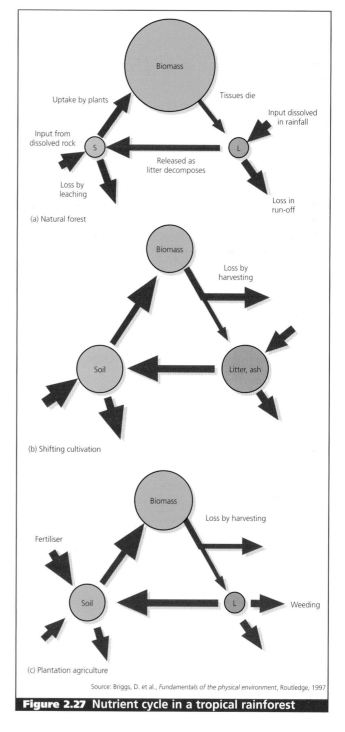

Source: Briggs, D. et al., *Fundamentals of the physical environment*, Routledge, 1997

Figure 2.27 Nutrient cycle in a tropical rainforest

Figure 2.28 Distribution of main kinds of soil in the humid tropics

Soil type	Million ha
Acid, infertile soils	938
Moderately fertile, well drained soils	223
Poorly drained soils	119
Very infertile sandy soils	104
Shallow soils	75
Total	1459

(Source Nortcliff, S., 'The clearance of the tropical rainforest', *Teaching Geography*, April 1987, 110–3), 1987.

continuing cropping the plot. Indeed, farmers try to replicate the rainforest environment by intercropping. This provides shelter for the soil and protects it from the direct attack of the intense rain (rain-splash erosion can otherwise be a serious problem). Compaction of the soil by heavy raindrops and the reduction of the infiltration capacity as a result will lead to overland flow and soil erosion.

Issues in the rainforest

Tropical rainforests are disappearing at an alarming rate, turning 'Green Jungles' into to 'Red Deserts'. The loss of rainforest is up to 200 000 km^{-2} per annum. It is claimed that there are no large-scale rainforest biomes left, only small-scale isolated refugia of a

few tens or hundreds of square kilometres.

To those who live within or close to the rainforest, the forest is a resource which they are eager to exploit. To many economically marginal households, clearing the land presently occupied by forest is seen as a way of improving their quality of life, and becoming self-sufficient in farming. Rainforests are areas of low population densities and, in some areas, relatively unexploited. However, in many cases new farmers have little experience of the tropical environment. Some are poor people from the cities while others are farmers familiar with very different environments.

Forest clearance in the tropics will continue in order to satisfy the demands of the growing population (see pages 212–3). The priority therefore should be to slow down the rate of clearance. This can be done by making the most effective use of the cleared land and also by limiting the need for further forest clearance to replace land cleared at an earlier stage. These aims can be achieved by:

- increasing the productivity of land already cleared by choosing suitable crop and land management combinations

- minimising the damage that results from forest clearing methods by adopting methods of clearance and timing of clearance which produce the least detrimental effects

- restoring eroded or degraded land by using more appropriate land management systems.

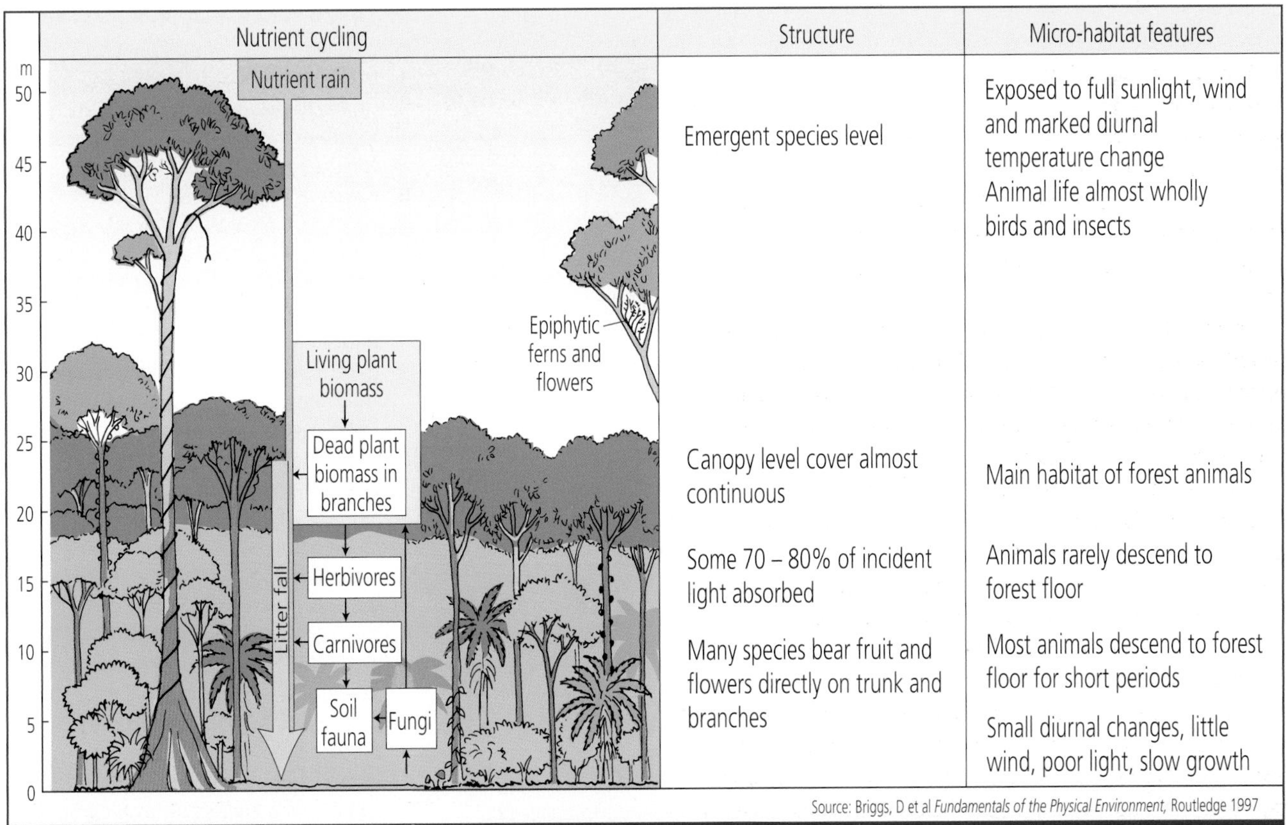

Figure 2.29 Vegetation in the rainforest

Source: Briggs, D et al *Fundamentals of the Physical Environment*, Routledge 1997

Vegetation in tropical rainforests

The tropical rainforest is the most diverse ecosystem in the world, yet it is also the most fragile. This is because the conditions of temperature and humidity are so constant that species specialise to a great extent. Their food sources are often limited to only a few species. Thus when this biome is subjected to stress by human activity it often fails to return to its original state.

The net primary productivity (NPP) of this ecosystem is high at over 2200 g/m² per annum. The hot, humid climate provides ideal conditions for plant growth and there are no real seasonal changes. Thus the plants are aseasonal and the trees shed their leaves throughout the year rather than in one season; the forest is turned evergreen as a result.

There is a great variety of plant species – in some parts of Brazil there are 300 tree species in an area of 2 km². The need for light means that only those trees that can grow rapidly and overshadow their competitors will succeed. Thus trees are notably tall and have long, thin trunks with a crown of leaves at the top; they also have buttress roots to support their height.

There are broadly speaking three main layers of tiers of trees (Figure 2.29):

■ **emergents** which extend up to 45–50 m

■ a **closed canopy** 25–30 m high which cuts out most of the light from the rest of the vegetation and restricts its growth

■ a limited **understorey** of trees, denser where the canopy is weaker; when the canopy is broken by trees falling, clearance or at rivers there is a much denser vegetation.

Trees are shallow rooted as they do not have problems getting water. Other layers include lianas and epiphytes, and the final layer is an incomplete field layer limited by the lack of light. The floor of the rainforests is littered with decaying vegetation, rapidly decomposing in the hot, humid conditions. Tree species include the rubber tree, wild banana and cocoa; pollination is not normally by wind due to the species diversity, but use insects, birds and bats which have restricted food sources.

Rain forests support a large number of epiphytes which are attached to the trees. Many of these are adapted to a system on the small intake of nutrients. There are also parasites taking nutrients from the host plants as they may kill their hosts. Those flora living on dead material are called saprophytres, an important part of the decomposing unit. The fauna are as diverse as the flora.

Figure 2.30 A comparison of a brown earth soil under deciduous woodland in southern England and a rainforest in soil in Amazonas, Brazil

Brown Earth

Depth cm	% Org. matter	% Sand	% Silt	% Clay	pH (H₂0)
0–6	10.4	39	37	14	4.8
6–18	5.0	49	36	15	4.5
18–30	0.6	31	9	60	5.3
30–64	0.3	30	6	64	5.8
64–93	0.1	7	6	87	6.3
93–108	0.2	5	4	91	6.5

Oxisol

Depth cm	% Org. matter	% Sand	% Silt	% Clay	pH (H₂0)
0–8	3.4	10	15	75	3.4
8–18	0.6	11	11	78	3.7
18–50	0.5	7	8	85	4.2
50–90	0.3	7	4	89	4.5
90–150	0.2	7	3	90	4.7
150–170	0.2	5	3	92	4.9

(Source Nortcliff, S., 'The clearance of tropical rainforest', *Teaching Geography*, April, 110–13.), 1987.

Questions

1 Describe the main types of soil found under deciduous woodlands, such as Wytham Woods. How do you account for their formation?

2 Figure 2.3 and Figure 2.4 show energy flows and productivity in Wytham Woods. Describe the trophic structure of Wytham Woods. What are the top carnivores in Figure 2.3? What effects might changes in the populations of mice, voles and caterpillars have on the populations of owls and titmice? Explain your answer.

3 Study Figure 2.28 and state what proportion of tropical soils are fertile. Explain why tropical rainforests have some of the world's most luxuriant vegetation and yet some of the world's least fertile soils.

4 Compare and contrast the characteristics of a brown earth with those of a ferruginous soil (oxisol). Use Figure 2.30 to support your answer.

5 Using examples, examine the effects of human activities on tropical soils.

Global distribution of major ecosystems

Savanna ecosystems

Savannas are areas of tropical grasslands that can occur with or without trees and shrubs. Savannas cover about one-quarter of the world's land surface and are found between the tropical rain forests and the world's great deserts. The development and maintenance of savannas has occurred as a result of a variety of factors including climate, soils, geomorphology, fire and the grazing of human herds of cattle.

The climate that characterises savanna areas is a tropical wet-dry climate (Figure 2.31). However, different savanna areas experience great variations in climate. The wet season occurs in summer: heavy convectional rain (monsoonal) replenishes the parched vegetation and soil. This rainfall, however, varies from as little as 500 mm to as much as 2000 mm, enough to support deciduous forest. On the other hand, all savanna areas have an annual drought: this can vary from as little as one month to as much as eight months. It is on account of the dry season that grasses predominate. Temperature remains high throughout the year ranging between 23 and 28°C. The high temperatures, causing high evapotranspiration rates, and seasonal nature of the rainfall cause a twofold division of the year into seasons of water surplus and water deficiency. This seasonal variation has a great effect on soil development (Figure 2.32).

The link between climate and soil could hardly be closer. During the wet season the excess of precipitation (P) over potential evapotranspiration (pEVT) means that leaching of soluble minerals and small particles will take place down through the soil. These are deposited at considerable depth. By contrast, in the dry season pEVT silica and iron compounds are carried up through the soil and precipitated

Figure 2.31 Climatic conditions in the tropical rainforest (humid tropics) and the savanna (tropical wet-dry)

Month	TRF (Amazon)		Savanna (Nigeria)	
	Ppt (mm)	Temp.	Ppt	Temp. (°C)
Jan	262	26	0	23
Feb	196	27	2	24
Mar	254	26	13	27
Apr	269	26	64	28
May	305	26	150	27
June	234	26	180	25
July	223	25	216	24
Aug	183	26	302	23
Sep	132	27	269	24
Oct	175	27	74	25
Nov	183	27	2	24
Dec	264	26	0	23
Total	2677 mm		1272 mm	

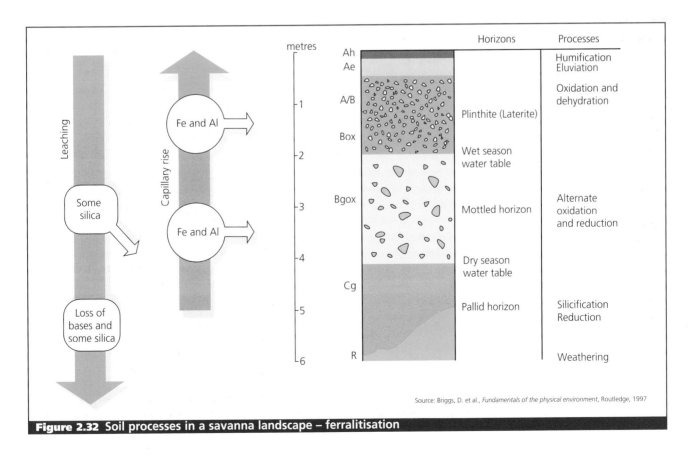

Source: Briggs, D. et al., *Fundamentals of the physical environment*, Routledge, 1997

Figure 2.32 Soil processes in a savanna landscape – ferralitisation

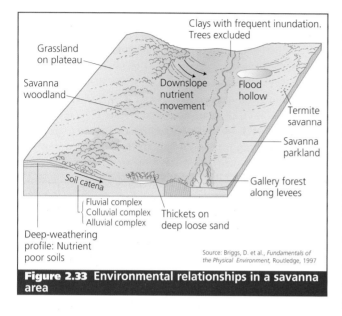

Source: Briggs, D. et al., *Fundamentals of the Physical Environment*, Routledge, 1997

Figure 2.33 Environmental relationships in a savanna area

close to the surface. Geomorphology plays an important role too. Some areas, notably the base of slopes and river valleys, are enriched by clay, minerals and humus that is deposited there. By contrast plateaus, plains and the tops of slopes may be depleted of nutrients by erosion. The local variation in soil leads to variations in vegetation: this control by the soil is known as **edaphic** control (Figure 2.33). For example, on the thicker clay-based soils there is frequently woodland whereas on the leached sandy soils, with poor water retention, grassland predominates.

Savanna vegetation is a mosaic including grasses, trees and shrubs. All, however, are **xerophytic** (adapted to drought), and therefore adapted to the savanna's dry season and **pyrophytic** (adapted to fire). Adaptations to drought include deep tap roots to reach the water table, partial or total loss of leaves and sunken stomata on the leaves to reduce moisture loss: those relating to fire include very thick barks and thick budding that can resist burning, the bulk of the biomass being below ground level and rapid regeneration after fire. The growth tissue in grasses is located at the base of the shoot close to the soil surface unlike shrubs, where growth occurs from the tips; hence burning, and even grazing, encourages the growth of grass relative to other plants.

The warm wet summers allow much photosynthesis and there is a large net primary productivity of 900g/m² per annum. This varies from about 1500g/m² per annum where it borders rain forest areas to only about 200g/m² per annum where it becomes savanna scrub. The biomass also varies considerably (depending on whether it is largely grass or wood) with an average of 4 kg/per metre. Typical species in Africa include the acacia, palm and baobab trees and elephant grass, which can grow to a height of over 5 m. Trees grow to a height of about 12 m and are characterised by flattened crowns and strong roots.

The role of fire, whether natural or man-made, is very important. It helps to maintain the savanna as a grass

community, mineralises the litter layer, kills off weeds, competitors and diseases and prevents any trees from colonising relatively wet areas.

The fauna associated with savannas is very diverse. The African savanna has the largest variety of grazers, over 40, including giraffe, zebra, gazelle, elephants and wildebeest (Figure 2.34). Selective grazing allows a great variety of herbivores: for example, the giraffe feeds off the tops of the trees, the rhinoceros the lower twigs and gazelle the grass beneath the trees. These animals are largely migratory searching out water and fresh pastures as the dry season sets in. A variety of carnivores including lions, cheetahs and hyenas are also supported.

Temperate grasslands

Temperate grasslands are found where rainfall is concentrated into one season in the year and there is insufficient to support forest and woodlands. True temperate grasslands are found in continental interiors. On their arid margins they are called **steppes** and where they are moister they are termed **prairies**.

In North America, the soils of the temperate grasslands are transitional, ranging between pedalfers (where E = pEVT and there is an overall movement of water down through the soil removing aluminium and iron from the upper soil horizons) and pedocals (where pEVT and there is an upward movement of water in the soil depositing calcium in the A horizons). This occurs because precipitation decreases from east to west.

Chernozems are associated with continental temperate climates, such as the Steppes of Russia or the Great Plains of North America. pEVT, therefore there is an upward movement of water through the soil. Hence the soil is a pedocal. They are also referred to as black earths, owing to the accumulation of organic matter throughout the soil. The soil is alkaline and has a crumb structure (owing to the humus). Thus they are very good for agriculture. The thickly

Figure 2.34 Rhinos

matted root system reduces the risk of soil erosion. The dominant process is the upward movement of water to the drying surface. Grasses also draw up calcium thus there is the accumulation of bases at the surface. Often there are no B or E horizons. Some leaching does occur, associated with snow melt in spring and periodic thunderstorms during the late summer.

There are two basic types of grass formation, turf forming grass such as wheat grass growing up to 60 cm, and tufted or tussock grass such as feather grass, which grows in compact clumps in drier areas.

The most extensive grasslands are found in continental interiors. The extent to which the grasslands represent a true climax vegetation system is debatable. Much temperate grassland has been destroyed by cultivation, the grazing of livestock and the use of fire. In fact, some argue that temperate grassland would not exist without human activity, although in areas of 'natural' grassland, large herbivores such as buffalo would prevent the growth of tree seedlings.

Some grasslands are a plagioclimax, that is a man-made climax vegetation, as opposed to one controlled by climate. In many parts of Britain, such as western Scotland, areas which were once wooded are now covered by grassland. Regrowth of the trees is prevented by the grazing of sheep (eating young shoots), and the strong winds. Leached brown soils and podzols have been formed, with a low nutrient content, thereby preventing the recolonisation of trees. Grasses can also occur where soils are too wet for trees.

Coniferous forests or taiga

Most coniferous forests consist of a large number of a few species (Figure 2.35). Common species include pine, fir and spruce. Trees are mostly in pure strands of one variety. Pines predominate in dry sandy soils, spruces on wetter soil. All trees become more stunted towards the poleward margin as the warm season decreases in length. The taiga is replaced polewards by the tundra. In coniferous forests there is little undergrowth. The layer of needle-shaped leaves decomposes very slowly to give a very acid podzol soil in which few plants will grow. Most trees have a conical shape, which reduces wind rocking and prevents extensive snow accumulation on the branches.

Typical climate is cold continental – over three months but less than six months with average temperatures over 6°C. Summer temperatures rise to just 10–15°C, but summers are very short and winters long and cold. Rainfall is concentrated in the short summer when it is most useful; precipitation is as snow in winter. 250 mm annual rainfall will support this type of forest.

The needle-shaped leaves reduce transpiration. The soil is often so cold that the tree cannot take in enough water to counteract transpiration loss which is increased by strong winds. The leaves are designed to conserve moisture.

Trees have to make the fullest use of the short summer for growth. Hence they are evergreen so the leaves are ready to begin work as soon as the temperature rises, without having to be grown as in the case of deciduous trees. Growth is, however, very slow. The summer is not long enough for flowers to be produced and pollinated, and the seed dispersed in the same season (as in the case of the deciduous trees). Coniferous trees, therefore, are pollinated one season and the seed is dispersed the next.

The main soils are podzols. Podzols are perhaps the most distinct soil types (Figure 2.21). They are found under coniferous vegetation and heathlands. Climatically, they are associated with cool temperate regions. These areas have a low annual precipitation (500–800 mm), low winter temperatures (<0°C) and summer temperatures over 10°C. Rapid leaching occurs in spring as a result of snowmelt. Under heathlands, permeable sands and gravels allow the free movement of water downwards.

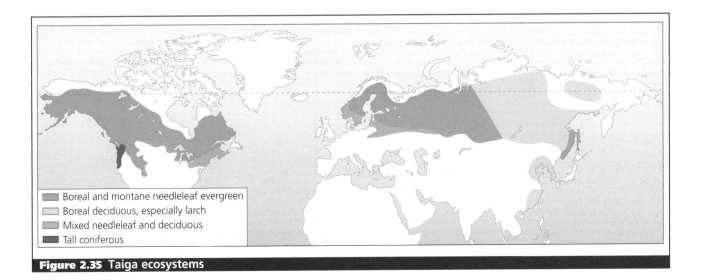

Boreal and montane needleleaf evergreen
Boreal deciduous, especially larch
Mixed needleleaf and deciduous
Tall coniferous

Figure 2.35 Taiga ecosystems

The effects of agriculture on ecosystems

Agriculture alters natural ecosystems so that biomass, in the form of crops and livestock, is used for human or animal consumption. Production can be increased by the use of fertilisers, genetic developments, irrigation and mechanisation. At the same time, crops losses to pests and weeds can be reduced by mechanical weeding, biological control, and the use of pesticides such as insecticides, fungicides and herbicides.

Water quality

The main types of pollution from agricultural activities are from nitrates, pesticides, runoff of silage effluent, and slurry (Figure 2.36). Rising nitrate levels threaten the quality of drinking water. In addition, levels of nitrates and phosphates, much lower than those resulting in groundwater contamination, can lead to eutrophication of fresh and coastal waters. Although the main source of phosphorus pollution is from sewage, animal manure is also an important source. Moreover,

Figure 2.36 Algal blooms in pond/river

once phosphorus is in groundwater or soil, it is more difficult to remove than nitrate).

The use of pesticides, especially herbicides, which are also used in non-agricultural activities such as on roads and railways, can lead to contamination of waterbodies and residues in drinking water supplies. Toxic pesticide residues can enter the food-chain via groundwater and surface waters, the soil, the crop itself or through direct contact. Accidental spills and leaks of materials high in organic matter, such as slurry and silage effluent into waterbodies, can deprive aquatic organisms of oxygen and lead to serious loss of aquatic life. Erosion of agricultural soils and associated runoff can lead to sedimentation of waterbodies.

Soil

Erosion of topsoil by surface runoff of water and, to a lesser extent, wind, can be exacerbated by a number of agricultural activities. Soil compaction occurs mostly in areas with highly productive soils and heavily mechanised agriculture. The degree of compaction depends on the type of soil, soil water content, the slope of land on which machinery is used, the size of machinery, weight of load, the timing and frequency of passage of machinery, and agricultural practices, such as irrigation.

Over-application of fertilisers can alter the chemical composition of the soil and local soil acidification may result. In some cases acidification causes increased leaching of nitrates and heavy metals to groundwater. Manure, artificial fertilisers and sewage sludge used on agricultural soils all contain heavy metals and other trace elements. Heavy metals can accumulate in the topsoil and may lead to reduced yields by reducing growth, or even enter the food-chain. They can also be directly ingested with soil by grazing animals.

In Europe about 7% (64 million ha) of agricultural land has been degraded by improper management, and about 5% (50 million ha) has been degraded by overgrazing. The problem is especially severe in southern Mediterranean areas such as Portugal, where over 20% of the agricultural area is highly erodable. This could lead to problems of desertification in the near future. Land degradation threatens over 60% of the total land in southern Europe.

Waterlogging may result from inadequate drainage, compaction by heavy machinery, heavy rainfall or overwatering. In addition, salinisation and alkalisation can result from irrigation with improper drainage, and from use of irrigation water with a saline content.

Many landscapes have been destroyed, in particular wetlands, peatlands and hedgerows. The channelisation of streams and small waterways running through agricultural land to improve land drainage has often been associated with a decrease in aquatic biodiversity in these waterways. Other landscapes been altered by afforestation, tourism and recreation, infrastructural developments such as rural road networks, urban sprawl and industrial development. All of these factors have played a role, but agricultural development has had the greatest influence on the reduction of biodiversity. The loss of natural habitats, combined with rising levels of toxic pesticide residues in the environment has had significant effects on wildlife and biodiversity. Pesticides also affect non-target species.

The loss of visual amenities by changes in the rural landscape is also a matter of grave concern. For example, hedgerow removal, clearing of woods, realignment of watercourses, the disappearance of meadows and riparian forest along waterways and the abandonment of terraces, all make mechanised farming easier.

Useful web sites include:

Ministry of Agriculture, Fisheries and Food on
http://www.open.gov.uk/maff/maffhome.htm
Environment Agency on http://www.environment-agency.gov.uk/

Questions

1 Describe and explain the relationship between climate, soils and vegetation in a savanna ecosystem.
2 Explain why tundra ecosystems have such low productivity.
3 How is vegetation adapted to survival in taiga forests?
4 Describe and explain the relationships between climate and soils in temperate grassland ecosystems.
5 In what ways have agricultural developments influenced ecosystems? Distinguish between positive and negative effects.
6 Briefly explain the impact of pesticides and fertilisers on ecosystems.
7 How does agriculture lead to a reduction in biodiversity?

Review

1 Examine the effects of climate and geology on soil formation.
2 For a small area that you have studied, describe and explain local variations in soil types and processes.
3 How does agriculture affect ecosystems?
4 With the use of examples, discuss the effects of human activity on ecosystems.

3 atmospheric *systems*

Climate and weather

The term **climate** refers to the state of the atmosphere over a period of not less than 30 years. It includes variables such as temperature, rainfall, winds, humidity, cloud cover and pressure. It refers not just to the averages of these variables but to the extremes as well. By contrast, **weather** refers to the state of the atmosphere at any particular moment in time. However, we usually look at the weather over a period of between a few days and a week. The same variables as for climate are considered.

Atmospheric composition

The atmosphere contains a mix of gases, liquids and solids (Figure 3.1). Atmospheric gases are held close to the earth by gravity. Close to the earth these gases are relatively constant. Nevertheless, there are important variations in atmospheric composition, and this causes variations in temperature, humidity and pressure between places and over time. The normal components of dry air include nitrogen (78.1%), oxygen (20.9%), argon (0.93%), and carbon dioxide (0.03%). In addition, there are other important gases such as helium, ozone, hydrogen and methane. These gases are crucial. For example, changes in the amount of carbon dioxide in the atmosphere is having an effect on global warming (see pages 87–90) and the destruction of ozone is having an important effect on the quality of radiation reaching the earth's surface (see pages 90–1).

The atmosphere also contains moisture. Most water vapour is held in the lower 10–15 km of atmosphere. Above this it is too cold and there is not enough turbulence or mixing to carry vapour upwards. The lower atmosphere also contains solids such as dust, ash, soot and salt. These allow condensation to occur which can cause cloud formation and precipitation.

Variations in composition with altitude

Turbulence and mixing in the lower 15 km of the atmosphere produces fairly similar 'air'. At high altitudes, in contrast, marked concentrations of certain gases occur. For example, between 15 km and 35 km there is a concentration of ozone. Although this forms only a small percentage of the atmospheric gas, it is significant enough to lead to an increase in atmospheric temperature in this region and has an important screening function.

The most significant concentrations of gases at altitude include:

- nitrogen between 100 and 200 km
- oxygen between 200 km and 1100 km
- helium between 1100 and 3500 km
- hydrogen above 3500 km.

Key Definitions 1

Albedo   The reflectivity of the earth's surfaces.

Atmosphere   The mixture of gases, predominantly nitrogen, oxygen, argon, carbon dioxide and water vapour, that surrounds the earth.

Climate   The average weather conditions of a place or an area over a period of years.

Meteorology   The study of the earth's atmosphere and weather processes.

Precipitation   Includes all forms of rainfall, snow, frost, hail and dew. It is the conversion and transfer of moisture in the atmosphere to the land.

Troposphere   One of the four layers of the atmosphere, which extends from the surface of the earth to an altitude of 10 to 16 km. Climate and weather take place in the troposphere.

Weather   The state of the atmosphere at a given time and place.

Figure 3.1 Average composition of dry air

Constituent gas	Percentage volume
Nitrogen	78.1
Oxygen	20.9
Argon	0.93
Carbon dioxide	0.03
Neon	0.0018
Helium	0.0005
Ozone	0.00006
Hydrogen	0.00005
Krypton Methane Xenon	Trace

Atmospheric energy

The atmosphere is an **open energy system** receiving energy from both sun and earth. Although the latter is very small, it has an important local effect, as in urban climates. Incoming solar radiation is referred to as **insolation**. In turn the earth radiates energy out to the atmosphere.

Atmospheric energy budget

The atmosphere constantly receives solar energy yet, until recently, the atmosphere was not getting any hotter. There has therefore been a balance between inputs (insolation) and outputs (reradiation) (Figure 3.3).

Of incoming radiation 17% is **absorbed** by atmospheric gases, especially oxygen and ozone at high altitudes, and carbon dioxide and water vapour at low altitudes. Scattering accounts for a net loss of 6%, and clouds and water droplets reflect 23%. In fact, clouds can reflect up to 80% of total insolation. **Reflection** from the earth's surface (known as the **albedo**) is generally about 7%. However, there are important local variations in albedo as shown in Figure 3.5. About 36% of insolation is reflected back to space and a further 17% is absorbed by atmospheric gases. Hence, only about 47% of the insolation at the top of the atmosphere actually gets through to the earth's surface.

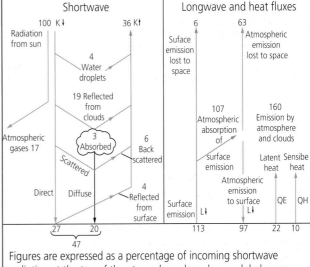

Figures are expressed as a percentage of incoming shortwave radiation at the top of the atmosphere, based on a global mean.

Source: Briggs, D. et al., *Fundamentals of the physical environment*, Routledge, 1997

Figure 3.3 The earth's energy budget

Key Definitions ②

Convection The transfer of heat by the movement of a gas or liquid.

Conduction The transfer of heat by contact.

Radiation The emission of electromagnetic waves such as X-ray, short- and long-wavelengths; as the sun is a very hot body, radiating at a temperature of about 5700°C, most of its radiation is in the form of very short-wavelengths such as ultraviolet and visible light.

Figure 3.2 Atmospheric processes

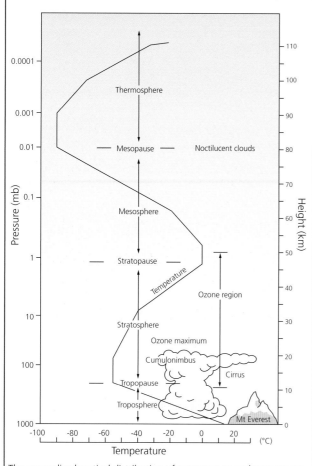

The generalised vertical distribution of temperature and pressure up to about 110 km. Note particularly the tropopause and the zone of maximum ozone concentration with the warm layer above.

Source: Barry, R. and Chorley, R., *Atmosphere, Weather and Climate*. Routledge, 1998

Figure 3.4 Changes in temperature through the atmosphere

Figure 3.5 Albedos for the short-wave part of the spectrum

Surface	Albedo (%)
Water (zenith angles above 40°)	2–4
Water (angles less than 40°)	6–80
Fresh snow	75–90
Old snow	40–70
Dry sand	35–45
Dark, wet soil	5–15
Dry concrete	17–27
Black road surface	5–10
Grass	20–30
Deciduous forest	10–20
Coniferous forest	5–15
Crops	15–25
Tundra	15–20

Source: Briggs, D et al, *Fundamentals of the physical environment*, Routledge, 1997

Questions

1 Study Figure 3.5.
  (a) What is meant by the term albedo?
  (b) Which surfaces have the highest albedo?
  (c) Which surfaces have the lowest albedo?
  (d) Why is albedo important?
2 **(a)** Why are light-coloured clothes often worn in summer?
  (b) Define the terms (i) conduction, (ii) convection and (iii) radiation.
3 Figure 3.6 shows annual temperature patterns.
  (a) Where are the highest temperatures found in January?
  (b) How do temperatures change northwards from the Tropic of Capricorn? Are there any exceptions to this pattern?
  (c) How do the patterns change in July?
  (d) Suggest reasons to explain the changing pattern of temperatures in January and July.

Energy received by the earth is reradiated at **long wavelength**. (Very hot bodies – such as the sun emit **short-wave radiation** – whereas cold bodies, such as the earth emit long-wave radiation.) Of this 8% is lost to space. Some energy is absorbed by clouds and reradiated back to earth. Evaporation and condensation account for a loss of heat of 23%. In addition, there is a small amount of condensation occuring (carried up by turbulence). Thus, heat gained by the atmosphere from the ground amounts to 39 units.

Hence, the atmosphere is largely heated from below. Most of the incoming short-wave radiation is let through, but the outgoing long-wave radiation is trapped by gases such as carbon dioxide. This is known as the **greenhouse principle**.

Factors affecting temperature

There are many factors which affect the temperature of a place. These include latitude, altitude, distance from the sea, the nature of nearby ocean currents, dominant winds, aspect, cloud cover, length of day, amount of dust in the atmosphere, and human impact.

Latitude

On a global scale latitude is the most important factor determining temperature (Figure 3.7). Two factors affect the temperature: the angle of the overhead sun and the thickness of the atmosphere. At the equator, the overhead sun is high in the sky, hence high intensity insolation is received. By contrast, at the poles, the overhead sun is low in the sky,

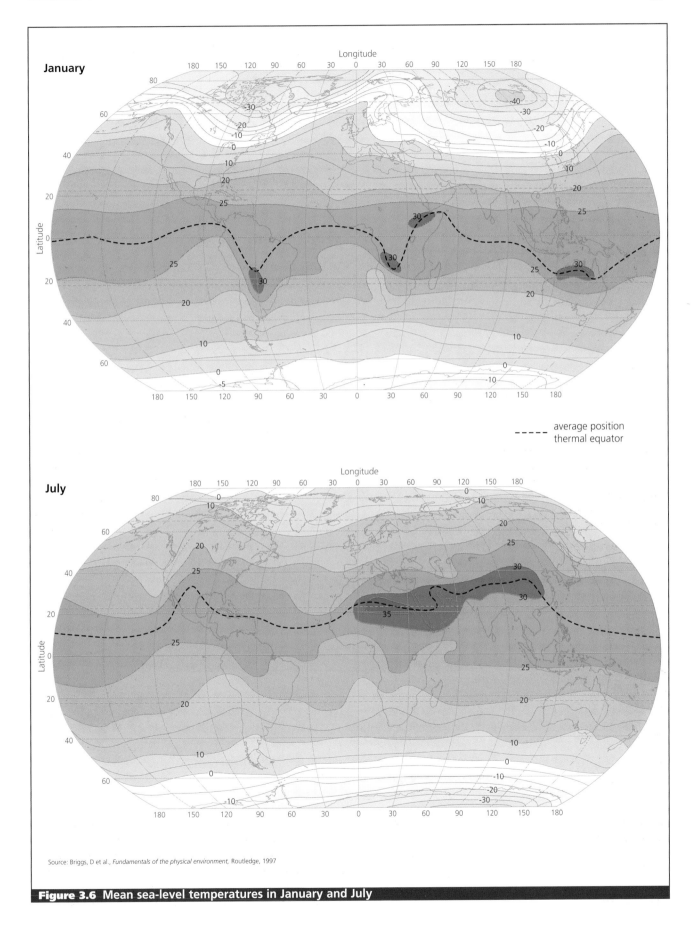

January

July

average position
thermal equator

Source: Briggs, D et al., *Fundamentals of the physical environment*, Routledge, 1997

Figure 3.6 Mean sea-level temperatures in January and July

Areas that are close to the equator receive more heat than areas that are close to the poles. This is due to two reasons:

1 incoming solar radiation (insolation) is concentrated near the equator, but dispersed near the poles.
2 insolation near the poles has to pass through a greater amount of atmosphere and there is more chance of it being reflected back out to space.

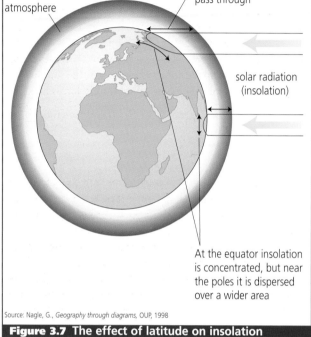

Near the poles insolation has more atmosphere to pass through

atmosphere

solar radiation (insolation)

At the equator insolation is concentrated, but near the poles it is dispersed over a wider area

Source: Nagle, G., *Geography through diagrams*, OUP, 1998

Figure 3.7 The effect of latitude on insolation

hence the quality of energy received is poor. Also, the thickness of atmosphere affects temperature. Radiation has more atmosphere to pass through near the poles, due to its low angle of approach. Hence more energy is lost, scattered or reflected than over equatorial areas, making temperatures lower over the poles.

In addition, the albedo (reflectivity) is higher in polar regions. This is because snow and ice are very reflective, and low angle sunlight is easily reflected from water surfaces. However, the lack of intensity in polar and arctic regions is partly offset by variations in the length of the day and season. The longer the sun shines the greater the amount of insolation is received, which may overcome in part, the lack of intensity of insolation in polar regions. (Alternatively, the long polar nights in winter lose vast amounts of energy.)

Because of the earth's axial tilt (23.5°) the planet's surface slopes away from the sun. This means that the incoming solar energy is very much concentrated in equatorial areas but is much more dispersed over polar areas as higher latitudes are much colder than lower latitudes. This is similar to the fact that the sun is at a low angle, in the morning and evening and tends to give us low amounts of heat at those times.

When the sun is overhead during the middle of the day, by contrast, we have the hottest conditions.

The result is a imbalance: positive budget in the tropics, negative one at the poles (Figure 3.8). Even so, neither region is getting progressively hotter or colder. To achieve this balance the horizontal transfer of energy from the equator to the poles takes place by winds and ocean currents. Thus there is a great imbalance in the heating of the earth's atmosphere. This gives rise to an an important second energy budget in the atmosphere – the horizontal transfer between low latitudes and high latitudes to compensate for differences in global insolation.

Proximity to the sea

Specific heat capacity (SHC) is the amount of heat needed to raise the temperature of a body by 1°C. There are important differences between the heating and cooling of water. Land heats and cools more quickly than water. It takes five times as much heat to raise the temperature of water by 2°C as it does to raise land temperatures.

Water heats more slowly because:

■ it is clear, hence the sun's rays penetrate to great depth (distributing energy over a wider area)

■ tides and currents cause the heat to be further distributed.

Since a larger volume of water is heated for every unit of energy than land, water takes longer to heat up.

Distance from the sea therefore has an important influence on temperature. Water takes up heat and gives it back much more slowly than the land. In winter, sea air of mid latitudes is much warmer than the land air; therefore on-shore winds bring heat to the coastal lands. By contrast, during the

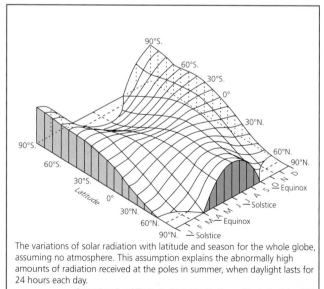

The variations of solar radiation with latitude and season for the whole globe, assuming no atmosphere. This assumption explains the abnormally high amounts of radiation received at the poles in summer, when daylight lasts for 24 hours each day.

Source: Barry, R. and Chorley, R., *Atmosphere, Weather and Climate*, Routledge, 1998

Figure 3.8 Latitudinal imbalance in insolation

summer coastal areas remain much cooler than inland sites. Areas with a coastal influence are termed **maritime** or **oceanic** whereas inland areas are called **continental**.

Ocean currents

The effect of ocean currents on temperatures depends upon whether the current is cold or warm (Figure 3.9). Warm currents from equatorial regions raise the temperatures of polar areas (with the aid of prevailing westerly winds). However, this effect is only noticeable in winter. For example, the Gulf Stream in particular transports heat northwards and then eastwards across the North Atlantic; the Gulf Stream is the main reason why the British Isles have mild winters and relatively cool summers. By contrast, there are other areas that are made colder by ocean currents. Cold currents, such as the Labrador Current off the north-east coast of North America, may reduce the summer temperature, but only if the wind blows from the sea to the land.

Altitude

In general temperatures decrease with altitude. At low altitudes heat escapes from the surface slowly because dense

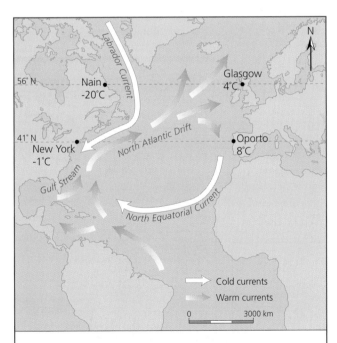

The effect of an ocean current depends upon whether it is a warm current or a cold current. Warm currents move away from the equator, whereas cold currents move towards it. The cold Labrador Current reduces the temperatures of the western side of the Atlantic, while the warm North Atlantic Drift raises temperatures on the eastern side.

Source: Nagle, G., *Geography through diagrams*, OUP, 1998

Figure 3.9 The effects of sea currents on temperatures

air contains dust and water vapour, which retain heat. By contrast, heat escapes rapidly from high altitudes because the thin air contains little water vapour or dust. The normal decrease of temperature with height is known as the **normal** or **environmental lapse rate**, on average 6.4°C per km. If temperatures increase with height then a **temperature inversion** is formed.

Winds

The effects of winds on temperature depends upon the initial characteristics of the wind. In temperate latitudes **prevailing** (dominant) winds from the land lower winter temperatures, but raise summer ones. This is because continental areas are very hot in summer but very cold in winter. By contrast, prevailing winds from the sea do the opposite: they lower the summer temperatures and raise the winter ones. This is due to the specific heat capacity of water which means that the sea is cooler then the land in summer but warmer than the land in winter.

Cloud cover

Cloud cover decreases the amount of insolation reaching the surface and the amount leaving it. If there is no cloud then incoming short-wave (s-w) radiation and outgoing long-wave (l-w) radiation are at a maximum. For example, rainforest areas with thick cloud cover experience days of about 30°C and nights of about 20°C, whereas deserts, without much cloud cover, might reach 38–40°C by day, while nights drop to around freezing. This is because humid air absorbs heat by day and retains it at night. The difference between daytime and night time temperatures is known as the diurnal range.

Aspect

Aspect is noticeable only in temperate latitudes. It refers to whether a slope faces to the south or the north. In the northern hemisphere, south facing slopes (**adret** slopes) are warmer than the north facing slopes (**ubac** slopes). The reverse is true in the southern hemisphere.

Length of day and season

Variations in the length of day and seasons are caused by the earth's revolution and rotation, and have a dramatic impact on the temperature of an area. The earth revolves around the sun every 365.25 days, and it rotates around its own axis every 24 hours. The earth's axis is inclined at an angle of 23.5°, thus the overhead sun appears to migrate from the Tropic of Cancer (23.5°N) on 21 June to the Tropic of Capricorn (23.5°S) on 22 December. The higher the overhead sun the greater the amount of heat energy received. Midsummer in the northern hemisphere (NH) is June, whereas in the southern hemisphere (SH) it is in December. March 21 and September 23 are the intermediate positions and are called the equinoxes.

Figure 3.10 Selected data for European sites

Shannon

	Jan.	Feb.	Mar.	Apr.	May	June	July	Aug.	Sep.	Oct.	Nov.	Dec.
Daily Max. °C	8	9	11	13	16	19	19	20	17	14	11	9
Daily Min. °C	2	2	4	5	7	10	12	12	10	7	5	3
Monthly ppt	94	67	56	53	61	57	77	79	86	86	96	117

London

	Jan.	Feb.	Mar.	Apr.	May	June	July	Aug.	Sep.	Oct.	Nov.	Dec.
Daily Max. °C	6	7	10	13	17	20	22	21	19	14	10	7
Daily Min. °C	2	2	3	5	8	1	13	13	11	8	5	3
Monthly ppt	52	47	40	48	48	44	56	54	50	52	56	53

Berlin

	Jan.	Feb.	Mar.	Apr.	May	June	July	Aug.	Sep.	Oct.	Nov.	Dec.
Daily Max. °C	2	3	8	13	19	22	24	23	19	13	7	3
Daily Min. °C	−3	−3	0	4	8	12	14	13	10	6	2	−1
Monthly ppt	43	40	31	41	46	62	70	68	46	47	46	41

Moscow

	Jan.	Feb.	Mar.	Apr.	May	June	July	Aug.	Sep.	Oct.	Nov.	Dec.
Daily Max. °C	−7	−6	0	9	17	22	24	22	16	8	0	−5
Daily Min. °C	−14	−13	−8	0	6	11	13	12	7	1	−4	−10
Monthly ppt	31	28	35	35	52	67	74	74	58	51	36	36

The length of daylight varies with the overhead sun. On the equinoxes, the sun is overhead the equator and all places receive 12 hours day and 12 hours night. On June 21, the summer solstice, the sun is overhead the Tropic of Cancer. Hence places in the northern hemisphere have their longest days and shortest nights. By contrast, 22 December is the winter solstice. The sun is overhead the Tropic of Capricorn, and places in the northern hemisphere have their longest nights and shortest days. By contrast, places south of the Antarctic Circle (66.5°S) receive 24 hours darkness on the same day.

Aerosols and human activity

Dust and other impurities such as volcanic fall-out may block insolation from reaching the earth and keep temperatures low. In addition, human impact may affect temperatures through agriculture, forestry, and urban lifestyle (see pages 87–90). Deforestation releases carbon, which allows insolation in, but does not allow outgoing long-wave radiation out, and therefore may lead to an increase in temperatures (see pages 87–90). Aerosols destroy ozone, which prevents harmful ultra-violet light from getting to the earth's surface (see pages 90–1).

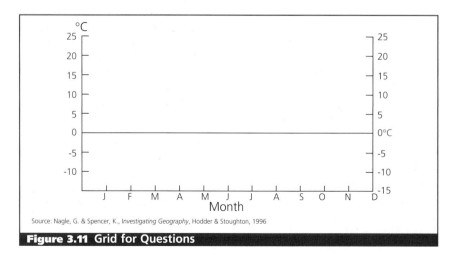

Source: Nagle, G. & Spencer, K., *Investigating Geography*, Hodder & Stoughton, 1996

Figure 3.11 Grid for Questions

Questions

1 **(a)** Make a copy of the grid provided (Figure 3.11). Using the data in Figure 3.10 show the variations in maximum and minimum temperatures between the four stations.
   (b) Using an atlas locate each of these areas in Europe. How do they vary in terms of continentality?
   (c) Work out (i) the average annual rainfall for each station; (ii) the average annual temperature for each station.
2 Define the term 'diurnal range of temperature'. How and why does the diurnal range vary between a hot desert area and a rainforest?
3 Describe how
   (a) maximum and
   (b) minimum temperature varies annually between the four sites.
4 Using an atlas, account for these differences in terms of:
   (a) proximity to the sea; **(b)** altitude;
   (c) latitude; **(d)** ocean currents;
   (e) cloud cover.

Key Definitions

Condensation level   The altitude at which relative humidity is 100% and hence clouds form.

Dew point   The temperature at which relative humidity is 100%.

Evaporation   The process by which a liquid is transformed into a gas.

Front   A boundary between a warm air mass and a cold air mass, resulting in frontal (depressional or cyclonic) rainfall.

Humidity   A measure of the amount of moisture in the air. Absolute humidity tells us how much moisture is in the air (g/m^3) whereas relative humidity expresses this amount as a percentage of the maximum air of a certain temperature could hold.

Instability   Unstable atmospheric conditions (rising air) likely to cause cloud formation and precipitation.

Precipitation   All forms of rainfall, snow, frost, hail and dew. It is the conversion and transfer of moisture in the atmosphere to the land.

Moisture in the atmosphere

Evaporation is the process by which a liquid is transformed into a gas. It depends on three major factors:

- the supply of **heat** – the hotter the air the more evaporation that takes place

- **wind** strength – in windy conditions saturated air is removed, and can continue whereas under calm conditions the air becomes saturated rapidly but is not removed

- the initial **humidity** of the air – if air is very dry then strong evaporation occurs, whereas if it is saturated then very little occurs.

Rain

For rain to occur, three factors must be satisfied:

- air is saturated, i.e. it has a relative humidity of 100%

- it contains particles of soot, dust, ash, ice, etc.

- its temperature is below dew point, i.e. the temperature at which the relative humidity is 100%, saturation is complete and clouds form.

Types of rainfall

There are three main types of rainfall (Figure 3.13):

- **cyclonic** – uplift of air within a low pressure area (warm air rises over cold air); it normally brings low-moderate intensity rain and may last for a few days

- **orographic** – a deep layer of moist air is forced to rise over a range of hills or mountains

- **convectional heating** causes pockets of air to rise and cool.

Fog

Fog is caused by clouds occurring at ground level. In fog visibility is less than a kilometre, whereas in mist, it is above 1 km. Fog is common in many areas, for example the North Sea coast of Britain in summer, the Grand Banks of Newfoundland and coastal Peru. Fog occurs when condensation of moist air cools below its dew point. The most common types are radiation fog and advection fog (Figure 3.14).

Advection fog is formed as warm moist air passes over a cold surface. As the air is chilled its temperature is reduced and it may reach **dew point** (temperature at which relative humidity is 100%). Thereafter condensation takes place. For example, air blowing from the North Atlantic Drift blowing over cold surfaces in Devon and Cornwall will often form a fog.

Radiation fog occurs at night when the ground loses heat through long-wave radiation. This occurs during high pressure conditions associated with clear skies (Figure 3.15).

Fog is a major environmental hazard – airports may be closed for many days and road transport is hazardous and slow. Freezing fog is particularly problematic. Fog can cause sizeable economic losses but the ability to do anything about it is limited. This is because it would require too much energy (and hence cost) to warm up the air or to dry out the air to prevent condensation.

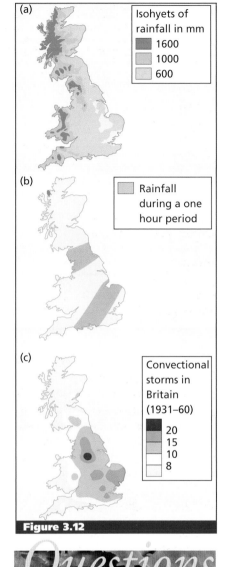

Figure 3.12

(a) Isohyets of rainfall in mm
- 1600
- 1000
- 600

(b) Rainfall during a one hour period

(c) Convectional storms in Britain (1931–60)
- 20
- 15
- 10
- 8

Questions

1 Describe the distribution of rainfall as shown in Figures 3.12 A, B, and C:
   (a) Explain the pattern of rainfall in Figure A.
   (b) Suggest why there is a distinct pattern of rainfall in Figure B.
   (c) Account for the pattern of convectional storms shown in Figure C.

2 Figure 3.13 shows the characteristics of fog.
   (a) What is fog?
   (b) What is the difference between radiation fog and advection fog?
   (c) Explain why fog only occurs during high pressure conditions. In what ways is human activity affecting the frequency distribution and nature of fog?

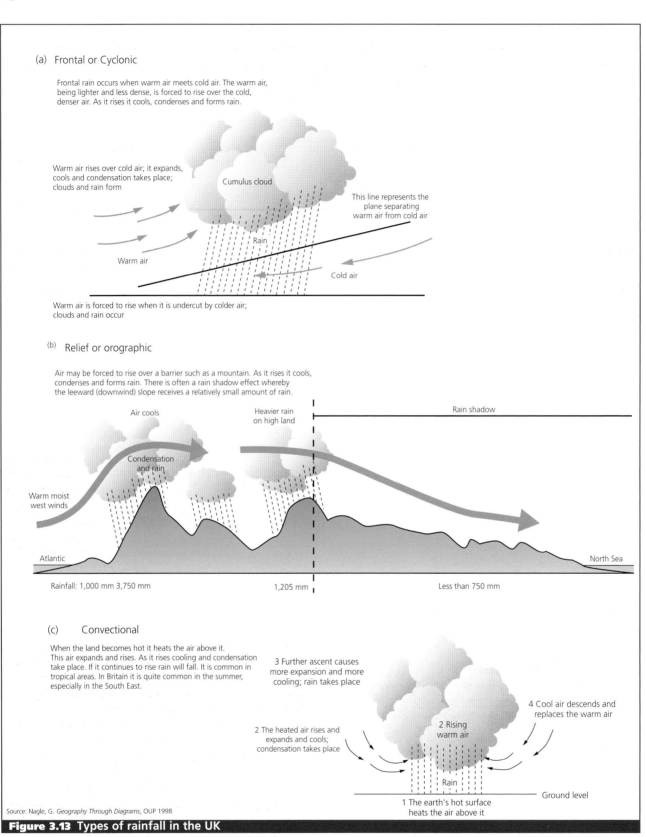

(a) Frontal or Cyclonic

Frontal rain occurs when warm air meets cold air. The warm air, being lighter and less dense, is forced to rise over the cold, denser air. As it rises it cools, condenses and forms rain.

Warm air rises over cold air; it expands, cools and condensation takes place; clouds and rain form

Cumulus cloud

This line represents the plane separating warm air from cold air

Rain

Warm air

Cold air

Warm air is forced to rise when it is undercut by colder air; clouds and rain occur

(b) Relief or orographic

Air may be forced to rise over a barrier such as a mountain. As it rises it cools, condenses and forms rain. There is often a rain shadow effect whereby the leeward (downwind) slope receives a relatively small amount of rain.

Air cools

Heavier rain on high land

Rain shadow

Condensation and rain

Warm moist west winds

Atlantic

North Sea

Rainfall: 1,000 mm 3,750 mm

1,205 mm

Less than 750 mm

(c) Convectional

When the land becomes hot it heats the air above it. This air expands and rises. As it rises cooling and condensation take place. If it continues to rise rain will fall. It is common in tropical areas. In Britain it is quite common in the summer, especially in the South East.

3 Further ascent causes more expansion and more cooling; rain takes place

2 The heated air rises and expands and cools; condensation takes place

2 Rising warm air

4 Cool air descends and replaces the warm air

Rain

1 The earth's hot surface heats the air above it

Ground level

Source: Nagle, G. *Geography Through Diagrams*, OUP 1998

Figure 3.13 Types of rainfall in the UK

Figure 3.14 Characteristics of fog in the British Isles

Type of fog	Season	Areas affected	Mode of formation	Mode of dispersal
radiation fog	October to March	inland areas, especially low-lying, moist ground	cooling due to radiation from the ground on clear nights when the wind is light	dispersed by the sun's radiation or by increased wind
advection fog (a) over land	winter or spring	often widespread inland	cooling of warm air by passage over cold ground	dispersed by a change in air mass or by gradual warming of the ground
(b) over sea and coastline	spring and early summer	sea and coasts, may penetrate a few miles	cooling of warm air by passage over cold sea	dispersed by a change in air mass and may be cleared on coast by the heat of the sun
frontal fog	all seasons	high ground	lowering of the cloud base along the line of the front	dispersed as the front moves and brings a change of air mass
smoke fog (smog)	winter	near industrial areas and large conurbations	similar to radiation fog	dispersed by wind increase or by convection

Figure 3.15 Radiation Fog

Air motion

The basic cause of air motion is the unequal heating of earth's surface. Variable heating of the earth causes variations in pressure and this in turn sets the air in motion. Most of the energy received by the earth is in the tropical areas whereas there is a loss of energy from more polar areas.

The major equalising factor is the transfer of heat by air movement.

Pressure variations

Pressure is measured in millibars (mb) and is represented by isobars, lines of equal pressure. On maps pressure is adjusted to mean sea level (MSL). MSL pressure is 1013 mb, although the mean range is from 1060 mb in the Siberian winter high pressure system to 940 mb (although some intense low pressure storms may be much lower). The trend of pressure change is of more importance than the actual reading itself. Decline in pressure indicates worse weather and rising pressure better weather.

Surface pressure belts

Sea-level pressure conditions show marked differences between the hemispheres. In the northern hemisphere there are greater seasonal contrasts, whereas in the southern hemisphere much simpler average conditions exist (Figure 3.16). The differences are largely related to unequal distribution of land and sea, because ocean areas are much more equable in terms of temperature and pressure variations.

One of the most permanent features is the subtropical high pressure belts, especially over ocean areas. In the southern hemisphere they are almost continuous at about 30° latitude, although in summer over South Africa and Australia they are somewhat broken. Generally pressure is about 1026 mb. In the northern hemisphere, by contrast, at 30° the belt is much more discontinuous because of the land. High pressure only occurs over the ocean as discrete cells such as the Azores and Pacific highs. Over continental areas such as south west USA, southern Asia and the Sahara, major

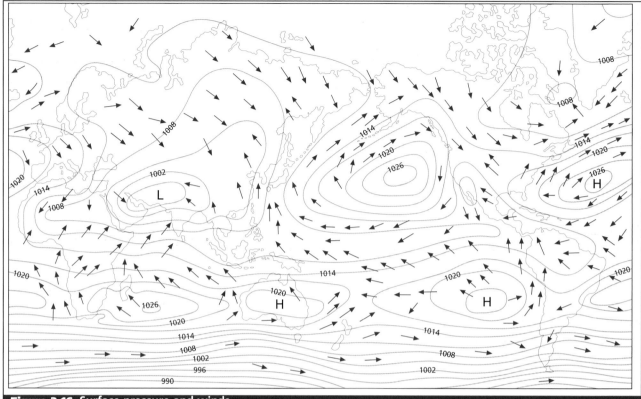

Figure 3.16 Surface pressure and winds

fluctuations occur: high pressure in winter and low pressure in summer because of overheating.

Over the equatorial region pressure is generally low. This is due the zone of maximum insolation. In July it is well north of the equator, whereas in January it is just south of the equator.

In temperate latitudes pressure is generally lower than in subtropical areas. The most unique feature is the large number of depressions (low pressure systems) and anticyclones (high pressure systems) which do not show up on a map of mean pressure. In polar areas pressure is relatively high throughout the year, especially over Antarctica, owing to the coldness of the land mass.

Factors affecting air movement

The driving force is the **pressure gradient**, the difference between pressure between any two points. Air blows from high pressure to low pressure. Globally very high pressure conditions exist over Asia in winter due to the low temperatures. By contrast the air pressure is low over continents in summer. High pressure dominates at around 25–30° latitude. The highs are centred over the oceans in summer and continents in winter, whichever is cooler.

The **Coriolis effect** is the deflection of moving objects caused by the easterly rotation of the earth (Figure 3.17). Air flowing from high pressure to low pressure is deflected to the right of its path in the northern hemisphere and to the left of its path in the southern hemisphere.

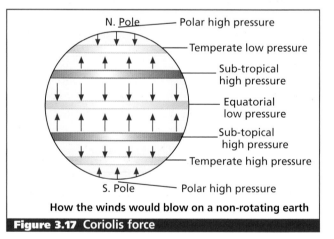

How the winds would blow on a non-rotating earth

Figure 3.17 Coriolis force

The General circulation

Surface winds

Air blows from high pressure to low pressure. The simplified map of world air pressure shows that the main areas of high pressure are at the subtropics and the poles whereas the main areas of low pressure are located near the equator and in the mid latitudes. Thus we would expect air to blow from the subtropics to the equator and to the mid latitudes, and air from the poles to blow to the mid latitudes. Without the Coriolis force this would produce southerly and northerly

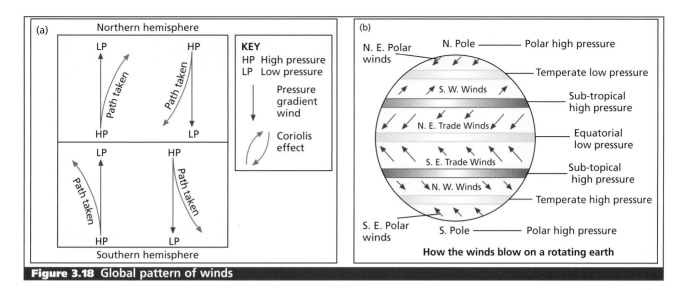

Figure 3.18 Global pattern of winds

winds (Figure 3.18). However, with the deflection of moving objects winds are deflected to the right of their paths in the northern hemisphere and to the left of their path in the southern hemisphere. These two forces, the pressure gradient and the Coriolis force, largely account for the surface easterly winds and westerly winds that characterise air motion (Figure 3.16).

The monsoon is caused by reversing wind systems. The monsoon is influenced by the reversal of land and sea temperatures between Asia and the Pacific during the summer and winter. In winter surface temperatures in Asia may be as low as $-20°C$. This creates very low pressure. By contrast the surrounding oceans have temperatures of 20°C. During the summer the land heats up quickly and may reach 40°C while the sea remains cooler at about 27°C. This initiates a land sea breeze blowing from the cooler sea (high pressure) in summer to the warmer land (low pressure), whereas in winter air flows out of the cold land mass (high pressure) to the warm water (low pressure). In addition, the presence of the Himalayan Plateau disrupts the strong winds of the upper atmosphere; this forces winds either to the north or south and consequently a deflection of surface winds.

General circulation model

In 1735, George Hadley described the operation of the Hadley Cell, created by the direct heating of the equator (Figure 3.19a). The air is forced to rise by convection, travels polewards then sinks at the subtropical anticyclone (high pressure belt). Hadley suggested that similar cells might exist in mid latitudes and high latitudes. William Ferrel suggested that Hadley Cells interlink with a mid latitude cell rotating it in reverse direction. These cells in turn rotate the polar cell (Figure 3.19b). The most recent models have refined the basic principles and include air motion in the upper atmosphere, in particular jet streams (very fast thermal winds) (Figure 3.19c).

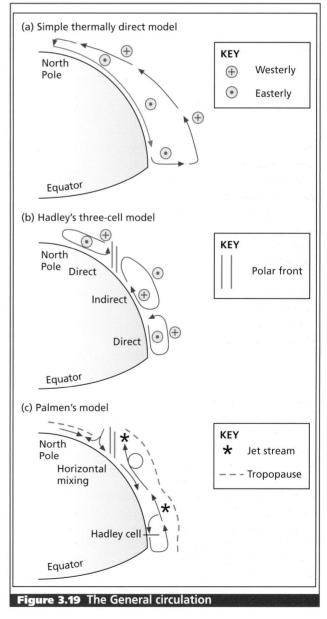

Figure 3.19 The General circulation

The El Niño Southern Oscillation

El Niño, which means the Christ Child, is an irregular occurrence of warm surface water in the Pacific off the coast of South America that affects global wind and rainfall patterns. In July 1997 the sea surface temperature in the eastern tropical Pacific was 2.0–2.5°C above normal, breaking all previous climate records.

In the 1920s Sir Gilbert Walker identified a characteristic of the southern hemisphere which became known as the Southern Oscillation (Figure 3.20). This consisted of a sequence of surface pressure changes within a regular time period of three to seven years, and was most easily observed in the Pacific Ocean and around Indonesia. During normal conditions dry air sinks over the cold waters of the eastern tropical Pacific and flows westwards along the equator as part of the trade winds. The air is moistened as it moves towards the warm waters of the western tropical Pacific. During El Niño periods the pattern is reversed: air rises over the western Pacific and sinks over near Indonesia.

The impact of El Niño is the most prominent phenomenon in year-to-year natural climate variability felt worldwide. During the 1982–83 El Niño, the most destructive this century, damage amounted to about $13 billion, and was blamed for droughts in countries from India to Australia, floods from Ecuador to New Zealand and fires in West Africa and Brazil.

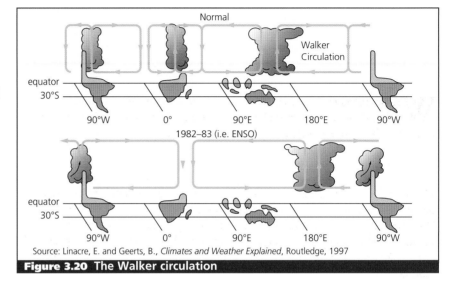

Source: Linacre, E. and Geerts, B., *Climates and Weather Explained*, Routledge, 1997

Figure 3.20 The Walker circulation

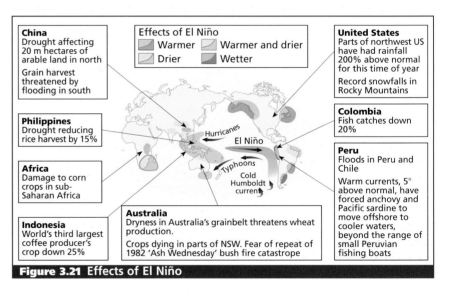

Figure 3.21 Effects of El Niño

Among the effects of the 1997–98 El Niño were:

- a stormy winter in California (the 1982–93 event claimed 160 lives and caused $2 billion damage in floods and mudslides)
- worsening drought in Australia, Indonesia, the Philippines, southern Africa and north-east Brazil
- drought and floods in China
- increased risk of malaria in South America
- lower rainfall in northern Europe
- higher rainfall in southern Europe.

Perception of the El Niño hazard has developed in a series of stages. Until the 1972–73 event it was perceived as only affecting local communities and industries along the eastern Pacific coast near Peru. Between 1972–73 and 1982–83 El Niño was recognised as a cause of natural disasters worldwide. Since 1982–83, however, countries have begun to realise that there is a need for national programmes that will use scientific information in policy planning and that an integrated approach from a number of countries is required to reduce the effect of El Niño.

Weather

Weather refers to daily changes in atmospheric conditions on a small scale, rather than anything global. One of the main factors which influences the weather is the nature and type of air mass affecting an area.

Air masses

The original concept of air masses was that they were bodies of air whose physical properties, especially temperature and humidity, were uniform over a large area. Nowadays, however, they are now redefined as large bodies of air where the variation of the main physical properties are fairly slack. It is generally applied only to the lower layers of the atmosphere, although air masses can cover areas tens of thousands of km² (Figure 3.22).

Air masses derive their temperature and humidity from the regions over which they lie. These regions are known as source regions. The principle ones are:

- areas of relative calm such as semi-permanent high pressure areas
- where the surface is relatively uniform including deserts, oceans and ice-fields.

Modification of air masses

As air masses move from their source regions they may be changed due to the effects of the surface which they move over. These changes create **secondary** air **masses**. For example, a warm air mass that travels over a cold surface is cooled and becomes more stable (Figure 3.23). Hence, it may form low cloud or fog but is unlikely to produce much rain. By contrast, a cold air mass that passes over a warm surface is warmed and becomes less stable. The rising air is likely to produce more rain. Air masses which have been warmed are given the suffix 'w' and those which have been cooled are given the suffix 'k' (*kalt*).

Principal air masses to affect the British Isles

The main air masses to affect the UK are polar (P) and tropical (T) air masses which meet at the polar front (Figure 3.24). Some polar air masses may originate in cool temperate or sub-Arctic areas. Sometimes the British Isles are affected by Equatorial (E) and Arctic (A) air masses. These are then generally divided into maritime (m) or continental (c) depending upon the humidity characteristics of the air, i.e. whether it originated over the ocean and so is moist (m) or over the land and thus dry (c).

Polar Continental (Pc) air masses originate over central Canada and Siberia, which can be extremely cold. The air is cold, dry and cloud free giving extremely cold weather in the UK. Pollution, fog and frost are common. The air mass is initially **stable**, but is warmed as it crosses the North Sea, picks up moisture and

Questions

1 (a) What is meant by the term pressure gradient?
 (b) Does air blow from high pressure to low pressure, or from low pressure to high pressure?
 (c) On a copy of Figure 3.16 mark on the areas of low pressure (with an L) and the areas of high pressure (with a H).
 (d) How do the areas of low and high pressure vary between January and July?
 (e) Suggest reasons to explain the differences you have noted in your answer.
2 (a) What is meant by the term 'El Niño'? Why is it called this?
 (b) What were the possible effects of the 1997–98 El Niño season.

Key Definitions 4

Air mass A large body of air with relatively similar temperature and humidity characteristics.
Anticyclone A high pressure system.
Cyclone An atmospheric low-pressure system that gives rise to roughly circular, inward-spiraling wind motion.
Front A boundary between a warm air mass and a cold air mass, resulting in frontal (depressional or cyclonic) rainfall.
Instability Unstable atmospheric conditions (rising air) likely to cause cloud formation and precipitation.
Weather The state of the atmosphere at a given time and place.

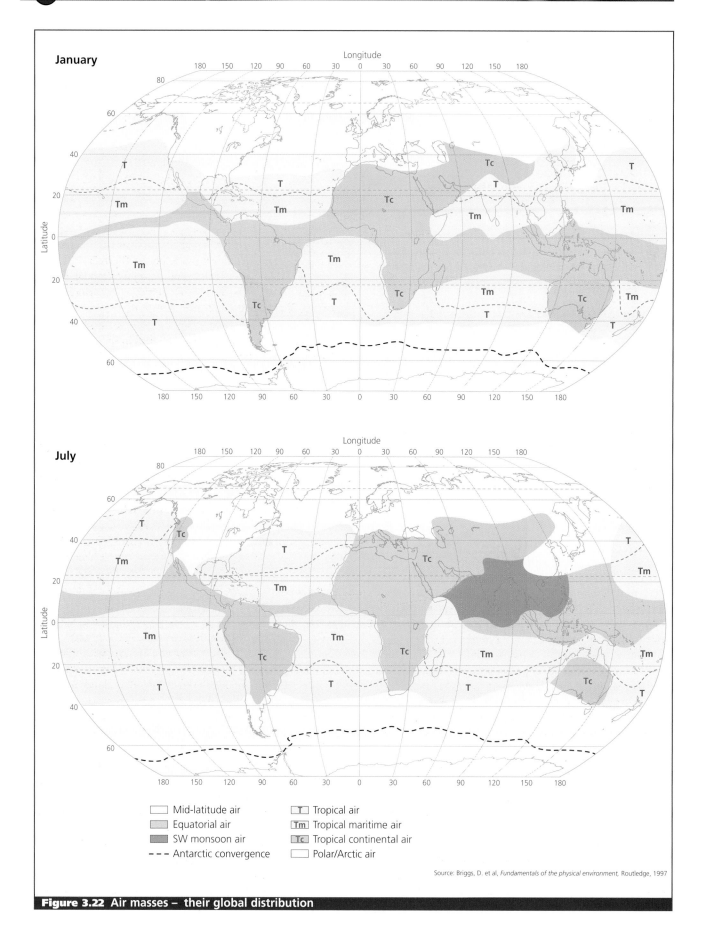

January

July

Mid-latitude air

Equatorial air

SW monsoon air

– – – Antarctic convergence

T Tropical air

Tm Tropical maritime air

Tc Tropical continental air

Polar/Arctic air

Source: Briggs, D. et al, *Fundamentals of the physical environment*, Routledge, 1997

Figure 3.22 Air masses – their global distribution

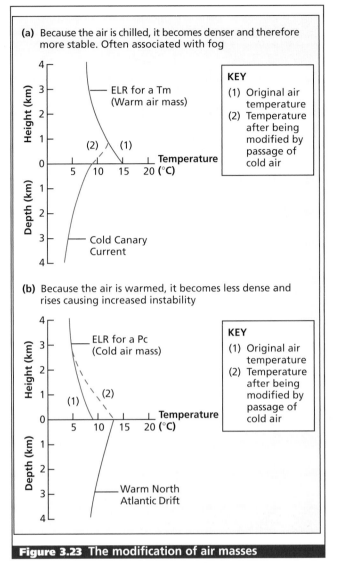

(a) Because the air is chilled, it becomes denser and therefore more stable. Often associated with fog

ELR for a Tm (Warm air mass)

Height (km)

Depth (km)

Cold Canary Current

KEY
(1) Original air temperature
(2) Temperature after being modified by passage of cold air

Temperature
5   10   15   20 (°C)

(b) Because the air is warmed, it becomes less dense and rises causing increased instability

ELR for a Pc (Cold air mass)

Height (km)

Depth (km)

Warm North Atlantic Drift

KEY
(1) Original air temperature
(2) Temperature after being modified by passage of cold air

Temperature
5   10   15   20 (°C)

Figure 3.23 The modification of air masses

may become **unstable** in lower layers. This produces heavy snow in eastern Britain, and bright, clear conditions on the west coast. Pc air often lasts several days. If it occurs in summer, it brings warm, cloudy weather.

Polar maritime (Pm) air comes from the high latitudes over the Atlantic Ocean and is essentially cool, moist and relatively unstable in the lower layers. It gives dull, wet weather with winds from the north-west or west. It is the most common air mass to affect the British Isles. It warms slightly as it crosses Atlantic, hence the instablity in its lower layers. Pm air masses are associated with cumulus clouds and strong winds, especially after the passing of cold front.

Tropical maritime (Tm) air masses are also common. Tm air comes from the oceans of the lower latitudes. It is warm, moist and unstable, especially in summer when convectional heating causing cooling of the air. It produces mild and wet weather in winter with thick cloud cover, while in summer it is warm rather than hot. The lower air is stable, but if forced to rise over hills, upper layers can become unstable to create thunder showers. Tm air masses often produce stratus cloud

producing hill and coastal fog. They are also associated with the warm sector of a depression. Winds are from the south-west and are usually moderate to fresh.

Tropical Continental (Tc) only occurs in summer when sub-tropical high pressure moves northwards. Tc comes from over the hot deserts. Tropical continental air masses are characterised by southerly winds from North Africa or southern Europe bringing hot dry air. They produce heatwave conditions, such as that in 1989. Tc air masses are very stable in lower layers, producing gentle winds and a dusty haze. Associated with high pressure, prolonged spells may cause drought whereas short episodes are associated with pollution and high pollen counts. Their effect varies over the UK. South-east England may suffer drought conditions whereas the north and west might be less affected.

Arctic maritime (Am) air brings northerly winds from Norway/Greenland with very cold air during the winter months. The air is dry and 'biting'. As it travels over the oceans it slowly heats up, picks up water and becomes unstable. It produces snow in winter in Scotland, hail in spring, and heavy showers at other times. Am air masses last several days.

Frontal weather

When two different air masses meet they form a front. For example, when a Pm and a Tm converge the temperature differences between them may be over 10°C. This creates differences in density and allows the warmer air mass to rise over the cooler one. The warmer, lighter air invades the colder, denser air to form the **warm sector**, while warm air rises over the cold air at the **warm front**. Where the cold air pushes the warm air up, a cold front is formed (Figure 3.25).

In general, the appearance of a warm front is heralded by high cirrus clouds. Gradually, the cloud thickens and the base of cloud lowers. Altostratus clouds may produce some drizzle while at the warm front nimbostratus clouds produce rain. A number of changes occur at the warm front (Figure 3.26). Winds reach a peak, are gusty and come from another direction; temperatures suddenly rise; pressure which had been falling remains more constant. The cold front is marked by a decrease in temperature; cumulonimbus clouds and heavy rain; increased wind speeds and gustiness and another change in wind direction, and a gradual increase in pressure. After the cold front has passed, the clouds begin to break up, and sunny periods are more frequent although there may be isolated scattered showers associated with unstable Pm air.

No two low pressure systems are the same. The weather that is found in any depression depends on the air masses involved. The greater the temperature difference between the air masses involved the more severe the weather. Depressions are divided into **ana** and **kata** depressions depending upon the vigour of the uplift of warm air. Where air masses of differing composition meet, an **anafront** is formed and this creates cloud systems of great height. By contrast **katafronts** occur when the air masses are fairly similar in composition (Figure 3.27).

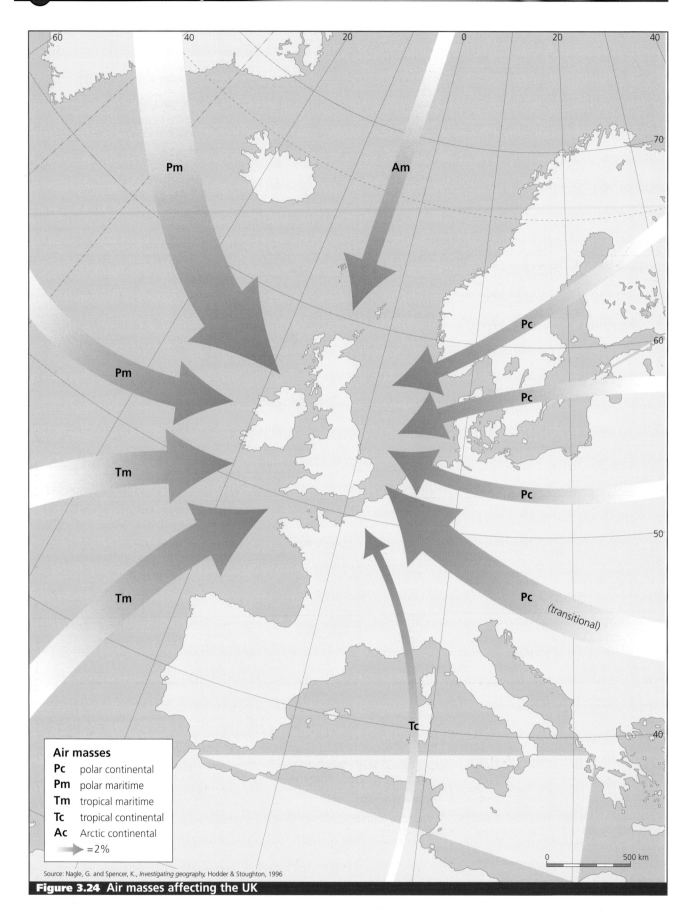

Air masses

Pc polar continental
Pm polar maritime
Tm tropical maritime
Tc tropical continental
Ac Arctic continental
→ = 2%

0   500 km

Source: Nagle, G. and Spencer, K., *Investigating geography*, Hodder & Stoughton, 1996

Figure 3.24 Air masses affecting the UK

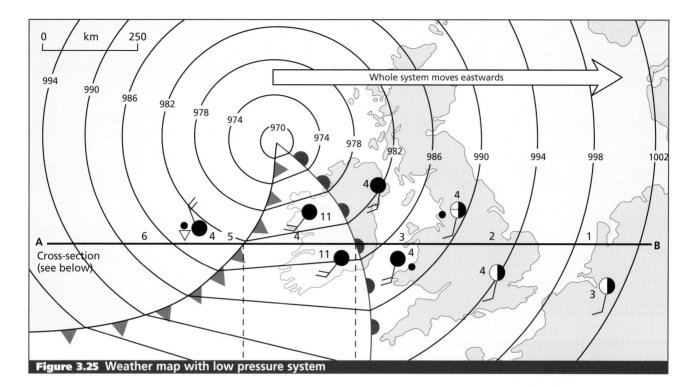

Figure 3.25 Weather map with low pressure system

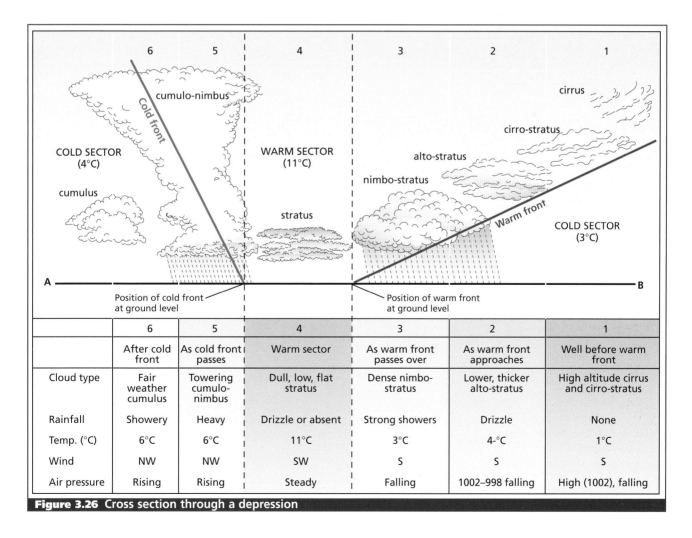

	6	5	4	3	2	1
	After cold front	As cold front passes	Warm sector	As warm front passes over	As warm front approaches	Well before warm front
Cloud type	Fair weather cumulus	Towering cumulo-nimbus	Dull, low, flat stratus	Dense nimbo-stratus	Lower, thicker alto-stratus	High altitude cirrus and cirro-stratus
Rainfall	Showery	Heavy	Drizzle or absent	Strong showers	Drizzle	None
Temp. (°C)	6°C	6°C	11°C	3°C	4-°C	1°C
Wind	NW	NW	SW	S	S	S
Air pressure	Rising	Rising	Steady	Falling	1002–998 falling	High (1002), falling

Figure 3.26 Cross section through a depression

Weather systems

High pressure or anticyclone

In a high pressure system the weather is usually settled and cloudless, but in winter clear skies and light winds can mean frost or fog. Winds blow clockwise around an anticyclone in the northern hemisphere. If high pressure persists over northern Europe in winter, then this can mean a spell of very cold east winds for Britain. In summer, however, high pressure over the British Isles or the continent usually brings warm, fine weather. Figure 3.28a shows a high pressure system off the south-west coast of Britain. The forecast for southern England stated that 'early cloud will clear to sunshine; the day will be crisp; maximum temperatures will reach 8°C but in most places it will be 4–5°C; wind speeds will be low to moderate but will increase during the night as the low pressure system E begins to affect the British Isles'.

Low pressure or depression

Low pressure systems are usually associated with variable weather – strong winds and rain belts, with perhaps snow in winter. Winds blow anticlockwise around a depression in the northern hemisphere. Figure 3.28b shows the weather for Britain just 24 hours after the forecast for Figure 3.28a was made. The forecast now stated that 'the weather will be wet and windy, with heavy bursts of rain; there will be a strong gale force north-westerly wind; maximum temperatures will be 12°C'.

Microclimates

Woodland microclimates

Trees and forests can have a marked effect on climate. For example, air movement is much less within a forest and temperatures are lower than in summer compared with open land. By contrast forests are more humid (less evaporation)

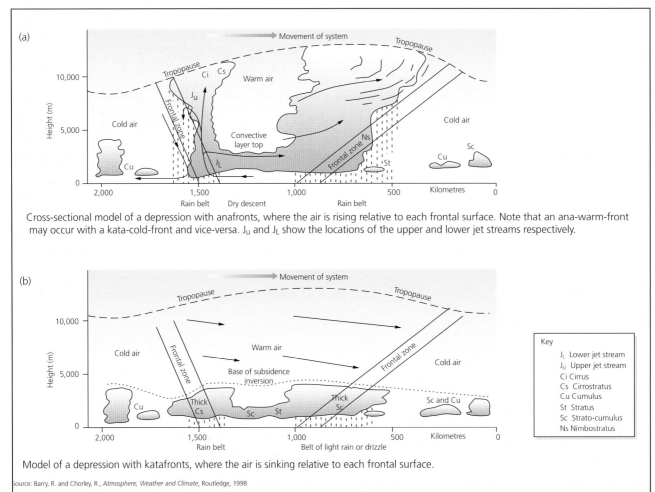

(a) Cross-sectional model of a depression with anafronts, where the air is rising relative to each frontal surface. Note that an ana-warm-front may occur with a kata-cold-front and vice-versa. J_u and J_L show the locations of the upper and lower jet streams respectively.

(b) Model of a depression with katafronts, where the air is sinking relative to each frontal surface.

Key

J_L Lower jet stream
J_u Upper jet stream
Ci Cirrus
Cs Cirrostratus
Cu Cumulus
St Stratus
Sc Strato-cumulus
Ns Nimbostratus

Source: Barry, R. and Chorley, R., *Atmosphere, Weather and Climate*, Routledge, 1998

Figure 3.27 Anafronts and katafronts

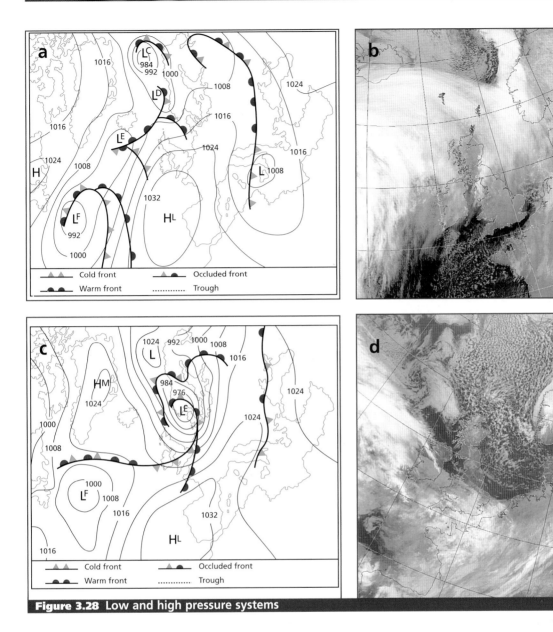

Figure 3.28 Low and high pressure systems

than open land. Most of the incoming radiation is absorbed by the canopy layer, although some energy is reflected. However, the albedo varies between species. On average it is about 15%, so only a small proportion reaches the ground. The direct sunlight is known as sun flex, hence forests do not heat up as much by day; by night vegetation traps and returns much of the outgoing long wave radiation. The presence of water vapour also helps absorb long wave radiation by night.

There are important seasonal differences, for instance deciduous forests lose their leaves in winter and there is then much less absorption and interception. There are also important differences between the type of forests:

■ **deciduous trees create larger seasonal differences than evergreens**

■ **large leaved trees such as sycamore absorb more energy than small leaved trees such as birch and oak**

■ **oak trees have a higher density of leaves than birch trees, hence more light reaches the ground in a birch wood.**

If a forest is layered there will be additional interception at each layer: In addition the outgoing long-wave radiation will be absorbed at each layer: in a tropical rainforest with as many as five layers up to 99.9% of the energy available is absorbed by the trees, and less than 0.1% reaches ground level. Trees also effect wind flow. Hedges have a similar, if less pronounced, effect on microclimate. A 2 m high hedge could result in reduced wind speeds as much as 56 m beyond it, with the maximum decrease – to 40% of the original speed – occurring some 8 m beyond the hedge. This dramatic effect makes hedges extremely important in providing shelter for stock. The planting of long, thin blocks of trees as shelter belts has a similar effect over even longer distances. Hedges have a number of other microclimatic effects. Soil moisture content, daytime air and soil temperatures can be increased

Figure 3.29 Difference of relative humidity (%) between the inside and outside of a forest

Forest	January	April	July	October	Year
Deciduous broad-leaf	3.4	3.2	−0.8	1.1	2.2
Needle tree (conifer)	4.8	4.8	6.5	9.5	6.8
Japanese cedar	1.6	−1.1	1.5	0.5	0.8

Note: Positive values indicate that inside the forest is more humid.

Source: Briggs, D et al, *Fundamentals of the physical environment*, Routledge 1997.

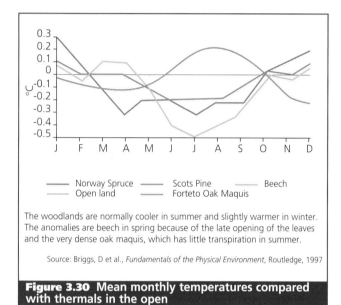

Norway Spruce ——— Scots Pine ——— Beech
Open land ——— Forteto Oak Maquis

The woodlands are normally cooler in summer and slightly warmer in winter. The anomalies are beech in spring because of the late opening of the leaves and the very dense oak maquis, which has little transpiration in summer.

Source: Briggs, D et al., *Fundamentals of the Physical Environment*, Routledge, 1997

Figure 3.30 Mean monthly temperatures compared with thermals in the open

by as much as 16–20% in the lee of a hedge, with the effect reaching as far as around 10 times the height of the hedge. Evaporation can be significantly decreased at a distance of around 15 times the height of the hedge.

Vapour pressure is higher in a forest than in open land. This is because of the presence of large amounts of moisture from the leaves and the low rates of evaporation due to cooler temperatures and low wind speeds.

Urban climates

Urban climates occur as a result of extra sources of heat released from industry, commercial and residential buildings as well as from vehicles. In addition, concrete, glass, bricks and tarmac, all act very differently from soil and vegetation. Some of these – notably dark bricks – absorb large quantities of heat and release them slowly by night. In addition, the release of pollutants helps trap radiation in urban areas. Consequently, urban microclimates can be very different from rural ones.

Questions

1 Study Figures 3.22 to 3.24 which deal with air masses. What effect does **(a)** latitude and **(b)** continentality (other word) have on the characteristics of an air mass. Give examples.

2 How would an air mass change if it where to pass over **(a)** a warm wet surface and **(b)** a cold dry surface. Give examples to back up your answer.

3 Why should **(a)** a high pollen count and high levels of pollution be associated with tropical continental air and **(b)** fog, frost and pollution with polar continental air?

4 Why are tropical maritime and polar maritime airstreams more common than others over the British Isles? How would the origin and frequency of air masses affecting the UK vary between January and July? Explain your answer.

5 Figures 3.28 A and C show the weather conditions over the British Isles on 16 and 17 December respectively. Figures 3.28 B and D show the satellite image for A and C respectively. Describe the weather conditions in southern England as shown on Figure A. How does this change by the 17 December?
Explain why the weather on the 16 and 17 December was so different.

Urban heat budgets differ from rural ones. In urban areas brick, concrete and stone have high heat capacities and a kilometre of an urban area may contain a greater surface area than a kilometre of countryside. The large surface area in urban areas allows a greater area to be heated.

In urban areas there is relative lack of moisture, this is due to:

■ lack of vegetation

■ high drainage density (sewers and drains) which remove water.

Little energy is therefore used for evapotranspiration, and more is available to heat the atmosphere. This is in addition to the man-made sources of heating such as industries, cars and the activities of people.

At night the ground radiates heat and cools, while in urban areas the release of heat by buildings offsets the cooling process. Moreover, some industries, commercial activities and transport networks continue to release heat throughout the night (Figure 3.32). The contrast between urban and rural is greatest under calm high pressure conditions. The typical heat profile of an urban heat island shows the maximum at the city centre, a plateau across the suburbs and a temperature cliff between the suburban and

Figure 3.31 Effects of urbanisation on climate: average urban climatic differences expressed as a percentage of rural conditions

Measure	Annual	Cold season	Warm season
Pollution	+500	+1000	+250
Solar radiation	−10	−15	−5
Temperature	+2	+3	+1
Humidity	−5	−2	+10
Visibility	−15	−20	−10
Fog	+10	+15	+5
Windspeed	−25	−20	−30
Cloudiness	+8	+5	+10
Rainfall	+5	0	+10
Thunderstorms	+15	+5	+30

*Note: Temperature is expresed as a difference only, not as a percentage.

Source: Briggs, D et al., *Fundamentals of the physical environment*, Routledge,1997.

Figures A to D show the sites where rain gauges and maximum-minimum thermometers were placed in order to assess the impact of vegetation on interception, and variations in temperature around a school.

The following table shows some of the data that was collected during the study.

Figure A Location: Close to a canal; Vegetation: Open grassland

Minimum temperature recorded

(°C)	7	5	6	8	6	7	3	3	5	6	6	4	3	3

Rainfall recorded

(mm)	12	7	0	0	2	5	10	8	5	0	0	3	10	8

Figure B Location: 250m from canal, 250m from main road; Vegetation: Deciduous woodland

(°C)	8	7	7	9	7	9	4	4	6	8	7	5	4	5
(mm)	8	3	0	0	1	3	8	5	3	0	0	2	6	4

Figure C Location: 250m from canal, 250m from main road; Vegetation: Coniferous woodland

(°C)	8	8	7	9	8	10	5	4	6	7	7	5	5	6
(mm)	2	1	0	0	0	1	3	2	1	0	0	0	1	1

Figure D Location: Next to kitchens and buildings

(°C)	9	8	9	10	8	9	7	7	7	8	8	7	7	7
(mm)	4	2	0	0	1	1	3	2	1	0	0	1	3	2

Questions

1 Define the terms interception; evapotranspiration; specific heat capacity.
2 Choose an appropriate method to show variations in the rainfall and temperature recorded at the four sites over the period of 14 days. For example you could use a dispersion diagram to show variations in rainfall, and a line graph to show variations in temperature.
  Comment on the results that you have produced.
3 Suggest ways in which **(a)** the presence of the canal **(b)** the buildings and **(c)** the vegetation influences microclimate.

Interception is high under evergreen shrubs

Open grassland and woodland

Close to the canal

Close to the kitchens during a flood

Coniferous forest

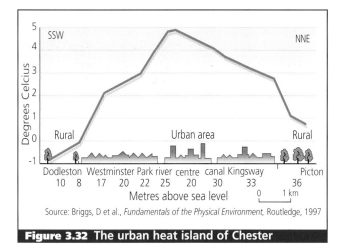

Source: Briggs, D et al., *Fundamentals of the Physical Environment*, Routledge, 1997

Figure 3.32 The urban heat island of Chester

rural area (Figure 3.32). Small-scale variations within the urban heat island occur due to the distribution of industries, open space, rivers, canals and so on.

The urban heat island effect is caused by at least four factors:

- heat produced by human activity, which can equal 50% of incoming energy in winter
- buildings have a high thermal capacity in comparison to rural areas – up to six times greater than agricultural land

- there are fewer bodies of open water, therefore less evaporation and fewer plants, therefore less transpiration
- the composition of the atmosphere, in which smog, smoke or haze traps heat in the urban atmosphere.

Air flow over an urban area is disrupted, winds are slow and deflected over buildings. Large buildings can produce eddying (Figure 3.33).

Nevertheless, the nature of urban climates is changing. For example, with the decline in coal as a source of energy, there is less sulphur dioxide pollution and thus less hygroscopic nuclei, hence less fog. However, an increase in cloud cover has occurred for a number of reasons:

- greater heating of the air
- increase in pollutants
- frictional and turbulent effective air flow
- changes in moisture.

1 Why are microclimates, such as urban heat islands, best observed during high pressure (anticyclonic) weather conditions?

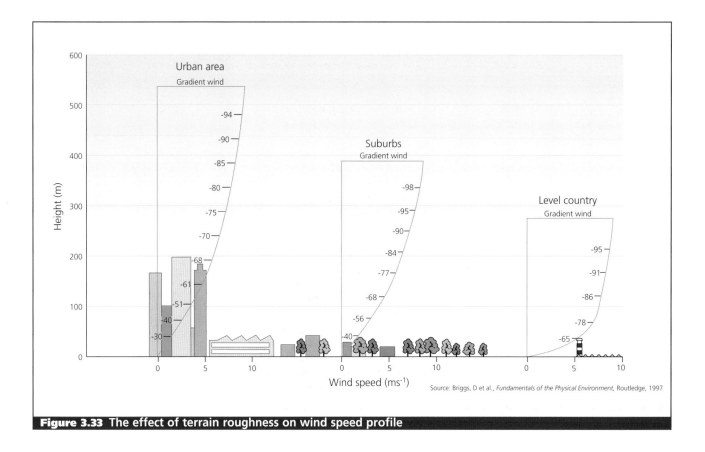

Source: Briggs, D et al., *Fundamentals of the Physical Environment*, Routledge, 1997

Figure 3.33 The effect of terrain roughness on wind speed profile

Weather systems: impact on human activity

Experiencing the weather is probably most people's most direct and frequent experience of the physical environment. This is especially true for those who live in urban areas who are less likely to experience other aspects of the physical environment. The impact on human activity ranges from the spectacular to the mundane. The weather affects what we wear, how we feel, and, in some cases, our levels of health (Figure 3.34).

Biometeorology is the study of the effects of weather and climate on people (animals and plants) (Figure 3.35). In addition to the natural environment, it is possible to study the microenvironment of the home, school or office on people's well-being.

An extreme view is that of **environmental determinism**. This states that if certain environmental conditions exist, the result can be predicted. According to Ellen Semple (1911) 'hot, moist equatorial climates encourage the growth of large forests which harbour abundant game and yield abundant fruits; they prolong the hunter gatherer stage of development and retard the advance to agriculture.'

Such an extreme view has been criticised on two main counts. First, similar environments do not always produce the same result. The Greek and Roman empires flourished in Mediterranean climates, but other Mediterranean areas such as California, South Africa and south-east Australia have not seen similar empires. Second, determinism fails to recognise the ways in human activity can affect the environment.

Figure 3.34 Applied meteorology: sectors and activities where climate has social, economic and environmental significance

Primary sector	General activities	Specific activities
Food	Agriculture	Land use, crop scheduling and operations, hazard control, productivity, livestock and irrigation, pests and diseases
	Fisheries	Management, operations, yield
Water	Water disasters	Flood-, drought-, pollution-abatement
	Water resources	Engineering design, supply, operations
Health and community	Human biometeorology	Heath, disease, morbidity and mortality
	Human comfort	Settlement design, heating and ventilation, clothing, acclimatisation
	Air pollution	Potential, dispersion, control
	Tourism and recreation	Sites, facilities, equipment marketing, sports activities
Energy	Fossil fuels	Distribution, utilisation, conservation
	Renewable resources	Solar-, wind-, water-power development
Industry and trade	Building and construction	Sites, design, performance, operations, safety
	Communications	Engineering design, construction
	Forestry	Regeneration, productivity, biological hazards, fire
	Transportation	Air, water and land facilities, scheduling, operations, safety
	Commerce	Plant operations, product design, storage of materials, sales planning, absenteeism, accidents
	Services	Finance, law, insurance, sales

Source: Goudie, A (ed.), *The Encyclopaedic Dictionary of Physical Geography*, Blackwell, 1994

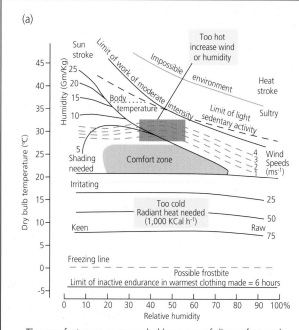

(a)

The comfort zone, surrounded by zones of discomfort and danger, within the range of climates experienced on the earth.

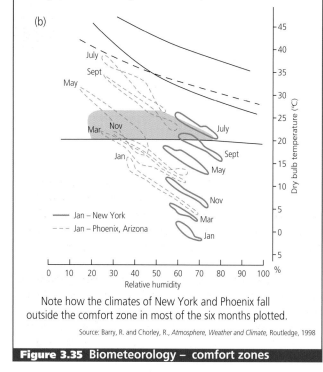

(b)

Note how the climates of New York and Phoenix fall outside the comfort zone in most of the six months plotted.

Source: Barry, R. and Chorley, R., *Atmosphere, Weather and Climate*, Routledge, 1998

Figure 3.35 Biometeorology – comfort zones

Rather than disregarding the influence of the environment entirely, however, the idea of **possibilism** was developed. This suggests that humans can act in a variety of ways in a given environment, but that some are more likely than others. This gives humans choice, and makes them active agents within the environment, but it also suggests that the environment sets limits within which human activity must take place.

It is not easy to measure the influence of the environment

on human activity, and increasingly, most people have little direct interaction with the environment apart from the weather. Nevertheless, there is evidence that climate and weather has an impact on a variety of features such as human comfort (Figure 3.35), patterns of disease, crime and suicide. Figure 3.36 shows how environments vary in terms of their suitability for human activity.

An important influence of weather and climate on human activity is through the effects of hazards. Climatic hazards such as tornadoes, hurricanes, storms and droughts are dealt with elsewhere in this chapter. However, weather-related losses are large, and increasing. Between 1950 and 1989 over 75% of insured losses in the USA were due to weather disasters.

Weather and climate have a direct impact on health and death rates (Figure 3.36). Death rates increase once a critical temperature has been reached. Mortality rates are higher, too, in cloudy, damp, snowy places. Certain diseases have a seasonal pattern. Even factors such as fog, poor driving conditions, wind chill and sunshine intensity can have an effect on our lives.

In addition, climate and weather have an important effect on economic activities. The most obvious example is that of agriculture, but there are increasingly important effects on the location of manufacturing industries and house building. The impact on farming systems is well known:

- **plants require water to survive and grow – too much may cause soil erosion, too little may cause plants to die**

- **most plants require temperatures of over 6°C for successful germination of seeds; accumulated temperatures are the total amount of heat required to produce optimum yield, for example, wheat needs about 1300°C**

- **wind can increase evapotranspiration rates and erode soil**

- **cloud cover may reduce light intensity and delay harvesting.**

Questions

1 Study Figure 3.36 which shows seasonal variations in deaths in London and in a rural part of South Africa.
  (a) Describe the seasonal pattern of respiratory illnesses in London
  (b) Describe seasonal variations in bowel (gastro-enteritis) complaints for London.
  (c) Describe the seasonal pattern of burns in South Africa. Suggest reasons for the patterns you have outlined above.

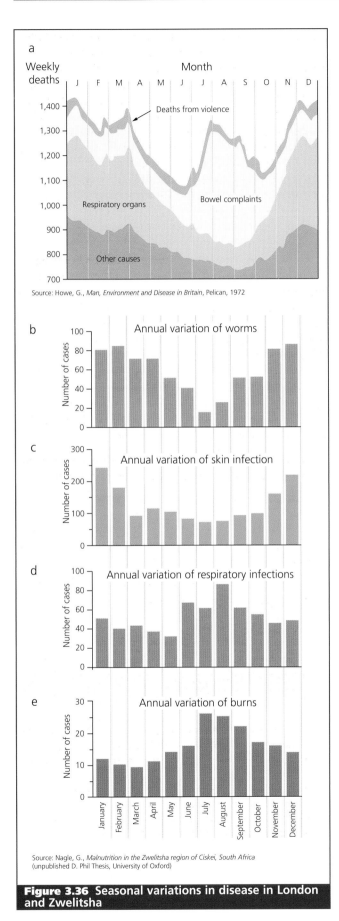

a

Weekly deaths

Month

J F M A M J J A S O N D

Deaths from violence

Respiratory organs

Bowel complaints

Other causes

1,400
1,300
1,200
1,100
1,000
900
800
700

Source: Howe, G., *Man, Environment and Disease in Britain*, Pelican, 1972

b

Annual variation of worms

Number of cases

0 20 40 60 80 100

c

Annual variation of skin infection

Number of cases

0 100 200 300

d

Annual variation of respiratory infections

Number of cases

0 20 40 60 80 100

e

Annual variation of burns

Number of cases

0 10 20 30

January
February
March
April
May
June
July
August
September
October
November
December

Source: Nagle, G., *Malnutrition in the Zwelitsha region of Ciskei, South Africa* (unpublished D. Phil Thesis, University of Oxford)

Figure 3.36 Seasonal variations in disease in London and Zwelitsha

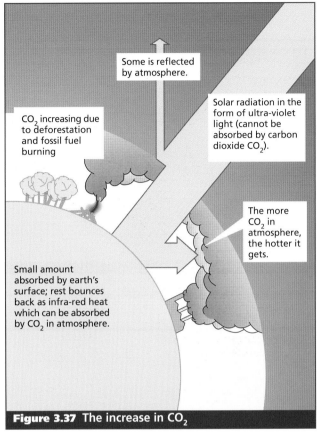

Some is reflected by atmosphere.

Solar radiation in the form of ultra-violet light (cannot be absorbed by carbon dioxide CO_2).

CO_2 increasing due to deforestation and fossil fuel burning

The more CO_2 in atmosphere, the hotter it gets.

Small amount absorbed by earth's surface; rest bounces back as infra-red heat which can be absorbed by CO_2 in atmosphere.

Figure 3.37 The increase in CO_2

The greenhouse effect and global warming

The earth's atmosphere is vital for life, and changes to it disrupt the natural balance of the earth's energy budget. The earth's temperature changes for a number of natural reasons. Recently, changes in atmospheric composition have been linked to an increase in global temperature (Figure 3.37).

Solar radiation is short-wave infra-red radiation and is able to pass through the atmosphere and warm the land and the sea. The infra-red heat that is re-radiated from the earth (as long-wave radiation) warms the lower part of the atmosphere. Carbon dioxide allows short-wave radiation to pass through and is able to absorb long-wave radiation. This is known as the greenhouse effect (Figure 3.38).

As long as the amount of carbon dioxide in the atmosphere and the amount of solar energy both remain the same then the temperature of the earth should also stay the same. However, human activities are upsetting the natural balance by increasing the amount of carbon dioxide in the atmosphere, as well as the other greenhouse gases.

There are a number of these greenhouse gases, such as methane, ozone, nitrous oxides and chlorofluorocarbons (CFCs) (Figure 3.39). They also absorb long-wave radiation.

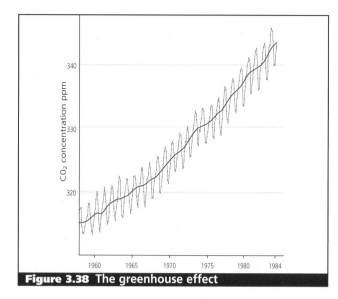

Figure 3.38 The greenhouse effect

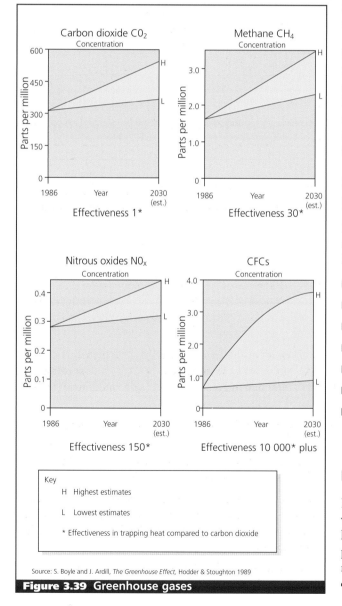

Key

H Highest estimates

L Lowest estimates

* Effectiveness in trapping heat compared to carbon dioxide

Source: S. Boyle and J. Ardill, *The Greenhouse Effect*, Hodder & Stoughton 1989

Figure 3.39 Greenhouse gases

Some are more powerful than others. One molecule of **CFC**, for example, has the same greenhouse impact as 10 000 molecules of carbon dioxide. **Methane** is another greenhouse gas. At present it has an atmospheric concentration of about 1.7 ppm and is increasing at a rate of about 1.2% per annum. This is largely due to the biological activity of bacteria in paddy fields and also due to the release of gas from oil and gas fields. The amount of **nitrous oxide** (NOx) is increasing from a concentration 0.3 ppm at an annual rate of 0.3% thanks to an increase in nitrogen-based fertilisers. **Ozone** near the ground in the troposphere is increasing also as a result of human activities. By 2030 increases in these minor greenhouse gases will probably have the same impact as the doubling of carbon dioxide from 270 ppm to 540 ppm.

The trend in carbon dioxide levels shows a clear annual pattern. This is associated with seasonal changes in vegetation, especially over the northern hemisphere. In addition, by the 1970s there was a second trend, one of a long term increase in carbon dioxide levels, superimposed upon the annual trends.

Studies of cores taken from ice packs in Antarctica and Greenland show that the level of carbon dioxide between 10 000 years ago and the mid-19th century was stable at about 270 ppm. By 1957 this had risen to 315 ppm and it has since increased to about 360 ppm. Most of the extra carbon dioxide has come from the burning of fossil fuels, especially coal, although some of the increase may be due to the disruption of the rainforests.

Some effects of the rise in greenhouse gases

Researchers have considered the effects of a doubling of carbon dioxide from 270 ppm to 540 ppm. Such a rise would lead to:

- an increase of temperatures by about 2°C
- greater warming at the poles rather than at the equator
- changes in prevailing winds
- changes in precipitation
- continental areas becoming drier
- sea level rising by as much as 60 cm
- ice caps changing in size (in fact they may increase due to more evaporation in lower latitudes and increase snowfall at higher latitudes) (Figure 3.40).

Bangladesh

Bangladesh is one of the least developed countries in the world. It has an average GNP of less than US$200 and a lower per capita use of fossil fuels. Less than one-third of the population is literate. The population, approximately 120 million in 1991, is growing at an annual rate of 2.2%. The densely populated low-lying country of about 144 000 km²

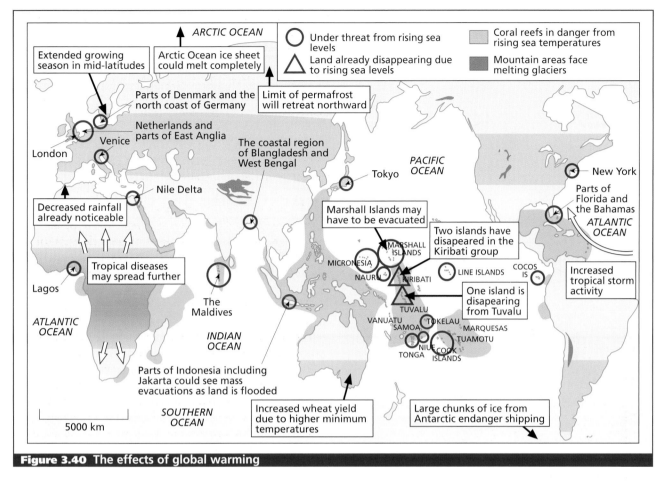

Figure 3.40 The effects of global warming

consists largely of the delta of three of the world's main rivers: the Ganges, the Bramaphutra and the Meghna. Bangladesh suffers recurrent climate related disasters: floods in 1987–88 and a cyclone in April 1989 are good examples. For these reasons Bangladesh may be one of the countries least responsible for the causes of climatic change but one of the most vulnerable to the effects of climatic change, especially changes in sea-level.

Sea-level change is not just a question of sea-levels rising. Instead it includes adverse environmental effects such as salt intrusion into fresh water and ground water, erosion of beaches or cliffs, and the flooding of farmland and homes.

The physical impacts on Bangladesh of a rise in sea-level are difficult to predict: this is because the coastal system responds dynamically as sea levels rise. Huge quantities of sediment, about 1 to 2.5 billion tonnes per year are carried by the rivers whose combined flood level can exceed 140 000 m^2 per second into Bangladesh from the whole of the Himalaya drainage system including the countries of Nepal, China and India. About two-thirds of this sediment goes into the Bay of Bengal and over the long-term causes land subsidence within the delta. Consequently the delta is changing. As a result it is debatable whether residents would even notice a rise of 6 mm a year or not. Indeed, if sedimentation rates keep pace with sea level rises the delta may remain very similar to what it is at present.

The main problem for Bangladesh is the increase from storms, especially extreme events which cause major storm surges.

Climatic change impacts in the UK

By 2030, winters in the UK could be approximately 1.5 to 2.1°C warmer than now while summer temperatures might rise by 1.5°C. As a result of these rises there could be a number of changes in Britain. This include:

- **movement of species northwards and to higher elevations; many sensitive species might be lost to the UK, although migration and invasion could increase the overall number of species**

- **sea-level could increase by 20 cm relative to today**

- **there might be an increased frequency of storm events and coastal flooding – areas particularly vulnerable to changes in sea-level are shown in Figure 3.40**

- **the water content of soils would be likely to decrease due to increased evaporation – this would have a major effect on the types of crops, trees and other land uses that soils can support**

- **many soils would shrink more than usual, with important implications for structural stability – the areas**

The potential impact of the greenhouse effect on Britain

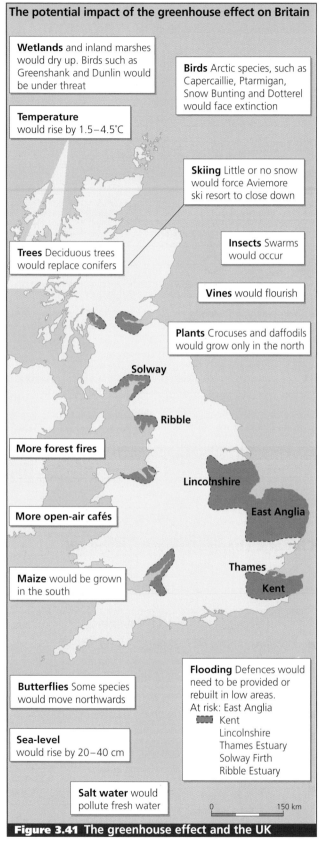

Wetlands and inland marshes would dry up. Birds such as Greenshank and Dunlin would be under threat

Birds Arctic species, such as Capercaillie, Ptarmigan, Snow Bunting and Dotterel would face extinction

Temperature would rise by 1.5–4.5°C

Skiing Little or no snow would force Aviemore ski resort to close down

Trees Deciduous trees would replace conifers

Insects Swarms would occur

Vines would flourish

Plants Crocuses and daffodils would grow only in the north

Solway

Ribble

More forest fires

Lincolnshire

East Anglia

More open-air cafés

Thames

Kent

Maize would be grown in the south

Butterflies Some species would move northwards

Flooding Defences would need to be provided or rebuilt in low areas.
At risk: East Anglia
Kent
Lincolnshire
Thames Estuary
Solway Firth
Ribble Estuary

Sea-level would rise by 20–40 cm

Salt water would pollute fresh water

0   150 km

Figure 3.41 The greenhouse effect and the UK

most affected would be central, eastern and southern England where there are clayey soils of large shrink-swell potential

- increases in the frequency of hot, dry periods would lead to decreases in water availability but increases in water demand

- groundwater levels (groundwater provides about 20% of the water supply to England and Wales).

The ozone layer

The amount of ozone in the atmosphere is a small but vital component of the atmosphe. Ozone is created by oxygen rising up from the top of the troposphere and reacting under sunlight to form ozone. Most ozone is created over the equator and the tropics because this is where solar radiation is strongest. However, winds within the upper atmosphere transport the ozone towards the polar regions where it tends to concentrate.

Ozone is constantly being produced and destroyed in the stratosphere in a natural dynamic balance. As well as being produced by sunlight it is also being destroyed by nitrogen oxides. However, this balance can be distorted by human activities. There is now clear evidence that human activities have led to the creation of a hole in the ozone layer over Antarctica (Figure 3.42).

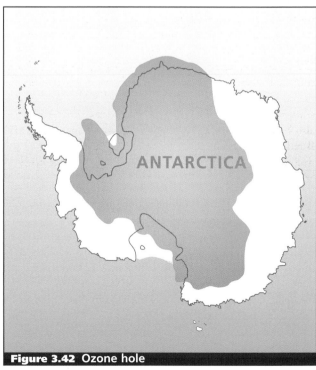

Figure 3.42 Ozone hole

The hole in the ozone layer over Antarctica was first discovered in 1982. It follows a very clear seasonal pattern – each spring time in Antarctica (between September and October) there is a huge reduction in the amount of ozone from the stratosphere. At the end of the long polar winter ozone is present in roughly the same quantities that were there in the 1960s and 1970s; as the summer develops the concentration of ozone recover – so what causes the depletion in ozone during the spring time?

During winter in the southern hemisphere the air over Antarctica is cut off from the rest of the atmosphere by strong winds – these winds block warm air from entering into Antarctica. The temperature over Antarctica therefore becomes very cold, often down as far as −90°C. This allows clouds of ice to form. Chemical reactions take place on this ice which include chlorine compounds resulting from pollution by human activities. These reactions release chlorine atoms. Once the sun returns during the spring the chlorine destroys ozone in a series of further chemical reactions. Hence the hole in the ozone layer occurs very rapidly in the spring. By summer however, the ice clouds have evaporated and the chlorine is converted to other compounds such as chlorine nitrate which remain dormant until the following winter.

The sources of chlorine atoms are called chlorofluorocarbons (CFCs) and include materials used in fridges, foamed plastic and aerosols. CFCs are particularly dangerous because they can be very long lived – over a hundred years – and they spread throughout the atmosphere. In the case of Antarctica, the build-up of chlorine appears to have very little impact until it reaches a critical threshold. Once this is reached only a small increase in chlorine will lead to a huge change in the ozone layer (Figure 3.43).

The hole in the ozone layer has major implications. This is because ultraviolet radiation could reach the ground in increased quantities. Some ultraviolet reaches the ground already – it is in the 290–320 wave band. This is known as UV-B, and can cause sunburn, skin cancer and eye problems such as cataracts. It is estimated that for every 1% decrease in the concentration of ozone there will be a 5% increase in the amount of skin cancers each year. Crops and animals have also been tested to see how they react with an increase in UV-B radiation – soya bean for example experiences a 25% decline in yield when UV-B increases by 25%, while cattle are affected by eye complaints including cancer of the eye.

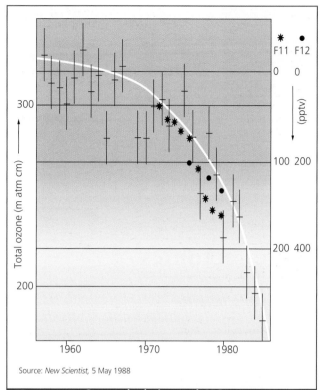

Source: *New Scientist*, 5 May 1988

Figure 3.43 Ozone depletion and CFCs

Questions

1 **(a)** What is meant by the term greenhouse effect?
  (b) What are the greenhouse gases? How do they vary in their contribution to the greenhouse effect?
  (c) With the use of examples, explain the consequences of the greenhouse effect.

2 **(a)** Why is ozone in the atmosphere important?
  (b) Under what natural conditions is ozone destroyed?
  (c) How has human activity led to a destruction of the ozone layer?
  (d) What are the consequences of a depletion of atmospheric ozone?

4 plate tectonics, earth processes and natural *hazards*

Plate tectonics

The earth's interior

The theory of plate tectonics states that the earth is made up of a number of layers (Figure 4.1). On the outside there is a very thin crust and underneath is a mantel which makes up 82% of the volume of the earth. Deeper still is a very dense and very hot core. In general these concentric layers become increasingly denser towards the centre. The density of these layers is controlled by temperature and pressure. Temperature softens or melts rocks.

Close to the surface rocks are mainly solid and brittle. This upper surface layer is known as the **lithosphere** and includes the **crust** and the upper **mantle**, and is about 70 km deep. The earth's crust is commonly divided up into two main types – **continental crust** and **oceanic crust** (Figure 4.2). In continental areas silica and aluminium are very common. When combined with oxygen they make up the most common rock, granite. By contrast, below the oceans the crust consists mainly of basaltic rock in which silica, iron and magnesium are most common.

The evidence for plate tectonics

The evidence of plate tectonics includes:

- the past and present distribution of earthquakes (Figure 4.3)
- changes in the earth's magnetic field
- the 'fit' of the continents: in 1620 Francis Bacon noted how the continents on either side of the Atlantic could be fitted together like a jigsaw
- glacial deposits in Brazil match those in West Africa
- the fossil remains in India matches that of Australia
- the geological sequence of sedimentary and igneous rocks in parts of Scotland match those in Newfoundland
- ancient mountains can be traced from east Brazil to West Africa, and from Scandinavia through Scotland to Newfoundland and the Appalachians (eastern USA)
- fossil remains of a small aquatic reptile, mesosaurus, which lived about 270 million years ago are found only in a restricted part of Brazil and the south-west of Africa. It is believed to be a poor swimmer!

In the early 20th century the American Harry Hess suggested that convection currents would force molten rock (magma) to well up in the interior and crack the

Key Definitions

Plate tectonics A group of theories that states that the earth's crust is separated into a number of plates which float on a semi-molten interior. It amounts for the distribution of volcanoes, earthquakes and fold mountains.

Volcano A mountain or plateau created by the eruption of lava and ash during volcanic activity.

Earthquate A sudden movement of the earth, usually caused when pressure at plate boundaries exceeds the ability of the rocks to withstand such pressure.

Hazards A natural event that threatens human life and/or property.

Epicentre The point on the earth's surface immediately above the centre (focus) of an earthquake.

Lahars Mudflows caused when water mixes with ash, dust and/or soot, often following a volcanic eruption.

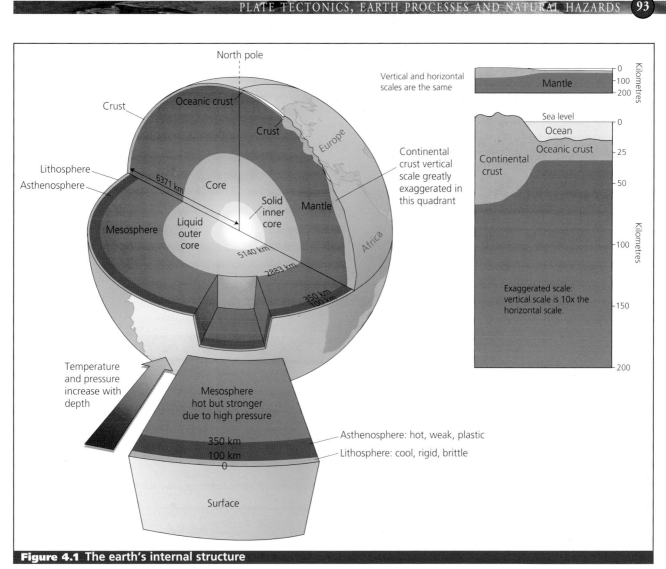

Figure 4.1 The earth's internal structure

Figure 4.2 A comparison of oceanic crust and continental crust

Examples	Continental crust	Oceanic crust
Thickness	35 to 70 km on average	6 to 10 km on average
Age of rocks	very old, mainly over 1500 million	very young, mainly under 200 million years
Sight of rocks	lighter with an average density of 2.6 light in colour	heavier with an average density of 3.0 dark in colour
Nature of rocks	numerous types, many contain silica and oxygen, granite is the most common	few types, mainly basalt

crust above and force it apart. In the 1960s research on rock magnetism supported Hess. The rocks of the Mid-Atlantic Ridge were magnetised in alternate directions in a series of identical bands on both sides of the ridge. This suggested that fresh magma had come up through the centre and forced the rocks apart. In addition, with increasing distance from the ridge the rocks were older. This supported the idea that new rocks were being created at the centre of the ridge and the older rocks were being pushed apart.

In 1965 a Canadian geologist J. Wilson linked together the ideas of continental drift and sea floor spreading into a concept of mobile belts and rigid plates, which formed the basis of plate tectonics (Figure 4.4).

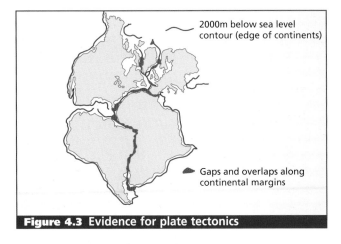

2000m below sea level contour (edge of continents)

Gaps and overlaps along continental margins

Figure 4.3 Evidence for plate tectonics

Plate boundaries

The zone of earthquakes around the world has helped to define six major plates and a number of minor ones (Figure 4.5). The boundaries between plates can be divided into two main types: spreading plates and colliding plates. Spreading ridges, where new crust is formed, are mostly in the middle of oceans (Figure 4.6a). These ridges are zones of shallow earthquakes (less than 50 km below the surface). Where two plates converge a deep sea trench may be formed and one of the plates is **subducted** (forced downwards) into

the mantle. In these areas fold mountains are created and chains of island arcs may be formed (Figure 4.6b). Deep earthquakes, up to 700 km below the surface, are common. Good examples include the trenches off the Andes and the Aleutian Islands that stretch out from Alaska. The partial melting of the descending ocean plate causes volcanoes to form in an arc shaped chain of islands, such as in the Caribbean.

On some plate boundaries the plates slide past one another to create a transform fault (fault zone) without colliding or separating (Figure 4.6c). These are also associated with shallow earthquakes. An example is the San Andreas Fault in California. Where continents embedded in the plates collide with each other there is no subduction but crushing and folding may create young fold mountains such as the Himalayas and the Andes (Figure 4.6d).

Questions

1 Briefly outline the evidence for plate tectonics.
2 What is a convection current? How does it help explain the theory of plate tectonics?
3 What happens at
   (a) a mid-ocean ridge
   (b) a subduction zone?

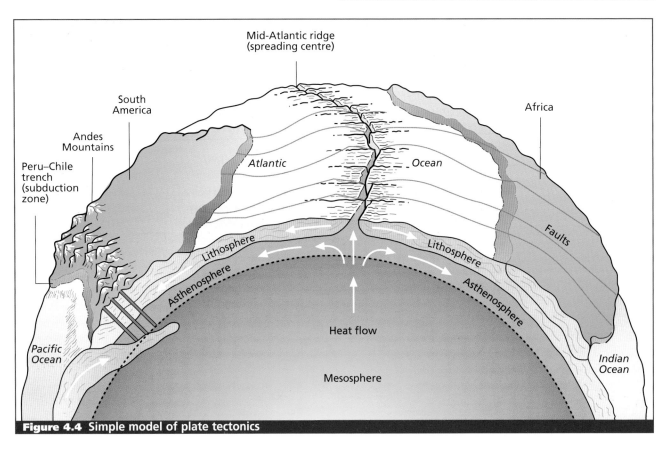

Figure 4.4 Simple model of plate tectonics

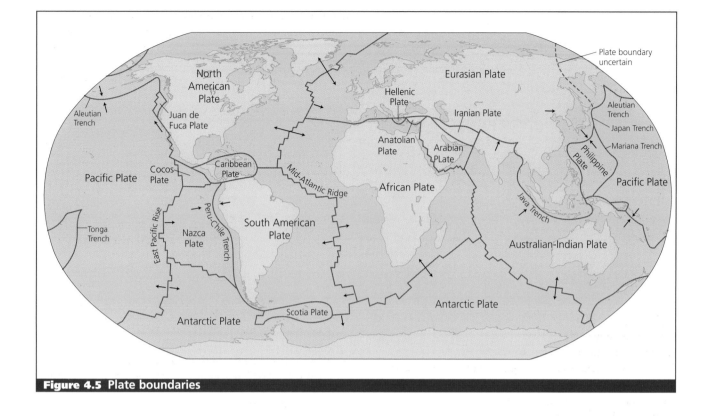

Figure 4.5 Plate boundaries

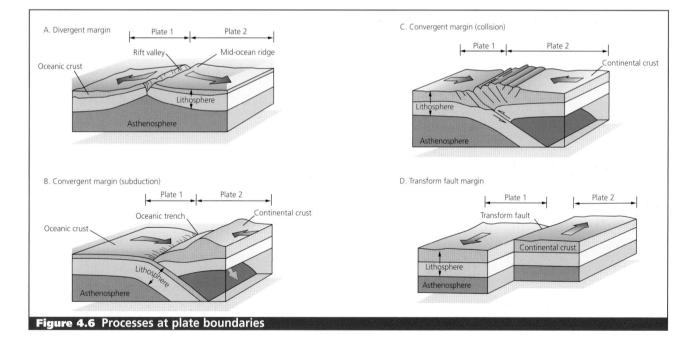

Figure 4.6 Processes at plate boundaries

Earthquakes

An earthquake is a sudden, violent movement of the earth. They occur after a build up of pressure causes rocks to give way. Most earthquakes are found at plate boundaries (Figure 4.7), but others are caused by human activity such as the weight of large dams, drilling for oil, coal mining and the testing of nuclear weapons.

The damage caused by an earthquake depends on a number of factors. These include the strength of the earthquake (Figure 4.8), the distance from the centre

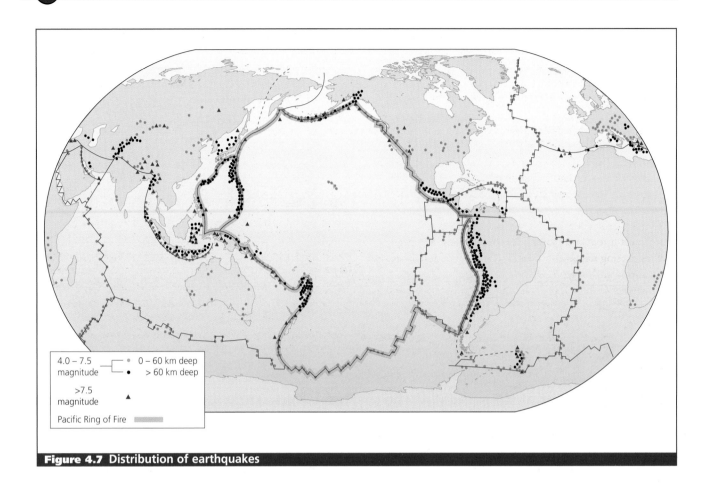

Figure 4.7 Distribution of earthquakes

Legend:
4.0 – 7.5 magnitude — 0 – 60 km deep / > 60 km deep
>7.5 magnitude ▲
Pacific Ring of Fire

| **Figure 4.8 The world's largest earthquakes** | | |
Place	Date	Strength
Kansu (Japan)	1920	8.6
Tokyo (Japan)	1923	8.3
Tangshen (China)	1976	8.0
Erzincan (Turkey)	1939	7.9
Mexico City	1985	7.8
North Peru	1970	7.7

(**epicentre**) of the earthquake, the nature and type of buildings, the type of rocks and sediments the buildings are on, population density and the time of day the earthquake occurred.

In areas of active earthquake activity the chances of an earthquake increase the longer ago the last earthquake occurred. Plates move at a rate of between 1.5 and 7.5 cm a year (the rate finger nails grow at). A large earthquake, however, can involve a movement of a few metres – these could occur every couple of hundred years rather than movements of a few centimetres each year. Many earthquakes are caused by the pressure created by moving plates. This increases the stress on rocks, the rocks deform somewhat and eventually give way and snap. The snapping is the release of energy, namely the earthquake. The strength of an earthquake is measured by the Richter Scale and the Mercalli Scale.

Earthquakes and seismic waves

Earthquakes release huge amounts of energy as shockwaves or seismic waves. These seismic waves radiate from the source or **focus** of the earthquake. The epicentre marks the point on the surface of the earth immediately above the focus of the earthquake. A large earthquake can be preceded by smaller tremors known as **foreshocks** and followed by numerous **aftershocks**. Aftershocks can be particularly devastating because they damage buildings that have already been weakend or damaged by the first main shock.

The nature of rock and sediment beneath the ground influences the pattern of shocks and vibrations during an earthquake. Unconsolidated sediments such as sand shake in a less predictable way than solid rock. Hence the damage is far greater to the foundations of buildings. Shock-waves from earthquakes can turn solid sediments into fluids like quicksand by disrupting sub-surface water conditions. This is known as **liquefaction** and can wreck foundations of large buildings and other structures.

Earthquake damage

Most earthquakes occur with little if any advance warning. Some places, such as California and Tokyo, with a history of coping with earthquakes, have developed 'earthquake action plans' and information programmes to increase public awareness about what to do in an earthquake.

Most problems are associated with damage to buildings, structures and transport systems. The collapse of buildings is the direct cause of many injuries and deaths, but it also reduces the effectiveness of the emergency services. In some cases more damage is caused by the aftershocks than the main earthquake itself, as they shake the already weakened structures. Aftershocks are more subdued but longer lasting and more frequent than the main tremor. Buildings partly damaged during the earthquake may be destroyed completely by the aftershocks.

Some earthquakes involve surface displacement, generally along fault lines. This may lead to the fracture of gas pipes, as well as causing damage to lines of communication. The costs of repairing such fractures is considerable.

Earthquakes may cause other geomorphological hazards such as landslides, liquefaction and tsunamis (tidal waves). For example, the Good Friday earthquake (magnitude 8.5) that shook Anchorage (Alaska) in March 1964 released twice as much energy as the 1906 San Francisco earthquake, and was felt over an area of nearly 1.3 million km^2. Over 130 people were killed, and over $500 million of damage was caused. It triggered large avalanches and landslides which caused extensive damage. It also caused a series of tsunamis throughout the Pacific as far as California, Hawaii and Japan.

Damage is often most serious where buildings are not designed to withstand shaking or ground movement. In the 1992 Cairo earthquake many poor people in villages and the inner city slums of Cairo were killed or injured when their old, mud-walled homes collapsed. By contrast, many wealthy people were killed or injured when modern high-rise concrete blocks collapsed. Indeed, some of these had been built without planning permission.

Human activity and earthquakes

In the Rocky Mountain Arsenal in Denver, Colorado, waste water was injected into underlying rocks during the 1960s (Figure 4.9). Disposal began in March 1962 and was followed soon afterwards a series of minor earthquakes, in an area previously free of earthquake activity. Between 1962 and 1965 over 700 minor earthquakes were monitored in the area.

The injection of the liquid waste into the bedrock lubricated and reactivated a series of deep underground faults which had been inactive for a long time. The more waste water was put down the well, the higher the number of minor earthquakes (Figure 4.9). When the link was uncovered, disposal stopped. The well was filled in 1966 and the number of minor earthquakes detected in the area fell sharply.

Underground nuclear testing

Underground nuclear testing has triggered earthquakes in a number of places. In 1968 underground testing of a series of

1200 tonnes bombs in Nevada set off over 30 minor earthquakes in the area over the following three days. Since 1966 the Polynesian island of Moruroa has been the site of over 80 underground nuclear explosion tests by France. More than 120 000 people live on the island. In 1966 a 120 000 tonnes nuclear device was detonated, producing radioactive fallout which was measured over 3000 km downwind.

Large dams

Earthquakes can be caused by adding increased loads on previously stable land surfaces. For example, the weight of water behind large reservoirs can trigger earthquakes. In 1935 the Colorado River was dammed by the Hoover Dam to form Lake Mead. As the lake filled, and the underlying rocks adjusted to the new increased load of over 40 km^3 of water, long-dormant faults in the area were reactivated, causing over 6000 minor earthquakes within the ten years. Over 10 000 events have been recorded up to 1973, about 10% of which were strong enough to be felt by residents. None caused damage.

Living with earthquakes

People cope with hazards in a number of ways. At an individual level there are at least three important factors which affect the actions a person takes. These include:

- **experience** – the more experience of hazards the greater the adjustment to the hazard
- **levels of wealth** – those who are better off have more choice
- **personality** – is the person a leader or a follower, a risk-taker or risk-minimiser?

Usually there are three choices in taking action:

- **do nothing and accept the hazard**
- **adjust to living in a hazardous environment**
- **leave the area.**

The first and last option are the most extreme and in most cases people follow the middle option. How people adjust to the hazard depends on:

- **the type of hazard**
- **the risk (probability) of the hazard**
- **the likely cost (loss) caused by the hazard.**

A number of factors influence the perception of risk.

The three main options for coping with hazards include:

1 **sharing the loss burden – through insurance or disaster relief**

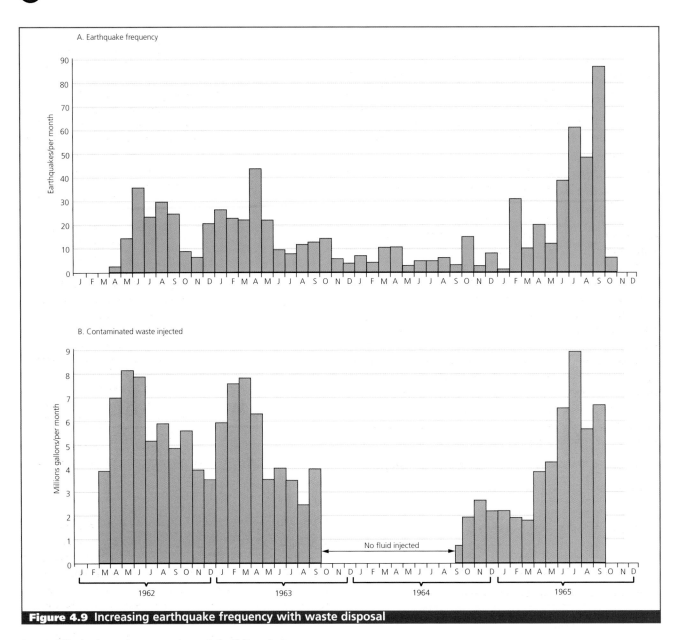

Figure 4.9 Increasing earthquake frequency with waste disposal

2 **modify the hazard event – through building design, building location, and emergency procedures**

3 **improved forecasting and warning.**

Building design

A single-storey building has a quick response to earthquake forces, whereas a high-rise building responds slowly, and shock waves are increased as they move up the building. If the buildings are too close together, vibrations may be amplified between buildings and increase damage. A stepped building offers greater stability, whereas complex buildings are often twisted during an earthquake. If the individual parts of the complex building are not joined very well they will pull apart during the earthquake. Buildings with a car park or pedestrian access at the base are often very unstable.

The weakest part of the building is where different elements meet. This is why elevated motorways are so much at risk from earthquakes. They have many connecting parts. Certain areas are very much at risk from earthquake damage such as areas with weak rocks, faulted (broken) rocks and soft soils. Earthquake frames and deep foundations are best to provide for a stable building.

Controlling earthquakes

In theory, earthquakes can be controlled by altering the fluid pressure deep underground at the point of greatest stress in the fault line. This can trigger a series of small and less damaging earthquake events that release the energy that would otherwise build up to create a major event. Additionally, a series of controlled underground nuclear explosions might relieve stress before it reached critical levels.

Prediction and risk assessment

There are a number of methods of detecting earthquakes: distortion of fences, roads and buildings, and changing levels of water in bore-holes. As strain can change the water holding capacity or porosity of rocks by closing and opening its tiny cracks, water levels in bore-holes will fluctuate with increased earthquake activity. Furthermore, satellites can now be used to measure the position of points on the surface of the earth to within a few centimetres. However, predicting earthquakes is not simple. Some earthquakes are very irregular in time pattern and may only occur less than once every one hundred years. Other earthquakes, by contrast, may continually slip and produce a large number of very small earthquakes. In addition, different parts of the fault line may behave differently. Areas which do not move are referred to as seismic gaps, whereas areas which move and have many mini earthquakes may be far less hazardous.

Earthquake prediction is only partly successful, although it offers a potentially valuable way of reducing the impact of earthquakes. Some bits are relatively easy. For example, the location of earthquakes is closely linked with the distribution of fault lines. However, the timing of earthquakes is difficult to predict. Previous patterns and frequencies of earthquake events offer some clues as to what is likely to happen in the future. Similarly the size of an earthquake event is difficult to predict. The most reliable predictions focus on:

- measurement of small-scale ground surface changes

- small-scale uplift or subsidence

- ground tilt

- changes in rock stress

- microearthquake activity (clusters of small quakes)

- anomalies in the earth's magnetic field

- changes in radon gas concentration

- changes in electrical resistivity of rocks.

Questions

1 Study Figure 4.10 which shows the size, location and impact of major earthquakes between 1980 and 1999.
   (a) What evidence is there that larger earthquakes cause more deaths? (Draw a graph with the size of earthquakes on the horizontal axis and the number of deaths on on the vertical axis.)
   (b) Is there any evidence to suggest that the number of large earthquakes, or their size, is increasing over time? (Draw a graph with the years on the horizontal axis and number and magnitude of earthquakes on the vertical axis.)

2 Account for the location of:
   (a) shallow focus earthquakes, and
   (b) deep focus earthquakes.

3 Study Figure 4.9 which shows the relationship between earthquake frequency and underground liquid waste disposal. Describe the relationship between the two variables. Suggest reasons to explain the relationship.

One particularly intensively studied site is Parkfield in California, on the San Andreas fault. Parkfield, with a population of less than 50 people, claims to be the earthquake capital of the world. Parkfield is heavily instrumented: strain meters measure deformation at a single point; two-colour laser geodometers measure the slightest movement between tectonic plates; and magnetometers detect alterations in the earth's magnetic field, caused by stress changes in the crust.

Figure 4.10 Deaths from earthquakes 1980–99

Place	Year	Magnitude	Number of deaths
Gilan and Zanjan, Iran	1990	7.7	35 000
Spitak and Leninakan, Armenia (CIS)	1988	6.9	25 000
Latur, India	1993	6.4	22 000
Mexico City, Mexico	1985	8.1	8000
Kobe, Japan	1995	7.2	6500
Rusaq, Afghanistan	1998	6.1	4500
Takhir, Afdghanistan	1998	6.9	3000
Dhamar, Yemen	1982	6.0	3000
Eboli, Italy	1980	7.2	2735
El Asnam, Algeria	1980	7.3	2590
East Nusa Tenggara, Indonesia	1992	6.8	2200
Qaen and Birjand, Iran	1997	7.1	2000
Cabantuan, Philippines	1990	7.7	2000
Sakhalin Island, Russia	1995	7.5	1989

Three earthquakes

The Colombian earthquake, 1999

On 25 January 1999, an earthquake of force 6 on the Richter Scale hit the city of Armenia in Columbia. Some of the aftershocks reached 5.6 and 5.5 on the Richter Scale. Armenia is the capital of Quindio Province where most of the deaths occurred. Many of its buildings were destroyed, including the fire station, and all 14 of the fire engines that should have been serving its 250 000 people. Doctors had to treat many of the injured on the floors of buses. Even though the hospitals were still standing, basic medical supplies ran short.

The earthquake hit the regional centre of Columbia's coffee industry, a pillar of the economy and a vital foreign currency earner bringing in some $2 billion each year. About 2000 people were killed and 180 000 people made homeless in the strongest earthquake for over a hundred years in the area. Most of the dead were from poor families in Armenia, although surrounding towns and villages were also badly affected. All the local services that would have been expected to cope were knocked out in the earthquake. Emergency food and water supplies were slow to reach the most needy. While Armenia's poorest residents had little to eat or drink, tonnes of supplies were tied up in warehouses. A few days after the earthquake desperate residents began looting supermarkets and shops to avoid starvation. At the same time bands of thieves descended on the city's population to steal what little possessions they had been able to take from their homes.

The Turkish earthquake, August 17 1999

In August 1999 more than 15 000 people were killed in a violent earthquake that devastated north-west Turkey (Figure 4.11). It was the worst earthquake in Turkey for over 50 years. The earthquake ripped through towns and cities in the night as people slept. The earthquake, measuring 7.4 on the Richter Scale, was centred on the industrial town of Izmit. Earth tremors were felt as far away as Ankara, Turkey's capital, some 340 km away.

The entire country of Turkey lies on a mini-'Ring of Fire', a band of fault lines, stretching across from the Pacific Ring of Fire through Central Asia to the Mediterranean. Earthquakes shake Turkey periodically, the last very destructive one being in 1939 when 33 000 people died.

What was surprising about the 1999 earthquake was the number of people who were killed. Partly this was because the earthquake occurred in the most densely populated part of the country where industries draw thousands of migrants from poorer regions. Partly it was also due to the time the earthquake occurred. Many people were asleep in their beds and were simply crushed by the falling masonry. Had the earthquate occurred during the day many more may have been in the streets and been able to escape the devastation.

Moreover, human negligence has been suggested as a contributory cause to the high death toll. In developing countries, such as Turkey, building regulations exist but are rarely enforced. The nature of the buildings therefore had a significant effect on the outcome of the earthquake.

It was not the older buildings that collapsed but the more recent buildings made of flimsy materials and built by 'cowboy' builders. In Istanbul hundreds of apartment blocks were destroyed in the earthquake.

Questions

1 Using an atlas locate Armenia (in South America).
   (a) What type of plate boundary is close to Colombia? Which two plates are involved in causing tectonic activity near Colombia?
   (b) Suggest reasons why looting took place a few days after the earthquake.
   (c) Why were the poor more affected by the earthquake than the rich?
   (d) Imagine that you are a newspaper reporter in Armenia just after the earthquake. Write a 200-word report on the causes and the affects of the earthquake, and how the local people are coping (or not).
2 Study Figure 4.8. Where have the worst earthquakes in the 20th century occurred? On a copy of Figure 4.5 mark the places mentioned in Figures 4.8. What conclusions can you make about the causes of earthquakes from this?
3 What else, other than what you have answered in Question 1, can cause earthquakes?

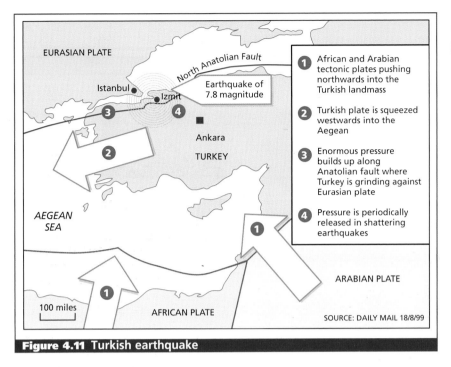

Figure 4.11 Turkish earthquake

Questions

1 Study Figure 4.11 which shows the location of the Turkish earthquake and some of the damage caused in the first few days.
   (a) Name the fault which separates the two plates that caused the earthquake.
   (c) Where was the epicentre of the earthquake.
   (d) Describe some of the impacts of the 1999 earthquake.
2 Study the text.
   (a) Why was the death rate from this particular earthquake so high. Give at least two contrasting reasons.
   (b) In what ways are human negligence (error) and corruption responsible for the damage and deaths caused by the earthquake?
3 Which crustal plates occur in the Taiwan region?
4 What causes earthquakes in the Taiwan region?
5 How many after shocks were there? Why are aftershocks important?
6 Why did the Taiwan earthquake kill fewer people than the Turkish earthquake a month earlier?

In addition, Turkey has no specialised rescue teams. Thus it was up to international rescue teams to bring with them equipment and sniffer dogs to locate and extract survivors. It was a mixed blessing, therefore, that the epicentre was close to Istanbul.

The Taiwan earthquake, September 1999

Taiwan lies at the edge of the small Philippine Plate which is slowly being forced underneath the large Eurasian Plate (Figure 4.12). Although Taiwan is on the Pacific Ring of Fire it has escaped major earthquakes for a long time. The last tremor to hit the island was an earthquake of 7.5 in 1986, but it happened hundreds of miles out to sea and only 15 people were killed.

The Taiwan earthquake was devastating for two main reasons:

■ it occurred in a densely populated region

■ the earthquake occurred quite close to the surface (about 30 km beneath the surface compared with the usual 600 km or so).

The Taiwan earthquake killed more than 1700 people, trapped more than 3000 and injured nearly 4000 people. Although the earthquake, which measured 7.6 on the Richter Scale, was more powerful than the one which hit Turkey the previous month (7.3) fewer people were killed. The devastation was less in Taiwan because most buildings were earthquake proof and were solidly built with concrete. However, over 1000 aftershocks severely damaged buildings, especially ones that had been weakened by the initial earthquake and subsequent aftershocks. Many of the aftershocks measured over 6.0 on the Richter Scale and so could be considered important earthquakes in their own right.

Even so, many of the buildings that were affected had tilted but remained intact. The result was that fewer people were crushed to death. Nevertheless, in some areas there were non-earthquake proof buildings. In central Nantou county, most of the buildings that did collapse were new high-rise buildings. The region had experienced an economic boom in recent years, and unsafe developments had been a problem in many areas despite strict building codes.

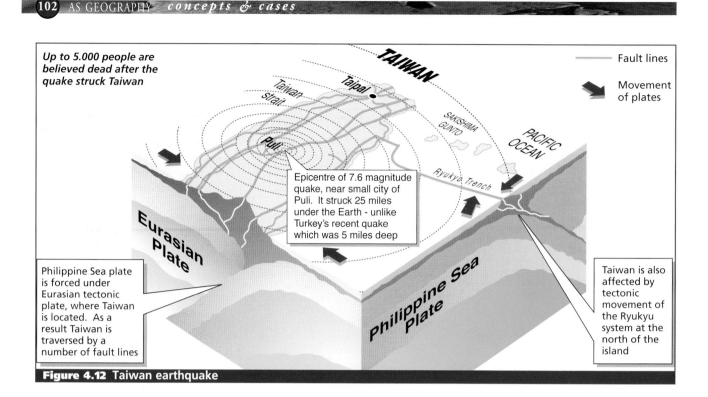

Up to 5.000 people are believed dead after the quake struck Taiwan

Philippine Sea plate is forced under Eurasian tectonic plate, where Taiwan is located. As a result Taiwan is traversed by a number of fault lines

Epicentre of 7.6 magnitude quake, near small city of Puli. It struck 25 miles under the Earth - unlike Turkey's recent quake which was 5 miles deep

Taiwan is also affected by tectonic movement of the Ryukyu system at the north of the island

Fault lines

Movement of plates

Eurasian Plate

Philippine Sea Plate

Figure 4.12 Taiwan earthquake

Taiwanese seismologists had warned that the country was due for a large earthquake. Because there had not been a large earthquake for some 30 years pressure was building up at the plate boundaries. Although about 15 000 earthquakes a year occur in the Taiwan region most of them are unnoticeable to all but the most sophisticated equipment. Such earthquakes do not release the energy built up by plate movement.

Volcanoes

Volcanoes are found along the boundaries of the earth's major plates where there is molten rock, or magma, to supply the volcanoes. Most of the world's volcanoes are found in the Pacific Rim or Ring of Fire (Figure 4.13). They are related to the subduction beneath either oceanic or continental crust. Subduction in the oceans creates chains of volcanic islands known as island arcs, such as the Aleutian Islands. Where the subduction of an oceanic crust occurs beneath the continental crust young fold mountains are formed such as the Andes.

Not all volcanoes are formed at plate boundaries. Those in Hawaii, for example, are found in the middle of the ocean (Figure 4.14). The Hawaiian Islands are a line of increasingly older volcanic islands which stretch north-west across the Pacific Ocean. These volcanoes can be related to the movement of plates above a hot part of the fluid mantle. A mantle **plume** or **hot spot**, or a jet of hot material rising from the deep within the mantle, is responsible for the volcanoes. Hot spots can also be found beneath continents and can produce isolated volcanoes. These hot spots can also play a

part in the break up of continents and the formation of new oceans.

At subduction zones volcanoes produce more viscous lava, tend to erupt explosively, and create much ash. Volcanoes that are found at mid-ocean ridges, or hot spots, by contrast, tend to produce relatively fluid basaltic lava, as in the case of Iceland and Hawaii. At mid-ocean ridges hot fluid rocks from deep in the mantle rise up due to convection currents. The upper parts of the mantle begin to melt and basaltic lava erupts forming new oceanic crusts. By contrast at subduction zones the slab of cold ocean floor slides down the subduction zone warming up slowly. Volatile compounds such as water and carbon dioxide leave the slab, and move upwards into the mantle so that it melts. The hot magma is thus able to rise.

Huge explosions occur wherever water meets hot rock. Water vaporises increasing the pressure until the rock explodes. Gases from within the molten rock can also build up high pressures. However, the likelihood of a big explosive eruption depends largely on the viscosity of the magma and hence its composition. Gases dissolve quite easily in molten rock deep underground due to the very high pressures. As magma rises to the surface the pressure drops and some of the gas may become insoluble and form bubbles. In relatively fluid magma the bubbles rise to the surface. Viscous magma, on the other hand, can trap enough gas so that it builds up enough pressure to create a volcanic eruption.

The strength of a volcano is sometimes measured by the Volcanic Explosive Index (Figure 4.15). This scale is based on the amount of material ejected in the explosion, the height of the cloud it causes, and the amount of damage caused. Any explosion over level 5 is considered to be very large and violent. As yet there has never been a level 8.

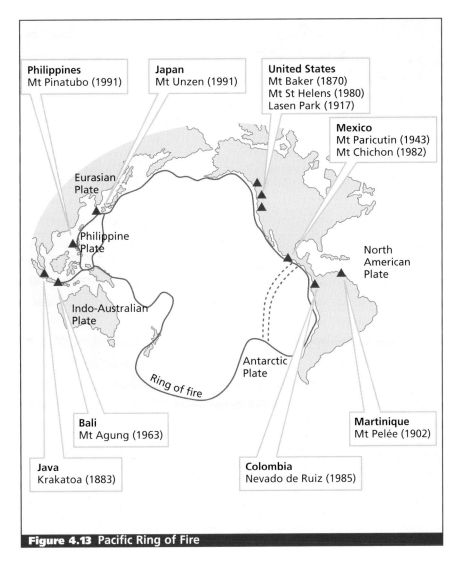

Philippines
Mt Pinatubo (1991)

Japan
Mt Unzen (1991)

United States
Mt Baker (1870)
Mt St Helens (1980)
Lasen Park (1917)

Mexico
Mt Paricutin (1943)
Mt Chichon (1982)

Eurasian
Plate

Philippine
Plate

North
American
Plate

Indo-Australian
Plate

Antarctic
Plate

Ring of fire

Bali
Mt Agung (1963)

Martinique
Mt Pelée (1902)

Java
Krakatoa (1883)

Colombia
Nevado de Ruiz (1985)

Figure 4.13 Pacific Ring of Fire

Questions

1 On a copy of Figure 4.13, and using an atlas, plot the following volcanoes listed in Figure 4.16: Mt Pinatubo (Philippines), Mt Unzen (Japan), Mt Baker (USA), Mt Popocatapetl (Mexico) and Nevado del Ruiz (Colombia).
Comment on the distribution of volcanoes that you have drawn. Which is the odd one out? Why? What do you think is meant by the term 'The Pacific Ring of Fire'?

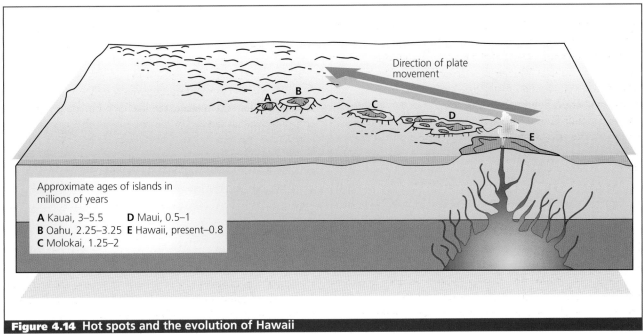

Direction of plate movement

A   B

C

D

E

Approximate ages of islands in millions of years

A Kauai, 3–5.5   **D** Maui, 0.5–1
B Oahu, 2.25–3.25 **E** Hawaii, present–0.8
C Molokai, 1.25–2

Figure 4.14 Hot spots and the evolution of Hawaii

Figure 4.15 The largest volcanic explosions

Crater Lake, Oregon, USA	c. 4895 BC	7
Towada, Honshu, Japan	915	5
Oraefajokull, Iceland	1362	6
Tambura, Indonesia	1815	7
Krakatoa, Indonesia	1883	6
Santa Maria, Guatemala	1902	6
Katmai, USA	1912	6
Mt St Helens	1980	5

Volcanic hazards

Ash and debris fall steadily from the volcanic cloud blanketing the ground with a deposit known as a pyroclastic fall. These can be very dangerous, especially as the fine ash particles can damage people's lungs. Ash is also fairly heavy –

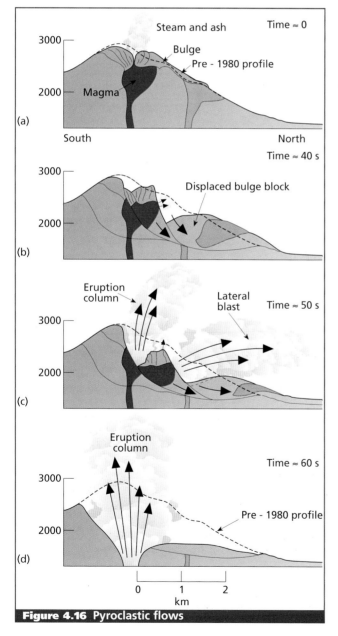

Figure 4.16 Pyroclastic flows

a small layer only a few centimetres thick can be enough to cause a building to collapse. Dust and aerosol also cause havoc with global climatic patterns. Pyroclastic falls are powerful enough to knock down trees and leave a trail of destruction. Some of them are extremely hot – up to 700°C and can travel at speeds of over 800 km per hour. Figure 4.16 shows the pyroclastic flows associated with Mount Pinatubo in 1991.

Lahars or volcanic mud flows are another hazard associated with volcanoes. A combination of heavy rain and unstable ash increase the hazard of lahars. The hazards associated with volcanic eruption vary spatially. Close to the volcano people are at risk of large fragments of debris, ash falls and poisonous gases. Further away pyroclastic flows may prove hazardous, and further still mud flows and debris flows may have an impact on more distant settlements. In addition, volcanoes can lead to tsunami and to famine. Although there is good evidence for the spatial distribution of volcanoes, there is little pattern in their temporal distribution (Figure 4.17).

Vulnerability and susceptibility

The concept of vulnerability encompasses not only the physical effects of a natural hazard but also the status of people and property in the area. Many factors can increase one's vulnerability to natural hazards (Figure 4.18), especially catastrophic events. Aside from the simple fact of living in a hazardous area, vulnerability depends on:

- **population density** – a large number of rapidly growing cities occur in hazardous areas
- **understanding of the area**
- **public education**
- **awareness of hazards**
- **the existence of an early-warning system**
- **effective lines of communication**
- **availability and readiness of emergency personnel**
- **insurance cover**
- **construction styles and building codes**
- **the nature of society**
- **cultural factors that influence public response to warnings.**

Many of these factors help explain the fact that less developed countries are much more vulnerable to natural hazards than are industrialised countries.

Volcano	Country	Year	Primary Cause of Fatalities			
			Pyroclastic Eruption	Mud flow	Tsunami	Famine
Mayon	Philippines	1814	1200			
Tambora	Indonesia	1815	12 000			80 000
Galunggung	Indonesia	1822	1500	4000		
Mayon	Philippines	1825		1500		
Awu	Indonesia	1826		3000		
Cotopaxi	Ecuador	1877		1000		
Krakatau	Indonesia	1833			36 417	
Awu	Indonesia	1892		1532		
Soufrière	St Vincent	1902	1565			
Mt Oelée	Martinique	1902	29 000			
Santa Maria	Guatemala	1902	6000			
Taal	Philippines	1911	1332			
Kelud	Indonesia	1919		5110		
Merapi	Indonesia	1930	1300			
Lamington	Papua-New Guinea	1951	2942			
Agung	Indonesia	1963	1900			
El Chichón	Mexico	1982	1700			
Nevado del Ruiz	Colombia	1985		23 000		

Figure 4.17 Hazards and deaths from volcanic explosions

Questions

1 What are the main hazards associated with volcanoes?

2 Study Figure 4.17 which shows volcanic disasters since 1800. Describe the location of these disasters. How do you account for this pattern?

Figure 4.18 The progression of vulnerability

1 Root causes	2 Dynamic pressures	3 Unsafe conditions	Disaster hazards
Limited access to ■ power ■ structures ■ resources	Lack of ■ local institutions ■ training ■ appropriate skills ■ local investments ■ local markets ■ press freedom ■ ethical standards in public life	Fragile physical environment ■ dangerous locations ■ unprotected buildings and infrastructure	Earthquakes High winds (cyclone, hurricane, typhoon) Flooding
Ideologies ■ Political systems ■ Economic systems		Fragile local economy ■ livelihoods at risk ■ low income levels	RISK = HAZARD + VULNERABILITY (R = H + V)
	Macro factors ■ rapid population growth ■ rapid urbanisation ■ arms expenditure ■ debt repayment schedules ■ deforestation ■ decline in soil productivity	Vulnerable society ■ special groups at risk ■ lack of local institutions	Volcanoes Landslides Drought Virus and pests
		Public actions ■ lack of disaster preparedness ■ prevalence of endemic disease	

Post-disaster response after the Nevado del Ruiz volcanic eruption

The volcano at Nevado del Ruiz, near Armero in Colombia was inactive between 1845 and l985. In 1985, however, a series of minor eruptions and earth tremors produced a cloud of hot pumice and ash. As a result, part of the ice cap of the 5400 metre volcano exploded and caused the Guali River to overflow. This in turn caused a natural dam to burst, releasing a torrent that travelled at speeds of about 70 km per hour and creating a massive mudflow that enveloped the town of Armero, population 29 000 (Figure 4.19). Because the volcano had been inactive for so long there had been limited evacuation drills, and the volcano had not been perceived as a threat to human life.

In 1988, three years after the disaster, a group of lawyers placed a notice in the local press of the small towns close to Armero. They invited anyone who had suffered injury, or the loss of relatives or property to sue the government of Colombia for gross negligence in not warning or evacuating them in time to avoid injury or property losses. Over 1000 claims were lodged amounting to a total claim about £40 million. It was expected that government lawyers would argue that the residents were aware of the risks in choosing to occupy a hazardous yet highly fertile area. Three expert vulcanologists were questioned as to whether the scale, location and timing of the mudflow could have been accurately forecast. They said it was impossible to do so, and on this basis the government was cleared of responsibility.

Following the legal challenge, the government created the Governmental Preparedness System which includes detailed warning and evacuation systems. However, while evacuation exists on paper, economic priorities may outweigh the needs for safety. For example, in 1993 the Galeras volcano erupted putting the nearby town of Pasto at risk. Although the government Disaster Preparedness Agency had repeatedly attempted to issue warnings to the public, the local authorities refused to authorise them, following an economic crash years earlier, when a volcano warning caused a financial crisis.

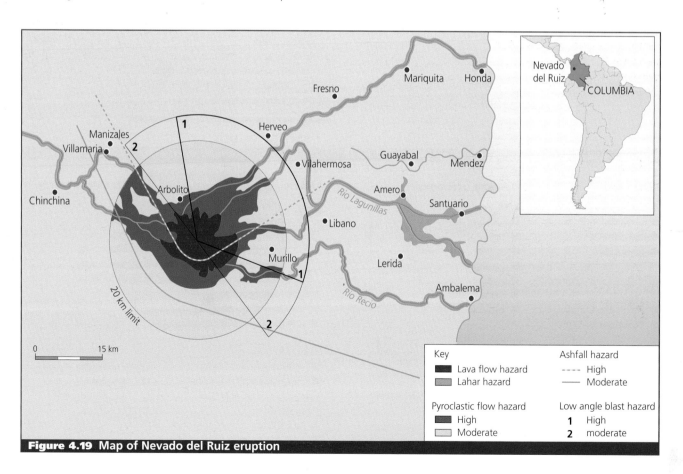

Figure 4.19 Map of Nevado del Ruiz eruption

Pre-disaster planning, Taal volcano, the Philippines

An excellent example of the conflicting demands of economic prosperity versus safety is illustrated by the example of Taal volcano in the Philippines. This is one of the world's deadliest volcanoes, and is located on an island in Lake Taal, about 60 km south of Manila (Figure 4.20). Taal has erupted 33 times since its earliest recorded explosion in 1572. The 1911 eruption resulted in 1334 deaths and covered an area of 2000 sq km with ash and volcanic debris which fell as far away as Manila.

The island is small. The population is less than 4000 people, yet they are relatively prosperous. The economy is based on fishing, fish-farming, agriculture, mining for scoria and tourism. The location of settlements on the island is closely related to the rich fertile soils suitable for sweet potatoes and corn. Alarmingly, population growth is rapid, increasing at 9.6% per year, more than three times the national average. Moreover, the island could not cope with a major eruption. There are only 215 boats, which could transport less than 2000 people. Hence, in the event of a very sudden eruption with limited warning, only about half of the population would be able to escape.

A Disaster Management Training Workshop in 1988 found that there was very little anxiety on the part of the population over the risks they faced. This was true even among survivors of the 1965 Taal eruption. The lack of escape boats was also of minimal concern. Islanders referred to a building set up by the Philippine Institute of Volcanology (PIV), as a form of 'volcanic eruption insurance policy'. They assumed that the PIV would look after them in the event of a disaster. The very presence of a warning station made some feel that the island was therefore safe for them to live on.

The view the local resident population took was one in which the Taal volcano was just one of many perceived risks that influenced their decision on where to live and work. By contrast, members of the Disaster Management Training Workshop sought to prevent residential occupation of the island. They adopted a narrow view of risk and vulnerability based on physical processes, and failed to acknowledge the advantages that the area offered. Each side has an entirely legitimate and logical response to the same hazard. However, their views differ on account of their different needs, priorities, perceptions and values.

Predicting volcanoes

Increasingly, scientists monitor volcanoes in an attempt to predict eruptions. There are a number of forces which give clues to an imminent eruption. The rise of magma beneath a volcano may fill a magma chamber and distort the shape of a volcano. This was certainly the case at Mount St Helen's in 1980. Seismometers monitoring earthquakes often pick up large clusters of earthquakes before a volcano and immediately after the eruption. In addition, gases may seep from fissures in the surface known as vents or fumaroles.

Managing volcanic eruptions

It is impossible to prevent volcanoes from erupting. However, there are a number of measures that can be taken to limit the damage:

- hazard zonation maps guide decisions regarding evacuation
- land-use planning
- monitoring of active volcanoes to provide early warning of likely eruptions.

The most reliable forecasts depend on detailed monitoring of microearthquake activity in the vicinity of the volcanic cone, which indicate that magma is working its way upwards. Recent years have also seen significant improvements in hazard assessment, volcano monitoring and eruption forecasting, particularly for less explosive eruptions, and lessons from disasters

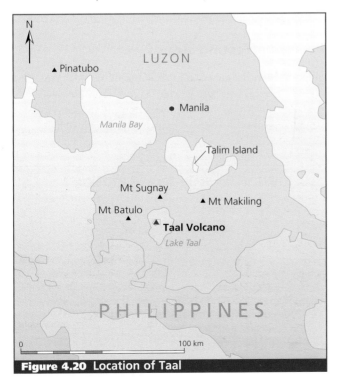

Figure 4.20 Location of Taal

like Mount St Helens (1980) are helping to improve volcanic hazard management and reduce risk. Other measures (including preparation of contingency plans) can be used to reduce the effects when vulnerable areas cannot be avoided. While it is virtually impossible to monitor all active volcanoes, satellites offer the prospect of global coverage from space and are being developed for remote warning systems.

One example of successful monitoring is the 1991 Mount Pinatubo eruption in the Philippines. This killed over 320 people, mostly due to collapse of ash-covered roofs. Many more lives were saved because early warnings were issued and at least 58 000 people were evacuated from the high-risk areas. Management of the 1991 eruption seems to have been well coordinated and effective because:

■ state-of-the-art volcano monitoring techniques and instruments were applied

■ the eruption was accurately predicted

■ hazard zonation maps were prepared and circulated a month before the violent explosions

■ an alert and warning system was designed and implemented

■ the disaster response machinery was mobilised on time.

There are also examples of lava flows having been diverted to stop a catastrophe. Lava flows on Mount Etna have been diverted by digging new channels for the lava to follow in. Just as a river takes the steeper gradient so too will lava. If an artificial channel is created which will allow the lava to flow more quickly it will take that route. In addition, lava flows can be slowed down by adding cool materials to it. Notable successes include Heimay, Iceland in 1975, where lava was cooled by pumping sea water onto it, thereby stopping a lava flow from engulfing the port. In Etna concrete blocks have been placed on lava flows to cool them. Success depends on the nature of the rock. Slow, viscous lava flows will often just carry the concrete blocks in their path and this only increases the impact of the lava flow.

Slopes

Slopes can be defined as any part of the solid land surface. They can be exposed (**sub-aerial**) or underwater (**sub-marine**), depositional (**aggradational**), eroded (**degraded**) or **transportational** or any mixture of these. Slope form refers to the shape of the slope in cross-section; slope processes the activities acting on the slopes, and slope **evolution** the development of slopes over time.

Slope processes

Soil creep

Individual soil particles are pushed or heaved to the surface by (a) wetting, (b) heating, or (c) freezing of water (Figure 4.21). They move at right angles to the surface (2) as it is the zone of least resistance. They fall (5) under the influence of gravity once the particles have (a) dried, (b) cooled, or (c) the

water has thawed. Net movement is downslope (6). Rates are slow: 1 mm per annum in the UK and up to 5 mm per annum in the tropical rainforest. They form terracettes such as those at Maiden Castle, Dorset and The Manger, Vale of the White Horse.

Rain-splash erosion

On flat surfaces (A) raindrops compact the soil and dislodge particles equally in all directions. On steep slopes (B) the downward component is more effective than the upward motion (a) due to gravity. Hence erosion downslope increases with slope angle (Figure 4.22).

Factors affecting slopes

Slopes vary with climate. In general, humid slopes are rounder, due to chemical weathering, whereas arid slopes are jagged or straight owing to mechanical weathering and overland runoff. Climatic geomorphology is a branch of geography which

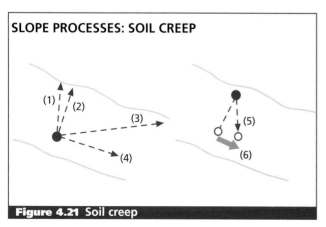

SLOPE PROCESSES: SOIL CREEP

Figure 4.21 Soil creep

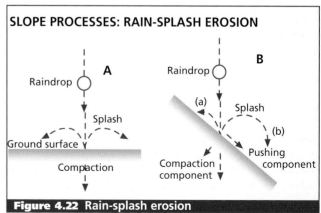

SLOPE PROCESSES: RAIN-SPLASH EROSION

Figure 4.22 Rain-splash erosion

studies how different processes operate in different climatic zones and thereby produce different slope forms or shapes.

Slopes are also influenced by rock type. The Tees–Exe line is an imaginary line running from the River Tees to the River Exe. It divides Britain into hard and soft rock. To the north and west are old, hard, resistant rocks, such as granite, basalt and carboniferous limestone forming upland rugged areas. To the south and east are younger softer rocks, such as chalk and clay, forming subdued low-lying landscapes.

Aspect is another important factor. In the UK north-facing slopes remain in the shade. During cold periglacial times temperatures rarely rose above freezing. By contrast, south-facing slopes, experienced many cycles of freeze-thaw. Solifluction and overland runoff lower the level of the slope, and streams remove the debris from the valley. The result is an asymmetric slope such as the River Exe in Devon and Clatford Bottom in Wiltshire.

Theories of slope evolution

Slope evolution refers to the change in slope form (shape) over time. Slopes can be divided into those that are (a) time independent, i.e. the slope retains a constant angle, although altitude may be lowered, and (b) time dependent slopes, in which slope angle and altitude decline progressively over time.

Slope decline: W. M. Davies

The main processes involved are soil creep, solution, overland runoff, weathering and fluvial transport at the base. Slopes decline progressively over time. The free face is changed by falls and slumps and develops a regolith. Weathered material is transported by overland runoff and surface wash, eventually producing a convex-concave profile.

Slope replacement: Walther Penck

Slope A is replaced by slope B, in turn replaced by C. Replacement is by lower angle slopes which extend upwards at a constant angle. The segments become increasingly longer as the slope develops. Some free faces may be completely removed. It is common in coastal areas.

Slope retreat: L. C. King

The key elements are a very hard lateritic cap rock controlling the rate of slope evolution. Mechanical weathering and sheet wash are the dominating processes in a semi-arid climate. All elements, except the pediment, retain a constant angle. Pediments vary from 1–2° on fine material to 3–5° on gravel and stony material. It is common in dry areas such as Monument Valley, Arizona.

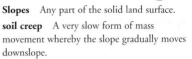

Key Definitions 2

Slopes Any part of the solid land surface.

soil creep A very slow form of mass movement whereby the slope gradually moves downslope.

Rain splash The compaction of the soil surface caused by falling rain, followed by the erosion of small particles.

Climatic geomorphology The processes which occur in certain environments are distinct from other environments, and lead to the formation of particular landforms.

Slope decline The gradual lowering of slope altitude and gradient over time.

Slope replacement The change in slope profile over time as high angled slopes are replaced by lower angle ones.

Parallel retreat A process of slope development whereby the slope retains its shape over time.

Mass movement The movement of material downslope due to gravity.

Weathering The wearing away of the land surface 'in situ' (on the spot).

Mass movements

Mass movements are any large scale movement of the earth's surface that are not accompanied by a moving agent such as a river, glacier or ocean wave. They include very small movements, such as soil creep, and fast movement, such as avalanches. They vary from dry movement, such as rock falls, to very fluid movements such as mud flows.

The sliding material maintains its shape and cohesion until it impacts at the bottom of a slope. They range from small-scale slides close to roads, to large-scale movements killing thousands of people, for example the Vaiont Dam in Italy where more than 2000 died on 9 October 1963.

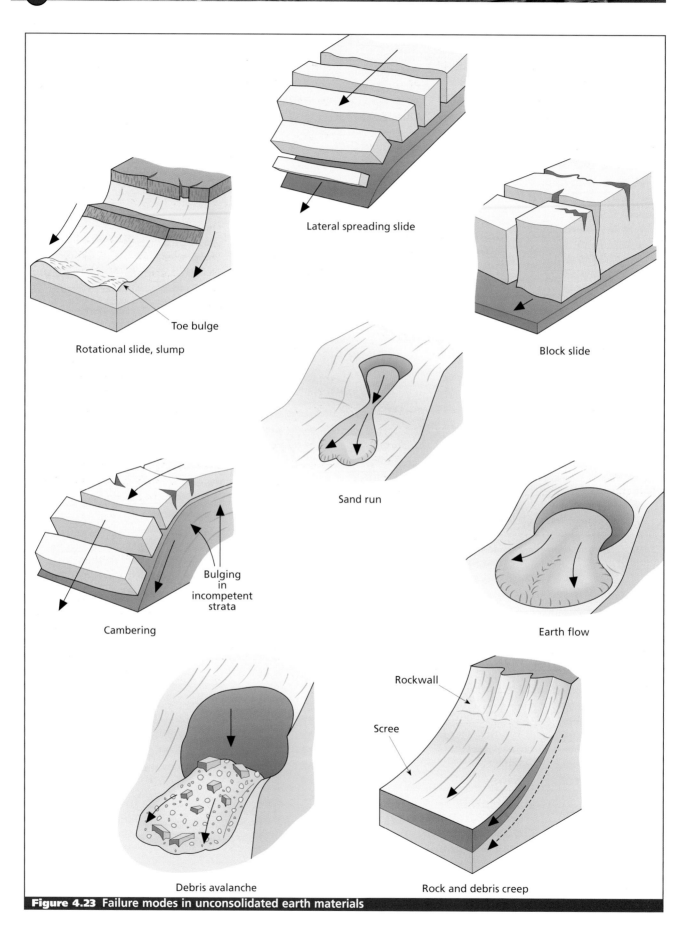

Lateral spreading slide

Toe bulge

Rotational slide, slump

Block slide

Sand run

Bulging in incompetent strata

Cambering

Earth flow

Rockwall

Scree

Debris avalanche

Rock and debris creep

Figure 4.23 Failure modes in unconsolidated earth materials

Falls

Falls occur on steep slopes (> 70°). The initial cause of the fall may be weathering, for example freeze-thaw or disintegration, or erosion prising open lines of weakness. Once the rocks are detached they fall under the influence of gravity. If the fall is short it produces a relatively straight scree; if it is long, it forms a concave scree. A good example of falls and scree is Wastwater in the Lake District.

Slumps

Slumps occur on weaker rocks, especially clay, and have a rotational movement along a curved slip plane. Clay absorbs water, becomes saturated and exceeds its liquid limit. It then flows along a slip plane. Frequently the base of a cliff has been undercut and weakened by erosion thereby reducing its strength. An example are the cliffs at Folkestone Warren. Human activity can also intensify the condition by causing increased pressure on the rocks such as at the Holbeck Hall Hotel, Scarborough.

Avalanches

Avalanches are rapid movements of snow, ice, rock or earth down a slope. They are common in mountainous areas: newly fallen snow may fall off older snow, especially in winter (a dry avalanche) while in spring partially melted snow moves (a wet avalanche), often triggered by skiing. Avalanches frequently occur on steep slopes over 22°, especially on north-facing slopes where the lack of sun inhibits the stabilisation of snow. Debris avalanches are a rapid mass movement of sediments, often associated with saturated ground conditions.

Weathering

Weathering is the **decomposition** and **disintegration** of rocks **in situ**. Decomposition refers to the chemical process and creates altered rock substances whereas disintegration or mechanical weathering produces smaller, angular fragments of the same rock. Weathering is important for landscape evolution as it breaks down rock and facilitates erosion and transport.

Mechanical (physical) weathering

There are four main types of mechanical weathering, freeze-thaw (ice crystal growth), salt crystal growth, disintegration and pressure release.

Freeze thaw occurs when water in joints and cracks freezes at 0°C and expands by 10% and exerts pressure up to 2100 kg/cm^2. Rocks can only withstand a maximum pressure of about 500 kg/cm^2. It is most effective in environments where moisture is plentiful and there are frequent fluctuations above and below freezing point such as periglacial and alpine regions.

Salt crystal growth occurs in two main ways: first in areas where temperatures fluctuate around 26–28°C, sodium sulphate (Na_2SO_4) and sodium carbonate (Na_2CO_3) expand by 300%. Secondly, when water evaporates, salt crystals may be left behind to attack the structure. Both mechanics are frequent in hot desert regions.

Disintegration is found in hot desert areas where there is a large diurnal temperature range. Rocks heat up by day and contract by night. As rock is a poor conductor of heat, stresses occur only in the outer layers and cause peeling or **exfoliation** to occur. Griggs (1936) showed that moisture is essential for this to happen.

Pressure release is the process whereby overlying rocks are removed by erosion thereby causing underlying ones to expand and fracture parallel to the surface. The removal of a great weight, such as a glacier, has the same effect.

Chemical weathering

There are four main types of chemical weathering, carbonation-solution, hydrolysis, hydration and oxidation.

Carbonation-solution occurs on rocks with calcium carbonate such as chalk and limestone. Rainfall and dissolved carbon dioxide form a weak carbonic acid. (Organic acids acidify water too.) Calcium carbonate reacts with an acid water and forms calcium bicarbonate, or calcium hydrogen carbonate, which is soluble and removed by percolating water.

Hydrolysis occurs on rocks with orthoclase feldspar such as granite. Orthoclase reacts with acid water and forms kaolinite (or kaolin or china clay), silicic acid and potassium hydroxyl. The acid and hydroxyl are removed in the solution leaving china clay behind as the end product. Other minerals in the granite, such as quartz and mica, remain in the kaolin.

Hydration is the process whereby certain minerals absorb water, expand and change, for example when gypsum becomes anhydrate.

Oxidation occurs when iron compounds react with oxygen to produce a reddish brown coating.

Factors affecting weathering

Climate

The rate of weathering varies with climate. Peltier's diagrams (Figure 4.24) show how weathering is related to moisture availability and average annual temperature. Frost shattering increases as the number of freeze-thaw cycles increases.

Chemical weathering increases with moisture and heat. According to Van't Hoff's Law, the rate of chemical weathering increases 2–3 times for every increase of temperature of 10°C (up to 60°C).

Geology

Rock type influences the rate and type of weathering in many ways due to:

- **chemical composition**
- **the nature of cements in sedimentary rock**
- **joints and bedding planes.**

For example, limestone consists of calcium carbonate and is therefore susceptible to carbonation-solution, whereas granite with orthoclase feldspar is prone to hydrolysis. In sedimentary rocks, the nature of the cement is crucial: iron-oxide based cements are prone to oxidation whereas quartz cements are very resistant.

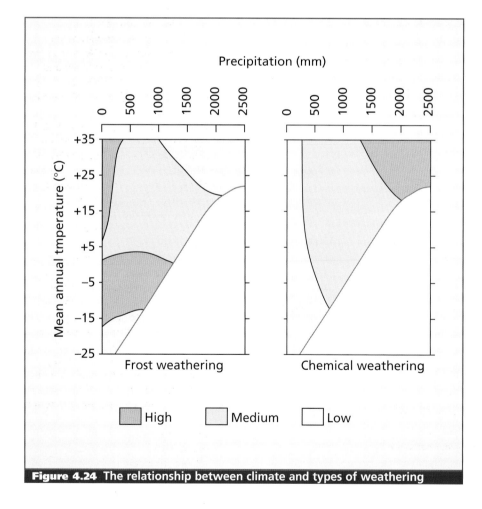

Figure 4.24 The relationship between climate and types of weathering

Gravestone index	Rock type	Lichen Cover (%)			Date	Rahn's
		Total	East	West	on stone	
1	Limestone	186	98	98	1848	5
2	Limestone	200	100	100	1865	5
3	Limestone	147	66	81	1896	5
4	Limestone	170	79	91	1891	5
5	Limestone	143	68	75	1909	4
6	Sandstone	140	61	79	1898	4
7	Sandstone	103	40	63	1937	3
8	Sandstone	119	55	64	1939	3
9	Sandstone	74	0	74	1949	2
10	Marble	74	35	39	1954	2
11	Marble	49	0	49	1965	2
12	Marble	27	11	19	1974	1

Simplified Rahn's Index

0	Unweathered	corners of letters sharp
1	Slightly weathered	rounding of corners of letters
2	Moderately weathered	lettering legible
3	Badly weathered	lettering difficult to read
4	Very badly weathered	lettering illegible
5	Extremely weathered	lettering disappeared

Figure 4.25

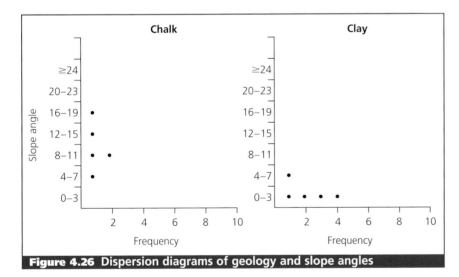

Figure 4.26 Dispersion diagrams of geology and slope angles

Useful websites

Global assessment of active volcanoes at http://www.geo.mtu.edu/eos/
Alaska volcanoes at http://www.avo.alaska.edu/
Hawaii volcanoes at http://www.soest.hawaii.edu/hvo
Internet resources in the earth sciences at:
http://www.lib.berkeley.edu/EART/EarthLinks.htmlArtwork summary
Volcano World at http://volcano.und.nodak.edu/
MTU Volcanoes Page at http://www.geo.mtu.edu/volcanoes/
World earthquakes at http://gldfs.cr.usgs.gov/
World quakes at http://www.civeng.carleton.ca/cgi-bin/quakes/

1 These results (Figure 4.25) were taken from an investigation into the rate of weathering on gravestones in an urban environment.
   (a) Choose an appropriate graph to show the relationship between the age of rock and the amount of lichen cover.
   (b) Describe the pattern you have found.
   (c) How do you explain this?
   (d) How do you account for the anomalies?
2 What is the relationship between Rahn's Index and the
   (a) age of rock
   (b) amount of lichen cover
   (c) rock type.
   How useful do you think Rahn's index is? Justify your answer.
3 The following figures were taken from slope surveys on the Vale of the White Horse (an area of chalk and clay).
   Chalk   7, 8, 8, 12, 16, 20, 24, 20, 20, 19, 17, 9
   Clay   3, 4, 3, 2, 1, 1, 0, 3, 2, 1, 2, 1.
   Figure 4.26 is a dispersion diagram showing slope angle frequencies for each of the rock types. Complete the diagram (the first five slope angles for both chalk and clay have been inserted on the diagram).
   For each rock type work out:
   (a) the mean or average
   (b) the mode (the most frequent slope angle)
   (c) the median (the middle value when all the angles are ranked from high to low)
   (d) the range (i.e. the maximum to the minimum).
   Illustrate the data in another way. Why is this way better than the ones illustrated here?
4 How do chalk and clay differ in terms of
   (a) permeability
   (b) porosity
   (c) rock hardness.
   How do these factors help to explain the variations in slope angles illustrated above?

5 coastal environments

Marine geomorphology

Coasts are complex landscapes (Figure 5.1), and there are a number of factors controlling coastal evolution. These include:

■ the work of waves and currents, including longshore drift

■ the degree of exposure to wave action – i.e. the 'trend' of the coast

■ variations in local geology such as rock type, structure and strength

■ long- and short-term changes in relative levels of land and sea

■ 'special' factors, such as volcanic activity and glaciation

■ the effect of vegetation and animals

■ human activities, particularly since the Coast Protection Act of 1949, and the use of improved marine engineering, and land reclamation.

Waves

Waves result from friction between wind and the sea surface. Waves in the open deep sea (waves of oscillation) are different from those breaking on shore. Waves of oscillation are forward surges of energy. Although the surface wave shape appears to move, the water particles only move in a roughly circular orbit within the wave (Figure 5.2).

The wave orbit is the shape of the wave. It varies between circular and elliptical. The orbit diameter decreases with depth, to a depth roughly equal to wavelength when no further movement occurs as related to wind energy – this point is called the **wave base**.

Wave height is an indication of wave energy. It is controlled by wind strength, **fetch**, and the depth of the sea. Waves of up to 12–15m are formed in open sea and can travel vast distances, reaching distant shores as swell waves. They are characterised by a low height and a long wavelength (Figure 5.3).

Figure 5.1 Isle of Wight aerial

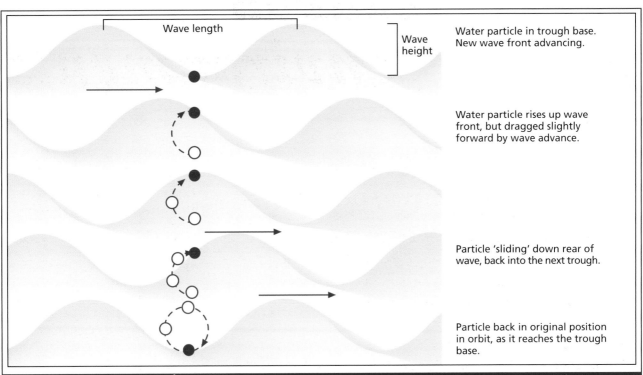

Figure 5.2 Water movement in a wave

Water particle in trough base. New wave front advancing.

Water particle rises up wave front, but dragged slightly forward by wave advance.

Particle 'sliding' down rear of wave, back into the next trough.

Particle back in original position in orbit, as it reaches the trough base.

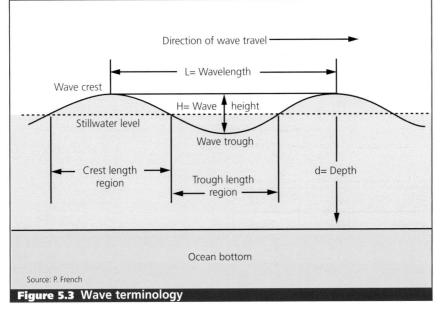

Source: P. French

Figure 5.3 Wave terminology

As waves move closer onshore, the wave base comes in contact with the sea bed. This slows down the wave advance, causing wavelengths to be reduced and the wave height to increase. Thus a breaker is formed. As the wave breaks the **swash** surges up the beach. Its speed gradually reduces due to friction and the uphill gradient. The water then drains back down the beach as **backwash** owing to the pull of gravity.

There are two main types of wave – constructive and destructive. Constructive waves occur when wave frequency is low (6–8 arriving onshore per minute), particularly when these waves advance over a gently shelving sea floor (Figure 5.4). These waves have been generated far offshore. The gentle offshore slope creates a gradual increase in friction, which causes a gradual steepening of the wave front. As the wave breaks, the powerful swash surges up the gentle gradient. Constructive waves cause deposition and lead to a build up of beach material.

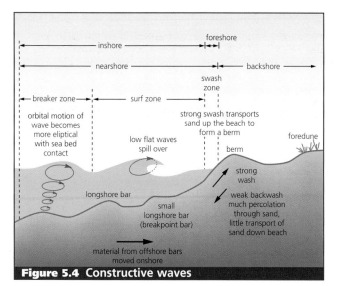

Figure 5.4 Constructive waves

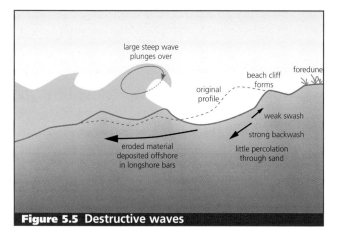

Figure 5.5 Destructive waves

Destructive waves are the result of locally generated winds, which create waves of high frequency (12–14 per minute) (Figure 5.5). The rapid approach of these waves, particularly if moving onshore up a steeply shelving coastline, creates a rapid increase in friction and thus a very steep wave. The wave becomes destructive, combing down the beach material.

Seasonal contrasts in wave energy are common. In the UK, for example, destructive waves are more frequent in winter, and lead to the removal of beach material. By contrast, constructive waves are more frequent in summer and material is moved back onshore and the beach is built up again (Figure 5.6).

Wave refraction

Wave fronts will usually approach the shore obliquely, due to wind direction (and therefore direction of wave advance) and the shape of the coast. As wave fronts approach the shore, their speed of approach will be reduced as the waves 'feel bottom'. This causes the wave fronts to bend and swing round in an attempt to break parallel to the shore. The change in speed and distortion of the wave fronts is called wave refraction (Figure 5.7). If refraction is completed, the fronts will break parallel to the shore. However, due primarily to the complexities of coastline shape, refraction is not always totally achieved – this causes longshore drift to occur, which is a major force in the transport of material along the coast.

Wave refraction also distributes wave energy along a stretch of coast. Along a complex coast with alternating headlands and bays, wave refraction will concentrate erosional activity on the flanks of headlands, while deposition will tend to occur in the bays (Figure 5.8).

If refraction is not complete long shore drift occurs (Figure 5.9). This is the gradual movement of sediment along the shore caused by differences in the direction of swash and backwash. The swash moves in the direction of the prevailing wind whereas the backwash moves straight down the beach

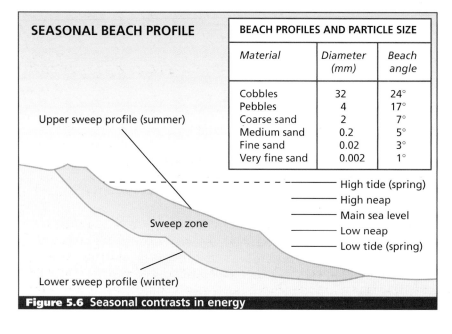

Figure 5.6 Seasonal contrasts in energy

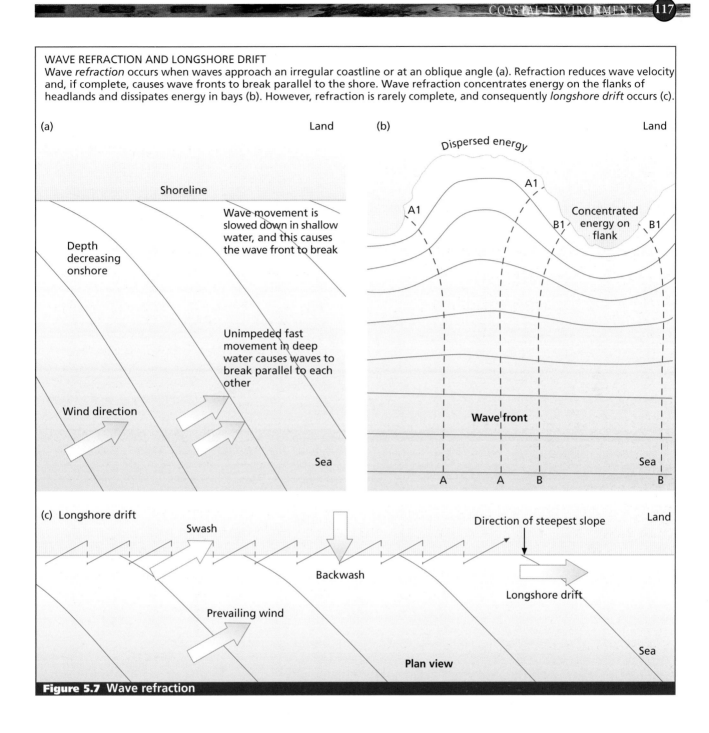

WAVE REFRACTION AND LONGSHORE DRIFT
Wave *refraction* occurs when waves approach an irregular coastline or at an oblique angle (a). Refraction reduces wave velocity and, if complete, causes wave fronts to break parallel to the shore. Wave refraction concentrates energy on the flanks of headlands and dissipates energy in bays (b). However, refraction is rarely complete, and consequently *longshore drift* occurs (c).

(a)

Land

Shoreline

Wave movement is slowed down in shallow water, and this causes the wave front to break

Depth decreasing onshore

Unimpeded fast movement in deep water causes waves to break parallel to each other

Wind direction

Sea

(b)

Land

Dispersed energy

A1

A1

Concentrated energy on flank

B1

B1

Wave front

Sea

A   A   B   B

(c) Longshore drift

Land

Swash

Direction of steepest slope

Backwash

Longshore drift

Prevailing wind

Sea

Plan view

Figure 5.7 Wave refraction

following the steepest gradient. Human activity can interrupt longshore drift by inserting artificial structures to reduce sediment movement. However, engineering structures on one part of the coast can have a serious impact further along the coast.

Human activity and longshore drift in West Africa

The Guinea Current is among the strongest in the world, and is removing approximately 1.5 million cubic metres of sand each year between the Ivory Coast and Nigeria (Figure 5.10).

The effect upon Ghana, Benin and Togo is especially catastrophic. The removal of the beach material has affected settlements, tourism and industry.

The cause of the coastal retreat is the Akosombo Dam on the Volta River in Ghana, built in 1961. It is just 110 km from the coast and disrupts the flow of sediment from the River Volta and prevents it from reaching the shore. Thus there is less sand to replace that which has already been washed away, and so the coastline retreats due to erosion by the Guinea Current. Towns such as Keta, 30 km east of the Volta estuary, have been destroyed as their protective beach has been removed.

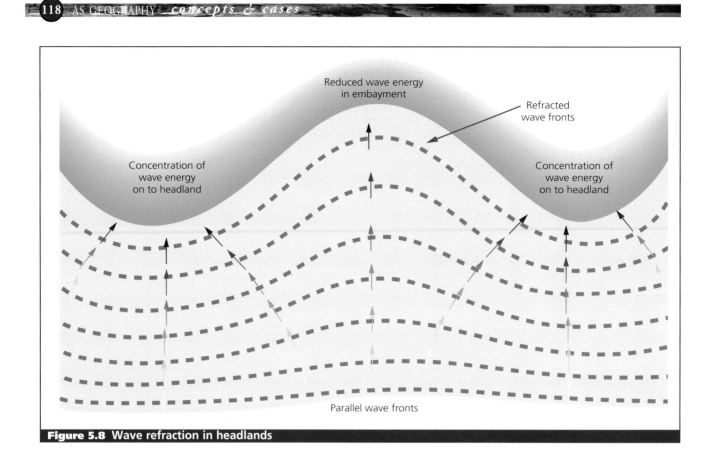

Figure 5.8 Wave refraction in headlands

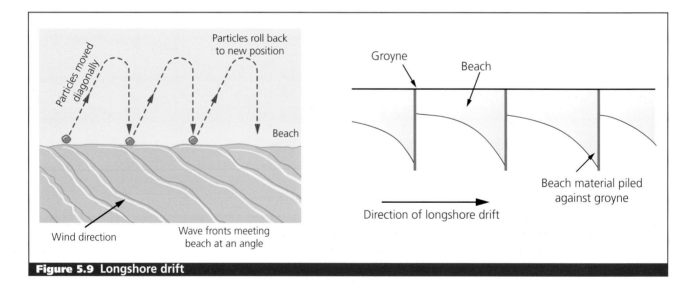

Figure 5.9 Longshore drift

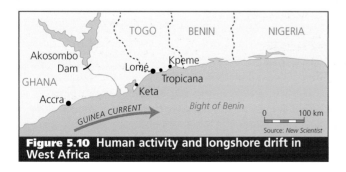

Figure 5.10 Human activity and longshore drift in West Africa

Figure 5.11 Brighton beach

Questions

1 Define the following terms: swash, backwash, trend, fetch.
2 Describe two changes that take place as a result of wave refraction.
3 Figure 5.11 shows a beach experiencing longshore drift. In which direction is drifting occurring? Suggest reasons to support your answer.
4 Figure 5.12 is a sketch of a beach in which groynes have been built. Suggest how the shape of the beach might look in 50 years' time. Give reasons for your answer.

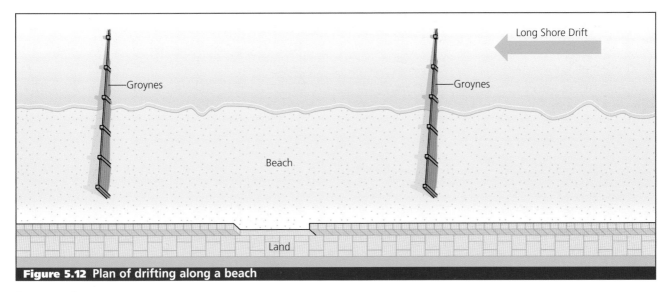

Long Shore Drift

Groynes

Groynes

Beach

Land

Figure 5.12 Plan of drifting along a beach

Erosion

There are a number of different types of wave erosion (Figure 5.13). **Hydraulic action** occurs as waves break onto cliffs. As the wave breaks, any air trapped in cracks, joints and bedding planes will be momentarily placed under very great pressure. As the wave retreats, the pressure is released with explosive force. Stresses weaken the rock, aiding erosion. This is particularly effective in well bedded and well-jointed rocks such as limestones and sandstones as well as in soft rocks such as clays, and glacial deposits. Hydraulic action is severe during times of storm wave activity – the average pressure of Atlantic storm waves is 11 000 kg per sq m.

Corrasion is the process whereby a breaking wave hurls pebbles and shingle against a coast, thereby eroding it. **Attrition** refers to the gradual reduction in the size of the load as it continuously crashes into itself. **Solution** is a form of chemical erosion. In areas of chalk and limestone, acidic water may help dissolve the rock.

In addition to erosion, there are a number of important processes of weathering and mass movement. Weathering processes include salt crystallisation, biological weathering and freeze-thaw, while mass movements include rock falls, slumping and flows.

Key Definitions

Abrasion   The wearing away of cliffs by the load carried by waves.

Attrition   The wearing away of the load carried by waves.

Hydraulic action   The force of air or water acting in joints and cracks.

Solution   The dissolving of the chemical load, especially on chalk and limestone.

Cliff   A steep slope with an angle of over 45°.

Wave cut platform   A horizontal layer of rock found at the base of a cliff caused by erosion and weathering.

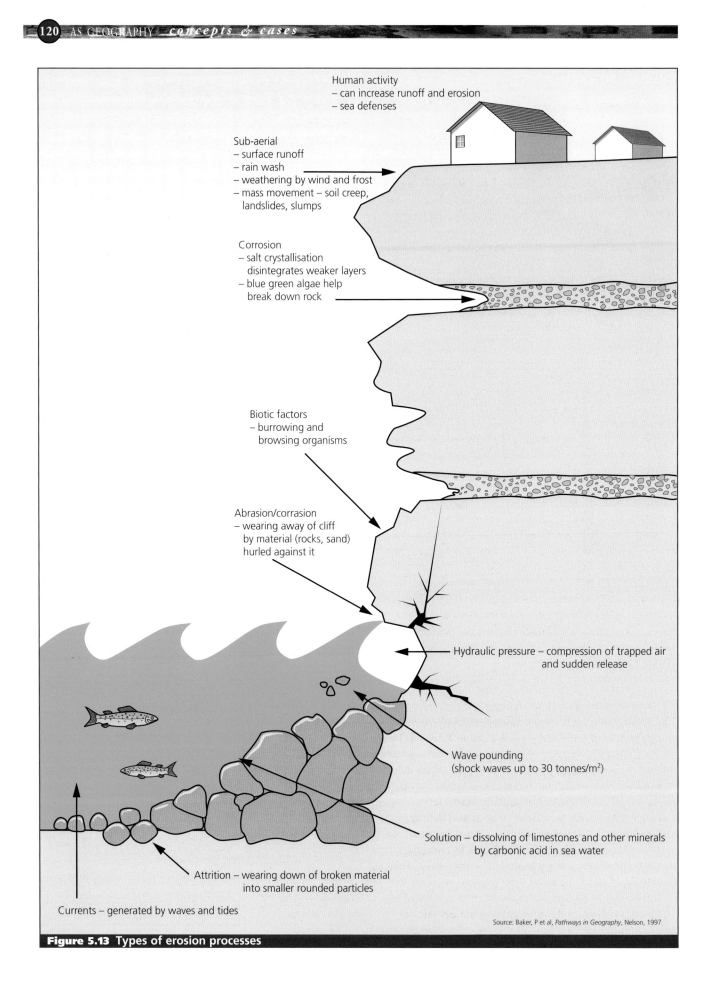

Human activity
– can increase runoff and erosion
– sea defenses

Sub-aerial
– surface runoff
– rain wash
– weathering by wind and frost
– mass movement – soil creep,
  landslides, slumps

Corrosion
– salt crystallisation
  disintegrates weaker layers
– blue green algae help
  break down rock

Biotic factors
– burrowing and
  browsing organisms

Abrasion/corrasion
– wearing away of cliff
  by material (rocks, sand)
  hurled against it

Hydraulic pressure – compression of trapped air
and sudden release

Wave pounding
(shock waves up to 30 tonnes/m^2)

Solution – dissolving of limestones and other minerals
by carbonic acid in sea water

Attrition – wearing down of broken material
into smaller rounded particles

Currents – generated by waves and tides

Source: Baker, P et al, *Pathways in Geography*, Nelson, 1997

Figure 5.13 Types of erosion processes

Cliffs and erosion

A **cliff** is a steep slope, generally with an angle of over 45°. Cliff profiles are very variable and depend on a number of factors such as rock type, rate of erosion, vegetation cover, and protection by a **wave cut platform**. One major factor is the influence of bedding and jointing. The well-developed jointing and bedding of certain harder limestones creates very geometric cliff profiles, with angular, steep cliff faces and a flat top (bedding plane). In other well-jointed and bedded rocks, a whole variety of features will be created by wave erosion, such as caves, geos, arches, stacks and stumps (Figure 5.14).

The dip of the bedding alone will create varying cliff profiles. For example, if the beds dip vertically, then a sheer cliff face will be found. By contrast, if the beds dip steeply seaward, then steep, shelving cliffs with landslips will be found (Figure 5.15).

Rate of cliff retreat are highly variable. In chalk, for example, it is over 100 m per century, whereas in hard limestone it is less than 1 m per century. This is due to the strength of the rock as well as the degree of exposure to erosive waves.

Each cliff profile, to some extent, is unique, but a model of cliff evolution or modification has been produced to take into account not only wave activity, but also subaerial weathering processes (Figure 5.16). As wave activity is constantly at work between high water mark (HWM) and low water mark (LWM), it causes undercutting of a cliff face, forming a notch and overhang. Breaking waves, especially during storms and spring tides, can erode the coast higher than HWM. As the undercutting continues, the notch becomes deeper and the overhang more pronounced. Ultimately the overhang will collapse, causing the cliff line to retreat. The base of the cliff will be left behind as a broadening **platform**, often covered with deposited material, with the coarsest near the cliff base, gradually becoming smaller towards the open sea.

The Seven Sisters

The Seven Sisters cliffs are one of the most impressive and recognisable landforms of the south coast of England. During the last glacial period, river systems would have flowed across the then impermeable chalk. These rivers flowed down to a sea-level that was much lower than now. At the end of the glacial period, as temperatures warmed, sea-levels rose again and the chalk became permeable once more. The rising sea drowned river valleys forming indented coastlines such as those in Devon and Cornwall. The chalk, however, was less resistant and the coastline was eroded by the waves, attacking first the headlands (due to wave refraction) and then planing the whole cliff line (Figure 5.17).

The dry valleys are short – the longest, Gap Bottom, is only 1.8 km, and the shortest just 400 m. The cliffs are eroding at a rate of between 50 m and 125 m per century.

Figure 5.14 Caves, geos and arches

Figure 5.15 Lulworth Cove

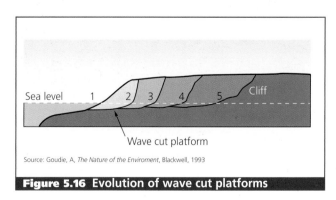

Sea level 1 2 3 4 5 Cliff

Wave cut platform

Source: Goudie, A, *The Nature of the Enviroment*, Blackwell, 1993

Figure 5.16 Evolution of wave cut platforms

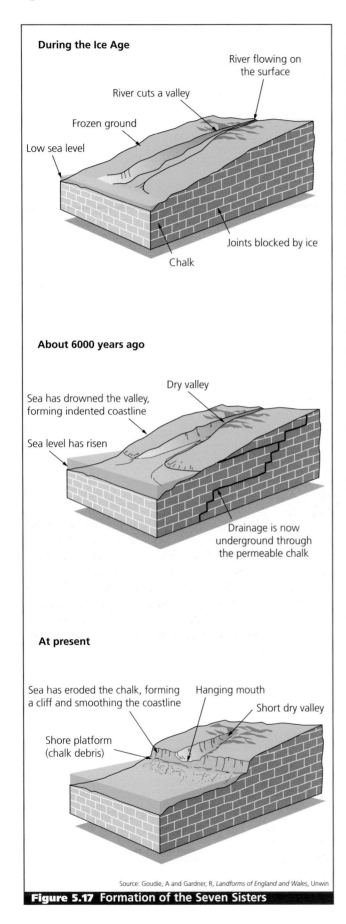

During the Ice Age

River flowing on the surface

River cuts a valley

Frozen ground

Low sea level

Joints blocked by ice

Chalk

About 6000 years ago

Dry valley

Sea has drowned the valley, forming indented coastline

Sea level has risen

Drainage is now underground through the permeable chalk

At present

Sea has eroded the chalk, forming a cliff and smoothing the coastline

Hanging mouth

Short dry valley

Shore platform (chalk debris)

Source: Goudie, A and Gardner, R, *Landforms of England and Wales*, Unwin

Figure 5.17 Formation of the Seven Sisters

Most retreat takes place in occasional pulses, such as in the early 1950s. Erosion and mass movement of the chalk is helped by the chalk beds dipping gently towards the sea. Erosion at the base of the cliff destabilises the cliff above. As the cliffs retreat they leave behind a lengthening shore platform.

In January 1999 a large section of Beachy Head fell into the sea (Figures 5.18 and 5.19). It was the biggest single loss of coastline in Britain in living memory. Hundreds of thousands of tonnes of chalk crashed 160 m into the sea lengthening the beach more than 30 m into the sea. The collapse was so large that a lifeboat spotted it some 7 km out at sea.

The fall may have been caused by water entering the chalk, freezing, expanding and causing the cliff to crumble. Another possibility is that waves along the south coast are getting stronger. Wave height in the Atlantic increased by 10% during the 1990s. Another theory is that following a series of dry years (in which the chalk dried out) 1998–99 was very wet. The chalk thus became increasingly wet and unstable. If a severe frost occurred when the chalk was moist a collapse would be possible.

Coastal platforms

As a result of cliff retreat, a platform along the coast is normally created (Figure 5.16). Traditionally, up to very recent times, this feature was described as a **wave-cut platform**, because it was believed that it was created entirely by wave action. However, there is now some controversy over the importance of other agents of weathering and erosion in the production of the coastal platform, especially the larger ones.

In high latitudes, **frost action** could be important in supporting wave activity. In other areas, **solution weathering**, **salt crystallisation** and **slaking** could also support wave activity, particularly in the tidal zone and splash zone. **Marine organisms**, especially algae, can accelerate weathering at low tide and in the area just above HWM. At night carbon dioxide is released by algae because photosynthesis does not occur. This carbon dioxide combines with the cool sea water to create an acidic environment, causing a solution of limestone and chalk.

Other organisms, such as limpets, secrete organic acids that can slowly rot the rock. Certain marine worms, molluscs and sea urchins can also actually 'bore' into rock surfaces, particularly chalk and limestone.

Coast erosion

Coast erosion rates vary greatly around the UK and depend on the geology of the coast and its exposure to wind, wave and current action. Rapid erosion is found along parts of the coast of southern and eastern England where relatively soft geological formations are being eroded (Figure 5.24). At the same time, however, parts of the same coast may be moving seawards as sediments build up.

Figure 5.18 Beachy Head before Jan 1999

Figure 5.19 Beachy Head after rock fall

Coast erosion is also a cause of landslides and rock fall. Of about 8500 landslides in Britain, about 15% were in the coastal zone. These latter include many of the largest landslide complexes such as The Undercliff on the Isle of Wight, Folkestone Warren and Black Ven in Dorset. On the Isle of Wight, it is almost impossible to obtain insurance with a post code of PO38, The Undercliff (Figure 5.25).

At Covehithe, in Suffolk, 6 m of coastline are eroded each year. This has carried on for a number of decades. A 40-km stretch of coast around Holderness, is eroded by about 2 m per year, depositing about 1.5 million cubic metres of sediment into the North Sea. Indeed, since Roman times the coast has moved inland by over 3 km, and over 30 villages have been lost to the sea. Rapid erosion is thus localised occurring most readily on soft rocks exposed to storm waves with a large fetch.

Coastal management on the Isle of Wight

The Isle of Wight contains almost 100 km of coastline. Much of this has been proclaimed as an Area of Outstanding Natural Beauty (AONB) and as a Heritage Coast. Natural forces operating around the coastline include marine erosion, weathering, mass movements, transport, longshore drift, as well as an array of human processes.

Over the last 150 years up to 25% of the Isle of Wight's coast has been subjected to intensive development. Much of this has been on the east side of the island whereas the west coast has been much less affected. Almost two-thirds of the 127,000 population live within 2 km of the coastline. The population on the island during the summer is over 250 000.

The Isle of Wight experienced severe storms in 1990, landslides in 1994 at Blackgang and Niton, pollution and dredging. To combat some of these effects major engineering works and cliff-stabilisation projects have occurred.

Coastal management on the Isle of Wight involves a wide range of issues:

■ planning

■ coastal protection

■ cliff stabilisation and ground movement studies

■ coastal infrastructure including seawalls, esplanades, car parks, paths

■ control of beaches and public safety

■ recreational activities and sport

■ beach cleaning

■ pollution and oil spills

■ offshore dredging

■ management of coastal land and property.

The coastal area has been divided into a number of natural cells, each of which contains its system of erosion, transport and deposition

Coastal defence

Coastal defence covers protection against coastal erosion (coast protection) and flooding by the sea. The coastal zone is a dynamic system which extends seawards and landwards from the shoreline. Its limits are defined by the extent to which natural processes and human activities occur. Coastal zone management is concerned with the whole range of activities that take place in the coastal zone and promotes integrated planning to manage them. Conflicting activities in coastal areas include housing, recreation, fishing, industry, mineral extraction, waste disposal and farming.

Hard engineering structures

The effectiveness of seawalls depends upon their cost and their performance. Their function is to prevent erosion and flooding but much depends on whether they are sloping or vertical, permeable or impermeable, rough or smooth, and of which material they are formed (clay, steel or rock for example.) In general, flatter, permeable, rougher walls perform better than vertical, impermeable smooth walls.

Cross-shore structures such as groynes, breakwaters, piers and strongpoints have been used for many decades. Their main function is to stop the drifting of material, as well as, in some cases, protecting the shore line from erosion. Traditionally groynes were constructed from timber, brushwood and wattle, and these have been very successful, especially on shingle coasts. However, modern cross-shore structures are often made from rock, and are part of a more complex form of management which includes beach nourishment and off-shore structures.

Sandown Bay

The cliffs at Shanklin on the south-east coast of the Isle of Wight are vertical, overhanging in places, and up to 40 m high. They are located in an area that is of prime importance for tourism and urban development (Figure 5.22). The cliffs, developed on a weak sandstone, are subject to mass movements, and have been managed and stabilised using a variety of methods such as:

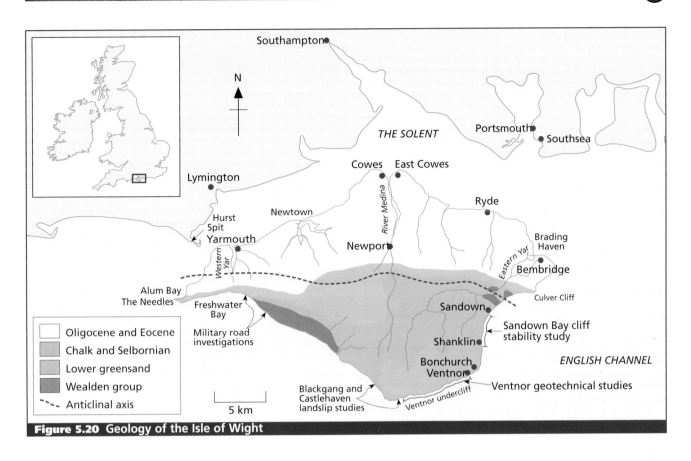

Figure 5.20 Geology of the Isle of Wight

Map labels: Southampton, THE SOLENT, Portsmouth, Southsea, Lymington, Cowes, East Cowes, Ryde, Newtown, Hurst Spit, Yarmouth, River Medina, Newport, Brading Haven, Bembridge, Eastern Yar, Western Yar, Alum Bay, The Needles, Freshwater Bay, Military road investigations, Culver Cliff, Sandown, Sandown Bay cliff stability study, Shanklin, ENGLISH CHANNEL, Bonchurch, Ventnor, Ventnor geotechnical studies, Blackgang and Castlehaven landslip studies, Ventnor undercliff

N

Legend:
- Oligocene and Eocene
- Chalk and Selbornian
- Lower greensand
- Wealden group
- Anticlinal axis

5 km

- realignment of the cliff top
- rock anchoring and rock bolting
- buttressing
- drainage and structural support to existing overhangs.

The 3.5 km length of weak sandstone cliffs between Shanklin and Sandown have a long history of cliff failure and recession. Cliff recession has made some cliff top paths and properties at risk of collapse. In addition, falling debris from mass movements and cliff erosion is a major threat to properties and businesses at the base of the cliff. In 1988 there was a major study into the stability of the cliffs at Sandown.

The Sandown Bay cliffs range in height from 10 m to 40 m, and are inclined at an angle of between 60° and 90°. The combination of steep cliff morphology, a weak and incompetent rock, and discontinuities (which act as lines of weakness that undermine the rock) leads to major instability.

Cliff recession occurs on the cliff through seepage erosion, mass movements, and weathering. The main types of weathering include freeze-thaw and the reduction of iron oxides. The debris at the base of the cliff, although endangering properties and businesses, does provide some protection to the cliff against further erosion.

Figure 5.21

Figure 5.22

Coastal erosion in England and Wales

Areas of rapid erosion

Source: *Independent*

Figure 5.23 Areas of rapid erosion in Britain

Figure 5.24

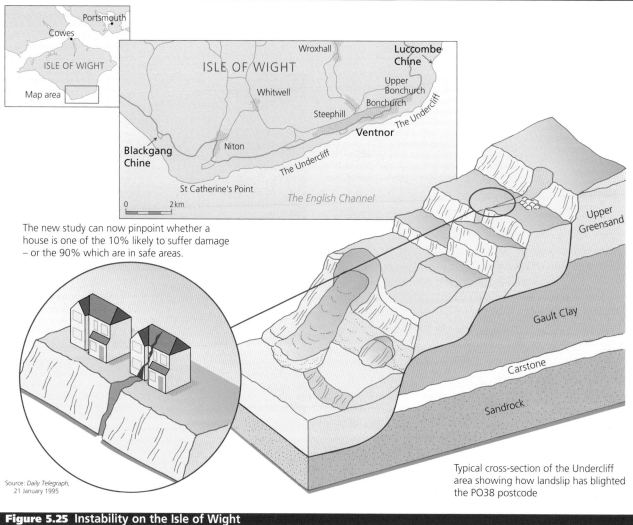

The new study can now pinpoint whether a house is one of the 10% likely to suffer damage – or the 90% which are in safe areas.

Source: *Daily Telegraph*, 21 January 1995

Typical cross-section of the Undercliff area showing how landslip has blighted the PO38 postcode

Figure 5.25 Instability on the Isle of Wight

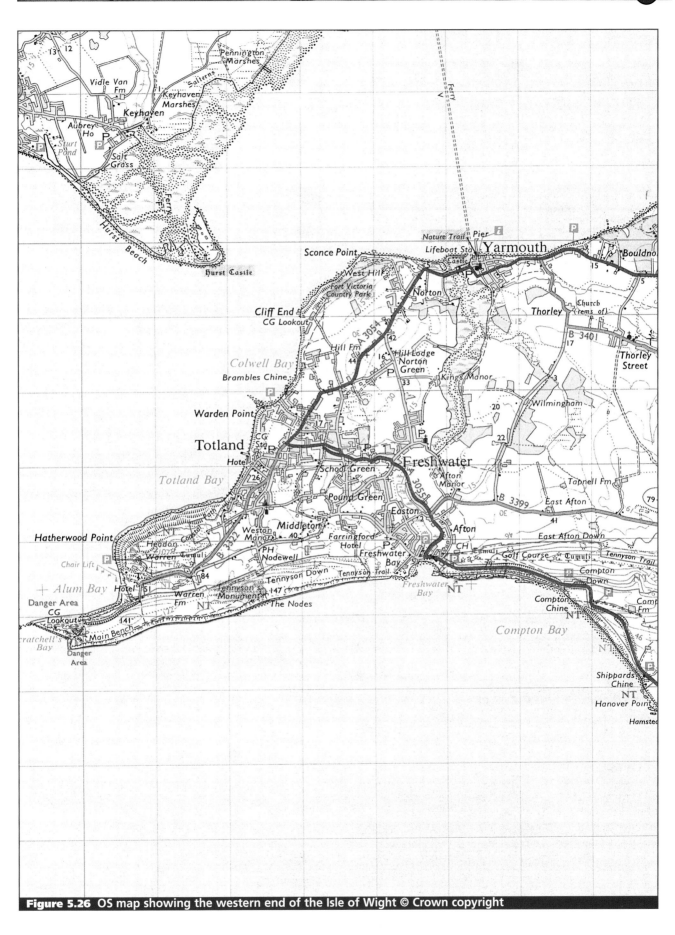

Approximately £1.5 million has been spent on cliff stabilisation works. This has included **reprofiling** the cliff top where possible (Figure 5.22), removal of superficial deposits, and removal of the unstable upper sections of the cliff. Engineers believe that an angle of about 60° is stable for the cliffs at Sandown. Cliffs at this angle could support vegetation which would reduce the amount of water getting into the rock. **Scaling** (the removal of loose materials) was also carried out. Anchors, bolts and dowels were used to provide **support** for the cliff beneath cliff top buildings (so that the buildings were protected). In addition, **meshing** was placed over large section of the cliff, in particular next to flights of steps where there is a risk to pedestrians from rock falls and rock slides. **Drainage holes** were inserted into the cliff to reduce excess amounts of water, and to drain seepage horizons (zones where water accumulates subsurface). Overhanging sections of cliff were **infilled** with timber shuttering and a polystyrene fill. To date the measures have prove effective.

Figure 5.27

Blackgang to Freshwater Bay

Protecting the south-west coast is problematic for a number of reasons:

- protective methods need to bear the full force of Atlantic gales

- much of the area is an Area of Outstanding Natural Beauty (AONB) and so engineering works could detract from the area's beauty

- much of the coastline is agricultural land, and not eligible for coastal protection grants

- cliff retreat is the result of weathering, mass movements and seepage, not just marine erosion.

The south-western part of the Isle of Wight includes some of the best coastal scenery in the British Isles. The coast contains a number of broad, sweeping bays such as Chale and Compton Bays, and headlands such as Atherfiels and Hanover Points. The cliffs are generally low and broken by a number of steep valleys (known as chines) such as Compton Grange.

The area contains some of the most important chalk downland in Britain, and much of the coastline has been designated as a Bird Reserve (under the West Wight Sanctuary Order of 1954). The area receives a considerable number of tourists each year, and the impacts of tourists is greatest where road access, car parking and facilities are concentrated. These areas include Freshwater Bay, Brighstone, Chilton Chine and Atherfield.

Pollution on this stretch of the coastline is due mainly to seabourne litter from Channel shipping. There is also a risk of oil spills from ships and from nearby oil explorations.

This area of coastline is completely exposed to the force of south-west winds, and so erosion is rapid especially where the cliffs are formed of soft Wessex Clay. Rates of erosion are between 0.3 and 0.5 m per annum, and this is aggravated by mass movements on the cliffs and by the seepage of water through the cliffs. The A3055 (Military Road) in particular is vulnerable at Brook and Compton (Figure 5.21). Coastal erosion could destroy parts of the road, in particular at Whiteways, Afton, Small Chine, Brook and Brook Bay. Although it is possible to move the road inland at Small Chine, Brook and Brook Bay it is not possible at Afton and Whiteways (Figure 5.21). On the southern side of Afton Down, there are a series of fissures in the rock which extend beneath the road towards the sea. These fissures or 'joints' are caused by pressure release, as weathering and erosion reduce the pressure on the adjacent cliffs. In some places these joints are up to 1 m wide and 13 m deep! some extend for over 150 m.

Overlooking the highest section of the Military Road the 13 m chalk cliffs are steeper than the cliffs below the road. This may make the chalk unstable and prone to mass movements such as toppling and rock falls.

To date this part of the island has received very little hard engineering. This is largely

Questions

1 Study the map extract of the western end of the Isle of Wight.
   (a) What coastal feature is located at 290848?
   (b) Describe the coastline between 305855 and 320865.
   (c) What map evidence is there to suggest mass movements are important in this area?
   (d) Suggest reasons for the bay in 3485.
   (e) What map evidence is there to suggest that human activity has affected the coastline of this region?
2 Describe the main characteristics of Hurst Beach in squares 2990, 3090, 3189, and 3190.
   (a) What type of feature is shown in these squares?
   (b) Explain how it might have been formed.
   (c) What feature is found in squares 3090 and 3191 behind Hurst Beach?
   (d) Explain how it is likely to have been formed.

due to the special ecological, environmental and scenic qualities of the area. The only coastal defenses are found at Freshwater Bay. However, the area is monitored closely. Although a policy of managed retreat has been adopted (i.e. allowing nature to take its course), the Military Road is protected both from marine erosion and mass movements on the upper cliffs.

Freshwater Bay to Norton

The chalk ridge reaches the sea at Compton Bay producing near vertical cliffs. These cliffs extend for about 5 km until they form the Needles.

Marine erosion is slow on the cliffs, about 10cm/year. Seepage erosion occurs along the coast, and mass movements are common, especially mudflows, earthflows, rotational slides and slips.

Beaches vary considerably. In Freshwater Bay and Alum Bay it is mostly shingle (the flint pebbles originating from the chalk), whereas north of Totland Bay it is sand, the material originating from local Eocene sands.

As the rate of erosion in Freshwater Bay is low, there is no need for any coastal protection schemes in this part of the shoreline. However, there are problems relating to sewage pollution at Yarmouth. Further along the coast is Alum Bay rockfalls and weathering have made some properties vulnerable. In other parts of the bay, cliff drainage and encasement have been used to remove excess water from the rock and to stabilise the slopes.

Questions

1 Suggest reasons why the Isle of Wight experiences high rates of coastal erosion.
2 Why is coastal erosion a bigger problem in the eastern side of the island rather than the western part?
3 Outline the potential conflicts between some of the users of the coastal area.
4 Explain what is meant by the following terms: reprofiling, scaling, meshing, drainage.
5 How and why has coastal protection been developed in the south east part of the Island?

Deposition

Beaches

Excellent beach development occurs on a lowland coast (constructive waves), with a sheltered aspect/trend, composed of 'soft' rocks, which provides a good supply of material, or where longshore drift supplied abundant material.

The term **beach** refers to the accumulation of material deposited between low spring tides and the highest point reached by storm waves at high spring tides. A typical beach will have three zones: backshore, foreshore and offshore. The backshore is marked by a line of dunes or a cliff. Above the high water mark there may be a **berm** or **shingle ridge**. This is coarse material pushed up the beach by spring tides and aided by storm waves flinging material well above the level of the waves themselves. These are often referred to as storm beaches. The seaward edge of the berm is often scalloped and irregular due to the creation of beach **cusps**.

The foreshore is exposed at low tide. The beach material may be undulating due to the creation of ridges, called fulls, running parallel to the water line, pushed up by constructive waves at varying heights of the tide. These are separated by troughs, called **swales**. Great stretches of sand too, may comprise the foreshore. In areas of complex coast sand beaches may only be exposed as small **bayhead beaches** in bays.

Offshore, the first material is deposited. In this zone, the waves touch the sea bed and so the material is usually disturbed, sometimes being pushed up as **offshore bars**, when the offshore gradient is very shallow.

Questions

1 Define the following terms: hydraulic action, solution, abrasion.
2 Under what conditions is erosion most intense?
3 Suggest reasons why rates of erosion are expressed in rates of m per century.
4 Study Figures 5.18 and 5.19 which show Beachy Head before and after the collapse of January 1999.
   (a) State two ways in which the new cliff has changed.
   (b) Draw a simple sketch section to show how the shape of the cliff has changed as a result of the cliff collapse.

Bars and spits

These more localised features will develop where:

- abundant material is available, particularly shingle and sand
- the coastline is irregular due, for example, local geological variety (transverse coast)
- where there are estuaries and major rivers.

Offshore bars are usually composed of coarse sand or shingle. They develop as bars offshore on a gently-shelving sea-bed (Figure 5.29). Waves touch the seabed far offshore and this causes deposition. Between the bar and shore, lagoons (often called **sounds**) develop. If the lagoonal water is calm, and fed by rivers, marshes and mud-flats can be formed. Bars can be driven onshore by storm winds and waves. A classic area is off the coast of the Carolinas in the south east of the USA.

Spits are common along an indented coast. For example, along a transverse coast where bays are common or near mouths (estuaries and rias), wave energy is reduced. The long, narrow ridges of sand and shingle are always joined at one end to the mainland. Orford Ness (Figure 5.30) has been deposited where the north–south longshore drift of the North Sea has been interrupted by the flow of the Rivers Alde and Butley. The spit, well-fed by the longshore drift, has grown rapidly – as recently as the 17th century, the town of Orford was a small port on the open sea.

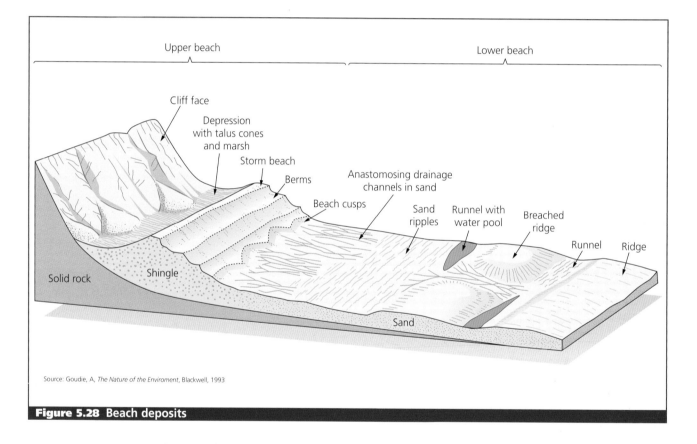

Source: Goudie, A, *The Nature of the Enviroment*, Blackwell, 1993

Figure 5.28 Beach deposits

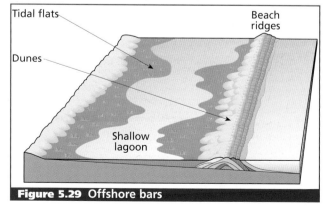

Figure 5.29 Offshore bars

Spits often become curved as waves undergo refraction. Cross currents or occasional storm waves may assist this hooked formation. A good example is Hurst Castle spit on the Dorset coast (Figure 5.26). The main body of the spit is curved but it has additional, smaller hooks, or **recurves**, and is 2 km long. Longshore drift moves sediment eastwards along the south coast. However, the coastline is very irregular and there is a sudden change in the trend of the coastline. Consequently, refraction occurs, causing the waves to bend around northwards. In addition, occasional storm waves come from the north-east down the Solent, building up the recurves.

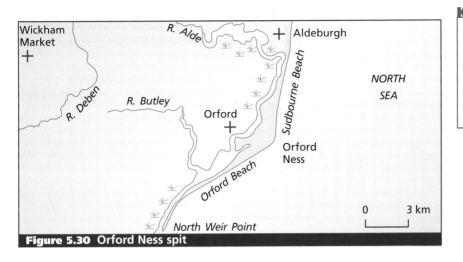

Figure 5.30 Orford Ness spit

On the seaward side, the slope to deeper water is very steep. Within the curve of the spit, the water is shallow and a considerable area of mud-flats and salt marsh (salting) is exposed at low water. These salt marshes are continuing to grow as mud is trapped by the marsh vegetation. The whole area of salt marsh is intersected by a complex network of channels, or creeks, which contain water even at low tide.

Managing a spit: the case of Dawlish Warren

The Dawlish Warren spit is located at the mouth of the Exe estuary (Figure 5.31). The sand spit receives its materials from a number of sources.

The main source is the red sandstone and breccia cliffs at Langstone Rock headland. Eroded material is drifted eastwards to Dawlish. Constructive waves and swell waves move material onto Dawlish from offshore bars. In addition, at low tide wind erosion blows sand from the beach onto the dunes where it is trapped by marram grass. Finally, the River Exe carries material down in suspension and solution. Much of this is deposited in the low energy environment behind the dunes. Furthermore, the mixing of fresh water and sea water causes clay particles to stick together and then, owing to their increased weight, they are deposited.

Dawlish Warren has been mapped for over 200 years. Over 200 m of the sand spit have been eroded since 1787, an annual average of 1 m (Figure 5.32). In addition, there has been sand lost from the dunes as a result of recreational trampling on the dunes. Moreover, the presence of a large breakwater at Langstone Rock has prevented sand from being drifted onto the spit and replenishing some of the sand lost by erosion. Hence the spit is getting narrower and there is a real danger that it could be breached in a major storm.

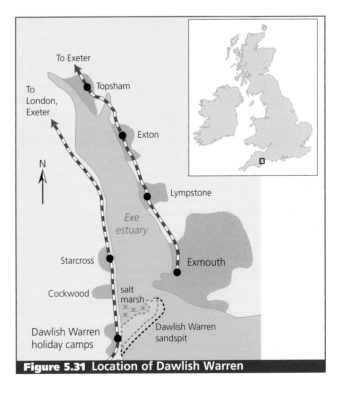

Figure 5.31 Location of Dawlish Warren

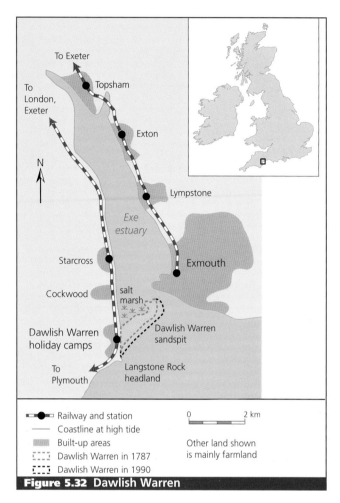

Figure 5.32 Dawlish Warren

Railway and station
Coastline at high tide
Built-up areas
Dawlish Warren in 1787
Dawlish Warren in 1990

0 — 2 km

Other land shown
is mainly farmland

In 1989–90 Dawlish Warren experienced severe storms. Initially, storm waves from the south built up sand and shingle against the breakwater at Langstone Rock, while removing sand from the western end of Dawlish. Later, storm winds from the south-east caused considerable damage to the rock armour. Boulders were dislodged, and unprotected sand was washed away, undermining the promenade. Other problems included:

■ boulders in the rock armour had moved position and were no longer interlocking

■ rocks that should have reduced the impact of backwash had been eroded

■ limestone boulders had been weathered and eroded

■ boulders had become more rounded making them liable to be rolled away during severe storms

■ boulders were now small enough to be lifted by storm waves.

There are a number of reasons why Dawlish deserves to be protected:

■ it is an internationally valuable habitat for plants and birds, and is managed as a nature reserve

■ it is an important holiday destination – over 20 000 people use the spit at peak times and the revenue from the car parks is a major source of income for Teignbridge District Council

■ there are over 50 businesses on the spit and the adjoining village providing services and employment for visitors and local residents

■ the spit is a natural flood defence, protecting the low-lying Exe estuary.

The new defences at Dawlish were built in 1992, costing £1.5 million (Figure 5.33). The main features were a new rock armour revetment with a curved sea wall. Rocks for the armour were imported from Norway, although local materials were used to 'face' the wall (Figure 5.34). Sea defences are not cheap (all 1992 prices):

■ large interlocked boulders cost £3500 per metre

■ sloping concrete walls cost £2000 per metre

■ gabions (stones in wire baskets) cost £100 per metre

■ offshore breakwaters cost £5000 per metre

■ beach nourishment costs £3 per cubic metre

■ stone walls cost £6000 per metre

■ groynes cost £10 000 each.

It is easy to see why some planning authorities are keen to allow nature to take its course, and why others have fallen behind in the maintenance of their defences. It is also clear that there has to be a good reason to protect land. So far, the sea defences have protected Dawlish Warren.

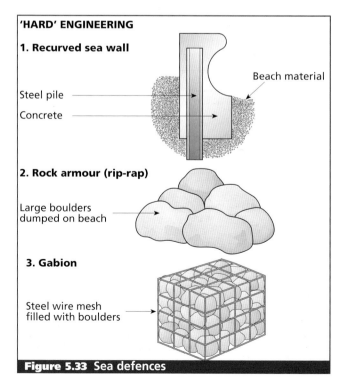

'HARD' ENGINEERING

1. Recurved sea wall

Steel pile
Concrete
Beach material

2. Rock armour (rip-rap)

Large boulders
dumped on beach

3. Gabion

Steel wire mesh
filled with boulders

Figure 5.33 Sea defences

Figure 5.34 Sea defences at Dawlish

Bars and barrier beaches

If a spit continues to grow lengthwise, it may ultimately link two headlands to form a **bay bar**. These are composed either of sand or shingle, as in the case of the Slapton Ley, south Devon.

Tombolo

A ridge of material that links an island with the mainland, is called a **tombolo**. The 'classic' example of a tombolo is Chesil Beach (Figure 5.35). Chesil Beach is 25 km long, connecting the Isle of Portland with the mainland Dorset coast at Burton Bradstock, near Abbotsbury. At its eastern end at Portland, the ridge is 13 m above sea-level and composed of pebbles about the size of a potato. At its western end near Abbotsbury, the ridge is lower, only 7 m above sea-level and built of smaller, pea-sized material.

The height of the ridge and the sizing of material suggests that dominant wave action occurred from east to west. Larger material is piled up at the eastern end, being the heavier and more difficult to transport. Smaller, lighter material is carried further west before being deposited. However, the dominant wave action comes from the south-west. The grading of the ridge should therefore be completely opposite to what it is.

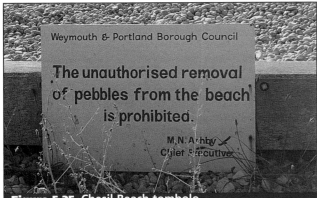

Weymouth & Portland Borough Council

The unauthorised removal of pebbles from the beach is prohibited.

M.N. Ashby
Chief Executive

Figure 5.35 Chesil Beach tombolo

An alternative explanation to longshore drift is that Chesil Beach is a very youthful feature dating from the Pleistocene (18 000–20 000 years BP) when sea levels fell at least 100 m below the present sea-level. As a result, much of the present English Channel was dry. During this period, vast amounts of debris could have been carried into the dry Channel area by meltwater. As sea-levels rose in early post-glacial times, this material could have been pushed onshore as an off-shore bar and was trapped by the Isle of Portland and Lyme Bay. Present-day wave action is gradually sorting this material.

Cuspate forelands

Cuspate forelands consist of shingle ridges deposited in a triangular shape, and are the result of two separate spits joining or the combined effects of two distinct sets of regular storm waves. The best example is at Dungeness, near Dover, where the foreland was deposited by longshore drift curling west from the North Sea and by the longshore drift flowing eastwards up the Channel.

Sand dunes

Sand dunes are young features. Sea-levels around the British Isles only reached their current position about 6000 years ago, so sand dunes have been formed since then. (There are fossil beaches and fossil dunes but live dunes are less than 600 years old.) Dunes are common around many parts of Britain, but especially so in north-east Scotland, north-east Norfolk and Cardigan Bay. Other good examples of sand dunes and their associated plant succession can be found at Formby, Lancashire (Figure 5.36), Studland Beach, Dorset, and Braunton Burrows, Devon.

Extensive sandy beaches are almost always backed by sand dunes because strong onshore winds can easily transport the sand which is exposed at low water. The sand grains are deposited against any obstacle on land to form dunes. Dunes can be blown inland and can therefore threaten coastal

Figure 5.36 Sand dunes

farmland and even villages. There are several methods to slow down the migration of dunes:

- special grasses, such as marram, with long and complex tap root systems bind the soil
- brushwood fencing reduces sand movement
- planting of conifers such as Scots and Corsican Pine which can stand the saline environment and poor soils.

Sand dune succession

Succession refers to the change in the sequence of plant communities over time. On sand dunes, and on salt marshes, there is a very noticeable change in species with increasing distance from the sea. These changes occur due to slight, but important changes in the micro-environment.

For sand dunes to form there needs to be a large supply of sand. Initially, sand is moved by the wind. Wind speed varies with height above a surface. As the belt of no wind is only 1 mm above the surface, most grains protrude above this height and they are moved by saltation. Irregularities cause increased wind speed and eddying, and more material is moved.

For stable dunes to occur vegetation is required. Plant succession and vegetation succession can be interpreted by the fact that the oldest dunes are furthest from the sea and the youngest ones are closest to the shore. On the shore, conditions are windy, arid and salty. The soil contains few nutrients and is mostly sand – hence the foredunes are referred to as yellow dunes. Few plants can survive, although sea couch and marram can tolerate these conditions. Once the vegetation is established it reduces the wind speed close to ground level. The belt of no wind may increase to a height of 10 mm. As grasses such as sea couch and marram need to be buried by fresh sand in order to grow, they keep pace with deposition. As the marram grows it traps more sand. As it is covered it grows more, and so on. Once established the dunes should continue to grow, as long as there is a supply of sand. However, once another younger dune, a foredune, becomes established the supply of sand, and so the growth of the dune, is reduced.

As the seaward dune gets higher the supply of fresh sand to inland dunes is reduced and marram and sea couch can no longer survive. In addition, as wind speeds are reduced, evapotranspiration losses are less, and the soil becomes more moist. The decaying marram adds some nutrients to the soil, which in turn becomes more acidic. In the slacks, the low points between the dunes, conditions are noticeably moister, and marsh vegetation may occur.

Towards the rear of the dune system 'grey' dunes are formed. The soil is 'grey' due to the presence of humus in the soil. The vegetation found depends largely upon the nature of the sand. Pine trees favour acid soils, whereas oak can be found on more neutral soils. Thus the vegetation at the rear of the sand dune complex is quite variable (Figure 5.37)

Mud-flats and salt marshes

The **intertidal zone**, the zone between high tide and low tide, experiences severe environmental changes in salinity, wetting and drying. Halophytic (salt tolerant) plants have adapted to the unstable, rapidly changing conditions (Figure 5.38).

Salt marshes are typically found in three locations – low energy coastlines such as the Norfolk coast, behind spits and barrier islands, such as Hurst Castle spit (Figure 5.26 map), and in estuaries and harbours, such as Poole Harbour.

Silt accumulates in these situations and, on reaching sea-level, forms mud-banks. With the appearance of vegetation a salt marsh is formed. The mud-banks are often intersected by creeks.

Questions

1 Under what conditions do beaches form?
2 What conditions are required for the formation of **(a)** spits and **(b)** bars?
3 What is a recurve on the end of a spit?
4 How are recurves formed?
5 **(a)** Why is it important to manage Dawlish Warren?
   (b) What are the physical pressures on Dawlish Warren?
   (c) How has human activity increased the pressures on Dawlish Warren?
6 How does the size of beach material vary along Chesil Beach?
   Suggest reasons to explain this pattern.

Key Definitions ④

Succession The change in the sequence of plant species over space and time.

Sand dunes Extensive areas of sand accumulation into regular dune (ridge) formation, each dune getting progressively smaller from the coast.

Salt marsh An area between high tide and low tide where mud is flooded by sea water during every tidal sequence.

Questions

1 **(a)** Define the term succession.
   (b) Suggest two factors necessary for sand dunes to form.
   (c) Describe the changes in vegetation across a typical sand dune.
   (d) Explain why these changes take place.
2 **(a)** What is a halophyte?
   (b) Describe how the environmental conditions at A and B change over a period of 24 hours.
3 In what ways are plants adapted to salt marsh environments?

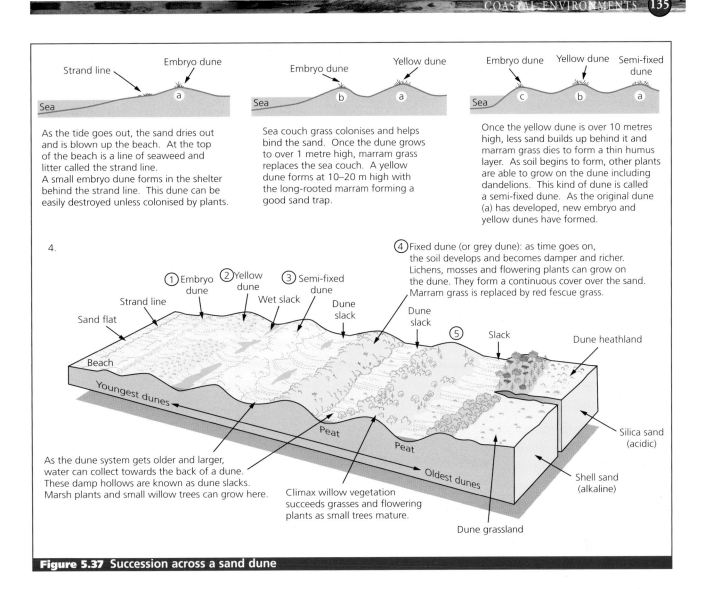

As the tide goes out, the sand dries out and is blown up the beach. At the top of the beach is a line of seaweed and litter called the strand line.
A small embryo dune forms in the shelter behind the strand line. This dune can be easily destroyed unless colonised by plants.

Sea couch grass colonises and helps bind the sand. Once the dune grows to over 1 metre high, marram grass replaces the sea couch. A yellow dune forms at 10–20 m high with the long-rooted marram forming a good sand trap.

Once the yellow dune is over 10 metres high, less sand builds up behind it and marram grass dies to form a thin humus layer. As soil begins to form, other plants are able to grow on the dune including dandelions. This kind of dune is called a semi-fixed dune. As the original dune (a) has developed, new embryo and yellow dunes have formed.

④ Fixed dune (or grey dune): as time goes on, the soil develops and becomes damper and richer. Lichens, mosses and flowering plants can grow on the dune. They form a continuous cover over the sand. Marram grass is replaced by red fescue grass.

As the dune system gets older and larger, water can collect towards the back of a dune. These damp hollows are known as dune slacks. Marsh plants and small willow trees can grow here.

Climax willow vegetation succeeds grasses and flowering plants as small trees mature.

Figure 5.37 Succession across a sand dune

Salt marsh succession

Once the bare marsh flat is formed, the first plants, such as green algae (enteromorpha), colonise it (Figure 5.38). The algae trap sediment from the sea and provide ideal conditions for the seeds of the salt-tolerant marsh samphire (*Salicornia*) and eel grass (*Zostera*) which then colonise the marsh. These plants are deep-rooted to survive the ebb and flow of the tide. They can survive being covered by sea water and can also tolerate drying out. These plants increase the rate of deposition by slowing down the water as it passes over the vegetation. This is known as **bioconstruction**.

Gradually, the clumps of vegetation become larger and the flow of tidal waters is restricted to specific channels, or creeks. The slightly increased height of the surface around the plants leads to more favourable conditions. Here plants are covered by sea water for shorter periods of time and this encourages other plants to colonise, such as sea aster, sea poa, and sea blite. These are even more efficient at trapping sediment and the height of the saltmarsh increases. New plants colonise as the marsh grows, including sea lavender, sea pink and sea purslane. As the height increases, tidal inundation of marsh becomes less frequent, and the rate of growth slows down. Sea rush (*Juncus*) and black saltwart become the most common type of plants. It takes about 200 years to progress from the marsh samphire (*Salicornia*) stage to the sea rush (*Juncus*) stage.

Key Definitions ⑤

Morphology The shape of the coastline.

Trend The direction or orientation of the coastline.

Isostatic Local changes in the level of the land.

Eustatic Global changes in the level of the sea.

Steric The rise of sea level due to the expansion of water as it gets warmer.

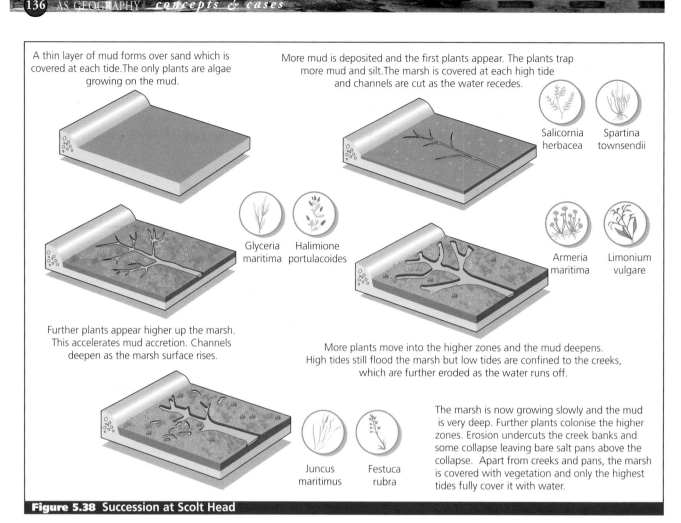

A thin layer of mud forms over sand which is covered at each tide. The only plants are algae growing on the mud.

More mud is deposited and the first plants appear. The plants trap more mud and silt. The marsh is covered at each high tide and channels are cut as the water recedes.

Salicornia herbacea

Spartina townsendii

Glyceria maritima

Halimione portulacoides

Armeria maritima

Limonium vulgare

Further plants appear higher up the marsh. This accelerates mud accretion. Channels deepen as the marsh surface rises.

More plants move into the higher zones and the mud deepens. High tides still flood the marsh but low tides are confined to the creeks, which are further eroded as the water runs off.

The marsh is now growing slowly and the mud is very deep. Further plants colonise the higher zones. Erosion undercuts the creek banks and some collapse leaving bare salt pans above the collapse. Apart from creeks and pans, the marsh is covered with vegetation and only the highest tides fully cover it with water.

Juncus maritimus

Festuca rubra

Figure 5.38 Succession at Scolt Head

Coastal morphology

Each stretch of coast is in its way, unique, dependent upon local geological differences (rock type and structures), exposure to wave action and types of wave action. However, it is possible to draw up certain classifications of coasts.

Coastal morphology due to differential erosion

Atlantic or discordant coastlines

Atlantic coastlines (or **discordant** or **transverse** coasts) develop where the rock strata run at right angles to the shoreline. The harder rocks will tend to form headlands and the softer rocks, bays. A clear example of this type of coast is the Dorset coast near Swanage – the local chalk and limestone form the headlands, while the bays are found in the clay belts.

Pacific (accordant) coastlines

Where the rock strata run parallel to the coast a Pacific (or **accordant**) coastline is formed (Figure 5.39). This is seen in the south coast of Dorset, for example.

Coastal morphology and changes in sea level

Changes in sea-level have occurred throughout geological time. The most recent changes (which are still affecting our present coastlines) have occurred in the last 20 000 years since the last glacial advance (Figure 5.40). Sea-level fell as water was locked up in the great ice sheets and glaciers that developed. During these cold phases, water was not returned to the oceans in the normal way but remained locked as ice. Hence the sea-level dropped. At the end of the glacial phase, the sea-level rose worldwide as the ice melted. These worldwide changes in sea level are known as **eustatic** changes.

In addition, the weight of ice caused the earth's crust to 'sag' locally causing a local fall in the level of the land. This had the same impact on landforms as a rise in sea-level. When the ice melted at the end of glaciation, the crust began to 'rebound' upwards, creating a local rise in the level of the land (and so a drop of sea-level) and the creation of certain 'raised' coastal landforms. These more localised changes, which are still occurring today, are called **isostatic** changes.

It is possible to classify coasts according to whether they have 'emerged' from the sea (by way of a fall in sea-level or a rise in land level), or have been 'submerged' (by a rise in sea-level or a drop in land level).

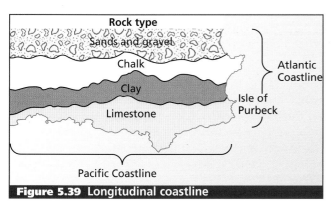

Figure 5.39 Longitudinal coastline

Submergence – drowned coasts

Submerged coasts are very common as a result of the rise of sea level of up to 100 metres in Britain over the last 18 000 years. **Rias** are drowned river valleys found in an upland area, where the river valleys were deep and narrow, and surrounded by undulating high ground (Figure 5.41). A good example is the Fowey Estuary in Cornwall. The Fowey, for example, is 300 m wide at its mouth, and extends inland for almost 8 km, varying in width from 300 m to 200 m. The surrounding land rises steeply from sea-level to 100m. By contrast, **fjords** are drowned glaciated U-shaped valleys (Figure 5.42). They are straighter, deeper and have steeper valley sides.

Emergence – raised coasts

Raised coasts are common in areas that were glaciated during the last glaciation – they are very well developed in Scotland and Northern England (the most intensely glaciated areas) as the pre-glacial beach, the 30 m beach, and the youngest 8 m raised beach.

The most typical raised coastlines are lowlands or former coastal plains that emerged, and as a result created a gently-shelving, smooth and wide coastal plain. This consists of the area of former offshore shallows, backed by the former beach (now raised) and/or cliff line. Coastal plains form rich farming land thanks to the thick marine deposits that cover it. An excellent example is the south-eastern coast of the USA (Figure 5.43).

Rising sea-levels

The earth's surface and lower atmosphere are expected to continue to warm over the next century as a result of increasing concentrations of greenhouse gases caused by emissions due to human activities. One of the major consequences of this warming will be a global rise in sea-levels. In addition, there may also be an increase in the frequency and intensity of coastal storms.

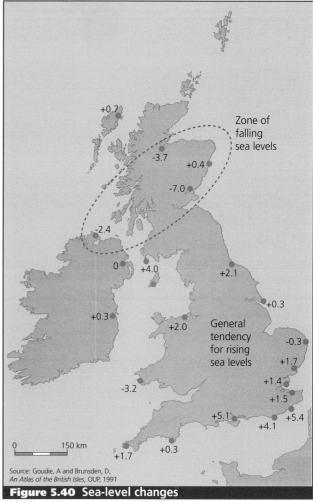

Source: Goudie, A and Brunsden, D,
An Atlas of the British Isles, OUP, 1991

Figure 5.40 Sea-level changes

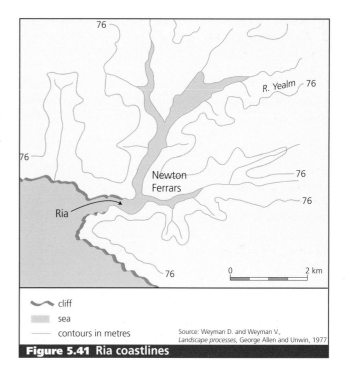

cliff
sea
contours in metres

Source: Weyman D. and Weyman V.,
Landscape processes, George Allen and Unwin, 1977

Figure 5.41 Ria coastlines

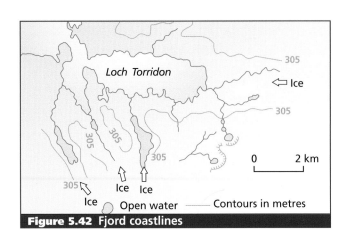

Figure 5.42 Fjord coastlines

Since 1900, sea-levels have risen by around 10 to 15 cm and this rate of rise is thought likely to increase over the next hundred years or so. (This figure is less than the local rising and subsidence of land in many areas, hence different areas will show different effects.) Estimates of the likely effects of global warming suggest additional rises in sea-levels of perhaps 20 cm by 2030, and 60 cm by 2100. Rising sea-levels could have many adverse effects, particularly coastal flooding and erosion. In certain parts of the UK (notably the South-east) the rise in the sea relative to the land may be made more owing to local subsidence. Estimates vary, but it could be of the order of 1 mm per year.

Figure 5.44 shows the areas of Britain where a rise in sea-level could have a significant effect (shaded areas indicate land which is less than 5 m above sea-level). Many of these include major conurbations or high grade agricultural land. Major road and rail links situated near the coast would also be at risk. Several power stations are also situated on low-lying land. In Northern Ireland, such areas are confined to narrow coastal strips and no significant areas exist inland.

The effects of sea-level rise may be worsened by possible increases in the frequency of storms and thus wave activity. The greatest impact is likely on the exposed western coast facing the Atlantic which has become notably rougher since 1970.

As well as direct effects such as coastal erosion or the flooding of coastal areas, higher mean sea-levels could also have an impact on underground water resources. The zone of mixing of sea water with freshwater in rivers is dynamic and a rise in sea-level can cause it to move upstream (Figure 5.46).

Much of the east coast of Britain is at serious risk of flooding due to inadequate sea defences and the willingness of planning authorities to allow development along low lying areas. Many of Britain's sea walls need substantial amounts of investment if they are to be effective. Many were constructed in the 1950s following the storm surge of 1953 which killed over 300 people, but since then the amount of money spent on sea defences has fallen (Figure 5.47).

In addition, natural flood defences, such as salt marshes, have been reclaimed for agricultural and leisure development. Existing sea defences were designed for certain return periods (the average-sized flow expected over a certain period, for example, every 50 years). However, changes in sea-levels, a

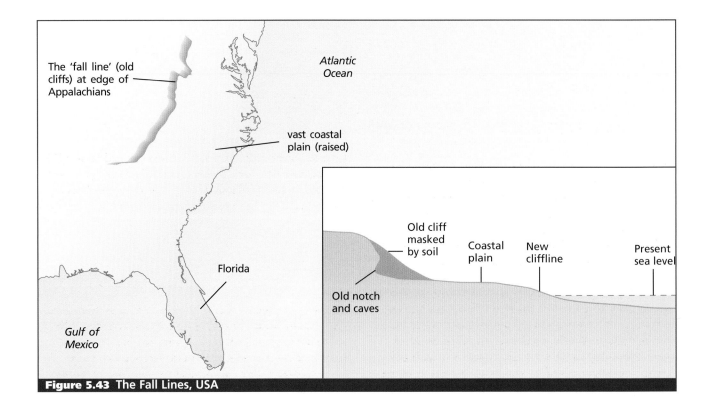

Figure 5.43 The Fall Lines, USA

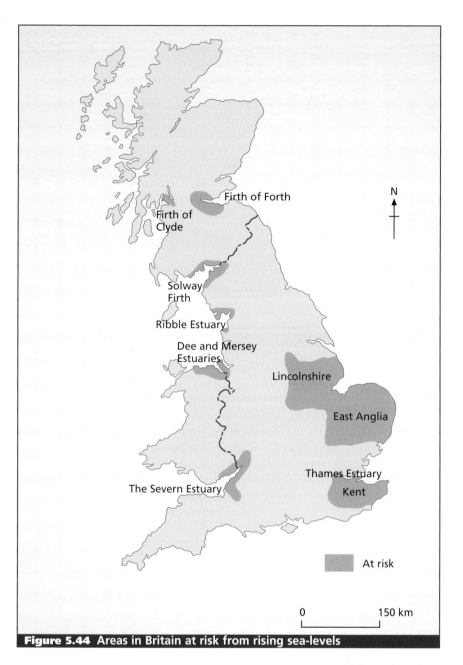

Firth of Forth

Firth of
Clyde

N

Solway
Firth

Ribble Estuary

Dee and Mersey
Estuaries

Lincolnshire

East Anglia

Thames Estuary

The Severn Estuary

Kent

At risk

0   150 km

Figure 5.44 Areas in Britain at risk from rising sea-levels

sinking land mass, and increased atmospheric storminess means that these return periods are over optimistic. Most of the risk is concentrated along the east coast between the River Humber and the Thames Estuary. This stretch of coast has a history of flooding, and tidal surges measured at London Bridge have been steadily increasing. This is in part due to global warming and in part the steric effect caused by water expanding as it heats up.

In 1994 the government introduced the Habitat Scheme in an attempt to persuade farmers to enhance the landscape and create natural habitats for plants and animals. In coastal areas this meant allowing farmland to become salt marsh again. The rising cost of sea defences and the impact of sea defences on other parts of the coast (i.e. the transfer of the erosion problem elsewhere), has meant that the role of sea

defences have been reevaluated. Farmers are paid up to £525 for each hectare of land that reverts to salt marsh. These salt marshes are therefore now the only defence mechanisms in some places.

Managing the coastline

The main concern of coast protection is to protect coastal areas by halting or reducing coast erosion and thereby saving homes, farmland, coastal paths, etc. However, other considerations need to be taken into account such as the possible adverse effects of remedial measures on the natural environment.

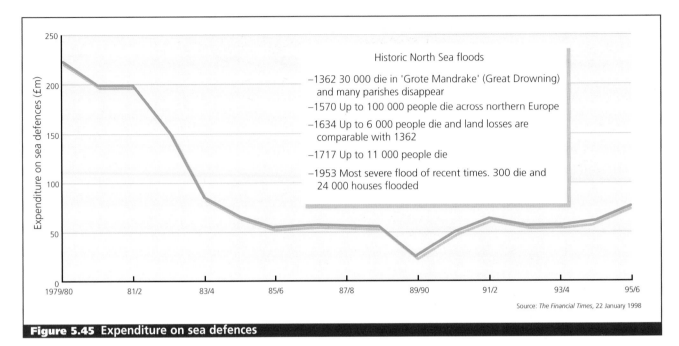

Historic North Sea floods

- –1362 30 000 die in 'Grote Mandrake' (Great Drowning) and many parishes disappear
- –1570 Up to 100 000 people die across northern Europe
- –1634 Up to 6 000 people die and land losses are comparable with 1362
- –1717 Up to 11 000 people die
- –1953 Most severe flood of recent times. 300 die and 24 000 houses flooded

Source: *The Financial Times*, 22 January 1998

Figure 5.45 Expenditure on sea defences

Flooding

Storm surges present the major flooding threat to low-lying coastal areas. They are caused by a combination of low atmospheric pressure and wind stress on the sea surface. A storm surge becomes stronger in shallowing water and converging coastlines, for example the North Sea from the north and the Bristol Channel from the south-west. The risk of flooding is greater in winter, with the most dangerous period around the fortnightly high waters ('spring' tides). In estuaries and tidal rivers the problem is worse after prolonged heavy rain. Although storm surges present the major threat, coastal defences may be overtopped or breached by wave action resulting in floods. Recent notable examples are Portland (1979) and Towyn (1989) where this was the primary cause.

Major surges in the North Sea result from depressions tracking north-eastwards across northern Scotland. These are exacerbated when the depression reaches the northern North Sea when the winds become northerly, helping the surge on its way, while the Coriolis force (the force generated by the earth's rotation) confines the surge to the east coast of Britain. Such conditions were responsible for the east coast floods of 1953, affecting areas from the Humber estuary to the Thames (Figure 5.47). In addition, high tides and high river levels (caused by melting snow) intensified the conditions. Following these floods, a national network of tide gauges and the Storm Tide Warning System (STWS) were established to obtain tidal information and to warn of possible recurrence.

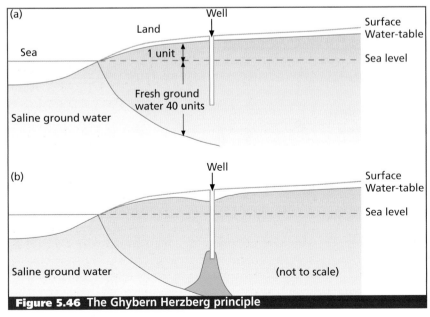

Figure 5.46 The Ghybern Herzberg principle

Questions

1. What are the natural conditions that cause sea-levels to **(a)** rise and **(b)** fall?
2. In what ways can human activity influence sea-level?
3. What is the difference between coastlines of emergence and coastlines of submergence?
4. State *two* landforms that are associated with emergence and *two* landforms that are associated with submergence. For each, explain how they were formed.
5. Distinguish between Atlantic and Pacific coastlines.

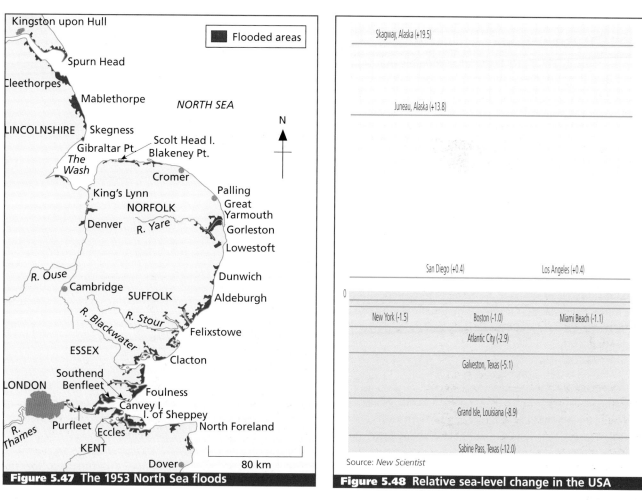

Figure 5.47 The 1953 North Sea floods

Figure 5.48 Relative sea-level change in the USA

Source: *New Scientist*

The USA's eastern seaboard

Along many parts of the USA's eastern seaboard coast, seawalls have protected buildings, but not beaches. Many beaches along the east coast have disappeared this century, such as Marshfield, Massachusetts, and Monmouth Beach, New Jersey. As sea-level rises, the beaches and barrier islands (barrier beaches) that line the coasts of the Atlantic Ocean and Gulf of Mexico from New York to the Mexican border, are in retreat. This natural retreat does not destroy the beaches or barrier islands, it just moves them inland.

The problem is that much of the shore cannot retreat naturally because there are industries and properties worth billions of dollar on them. Many important cities and tourist centres, such as Miami, Atlantic City and Galveston (Texas) are sited on barrier islands. Consequently, many shoreline communities have built seawalls and other protective structures to protect them from the power of destructive waves. Such fortifications, which can cost millions of dollars for a single kilometre, protect structures at least for the short term, but they accelerate erosion elsewhere. The first great seawall was built at Galveston after a hurricane devastated the city and killed more than 6000 people in 1900. The city survived a later hurricane, but lost its beach. Now, the rising sea level is making the seawall protection less effective. Much of the city is less than 3m above sea level.

Three factors put the east coast of the USA at particular risk. First, the flat topography of the coastal plains from New Jersey southward means that a small rise in sea level can make the ocean advance a long way inland. A rise of just a few millimetres each year in sea level could push the ocean a metre inland, while a rise of a few metres could threaten large areas such as southern Florida. Miami, in particular, faces severe problems as it is the lowest-lying US city facing the open ocean. Few places in metropolitan Miami are more than 3 m above sea level.

Barrier islands

Barrier islands are natural sandy breakwaters that form parallel to flat coastlines. By far the world's longest series is that of roughly 300 islands along the east and southern coasts of the USA (Figure 5.51). The distance between the barrier islands and the shore is variable. The islands are generally 200 to 400 metres wide but some are wider. Some Florida islands are so close to the shore that residents do not even realise they are on an island. Parts of Hatteras Island in North Carolina, on the other hand, are 20 km offshore.

Barrier islands form only under certain conditions and America's eastern seaboard provides the ideal conditions for barrier islands. First, there is a gently sloping and low-lying coast. Over the last 15 000 years, the sea level has risen by 120 m as glaciers and ice caps have melted. Wind and waves have formed sand dunes at the edge of the continental shelf. As the rising sea breaks over the dunes they form lagoons behind the sandy ridge of islands. Waves erode sand from the islands, depositing it further inland, forming new islands. Currents, flowing parallel to the coast, scour sand from the barrier islands and deposit it further up or down the coast to form new islands.

Second, much of the North American coast is sinking relative to the ocean, so local sea-levels are rising faster than global averages. The level of tides along the coasts shows that subsidence varies between 0.5 and 19.5 millimetres a year. By contrast, the west coast, in particular Alaska, is rising (Figure 5.50).

Third, extensive coastal development has accelerated erosion. Apartment blocks, resorts and second homes have developed rapidly along the shoreline. By 1990, 75% of Americans lived within 100 km of a coast (including the Great Lakes).

Until the late 1970s, most Americans assumed they could successfully protect their coastline against the rising sea. Now

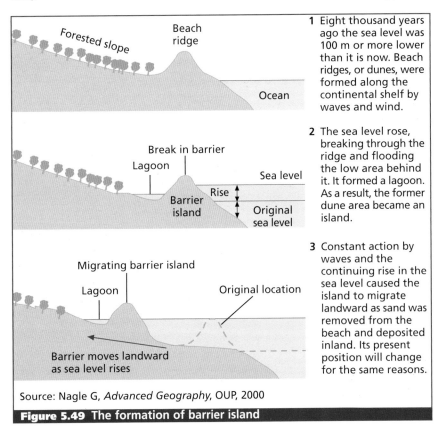

1 Eight thousand years ago the sea level was 100 m or more lower than it is now. Beach ridges, or dunes, were formed along the continental shelf by waves and wind.

2 The sea level rose, breaking through the ridge and flooding the low area behind it. It formed a lagoon. As a result, the former dune area became an island.

3 Constant action by waves and the continuing rise in the sea level caused the island to migrate landward as sand was removed from the beach and deposited inland. Its present position will change for the same reasons.

Source: Nagle G, *Advanced Geography*, OUP, 2000

Figure 5.49 The formation of barrier island

they are considering an alternative: strategic retreat. The term retreat does not mean abandoning the shore, but moving back from it. Instead of protecting the coast with seawalls, buildings are moved away from the rising sea, and new developments are prevented from being too close to the sea. Hence engineers have stopped challenging nature and have begun to work with natural coastal processes.

In the long term this makes the most economic sense. While it is impossible and impractical to abandon coastal cities such as Boston and New York, national and local governments are discouraging new coastal developments, especially in areas presently undeveloped.

The nature of erosion further complicates the issue. It is far from a uniform process. Most erosion occurs during coastal storms, especially at high tide. In addition, annual storm intensities are very variable, so coastal geologists try to plan for 'hundred year' storms, i.e. an intensity likely only once every century. This often makes their plans seem excessively cautious to coastal residents, especially in areas which have not experienced a severe storm for many years.

In many places there are observable cycles of erosion and deposition. During calm conditions, moderate currents often redeposit large quantities of sediment removed during a severe storm. This natural compensation reduces total erosion, but it can also disguise the real hazards of storms.

Erosion is evident at many places along the coast of the Atlantic and the Gulf of Mexico. Major resorts such as Miami Beach and Atlantic City have pumped in dredged sand to replenish eroded beaches. Erosion threatens islands to the north and south of Cape Canaveral, although the Cape itself appears safe. Resorts built on barrier beaches in Virginia, Maryland and New Jersey have also suffered major erosion.

Overall losses are not well known. Massachusetts loses about 26 ha a year to rising seas. Nearly 10% of that loss is from the island of Nantucket, south of Cape Cod. However, these losses pale into insignificance when compared with Louisiana, which is losing 40 ha of wetlands a day – about 15 500 ha a year (see pages 25–7).

Florida's extreme measures to combat erosion are well known (Figure 5.50). Intense development of Miami Beach in the 1920s started the widespread exploitation of coastal areas exposed to major storms and erosion. At the same time coastal towns in New Jersey, such as Sea Bright and Monmouth Beach, began building seawalls and groynes to prevent erosion (Figure 5.51). Since 1945 there have been many developments in coastal areas near large cities, especially for holiday homes and retirement communities.

Hard defences can cost millions of dollars a kilometre, and they require maintenance. Despite their cost, seawalls have failed at several places, including Texas, South Carolina and California. This is usually due to flaws in construction or poor maintenance.

Many US coastal geologists believe that the best compromise between building defences and leaving the shore to be eroded is pumping sand from other locations, usually offshore, to replace eroded sand. For example, between 1976 and 1980 the US Army Corps of Engineers spent $64 million on beach replenishment and

Figure 5.50 Hard defences along the US east coast

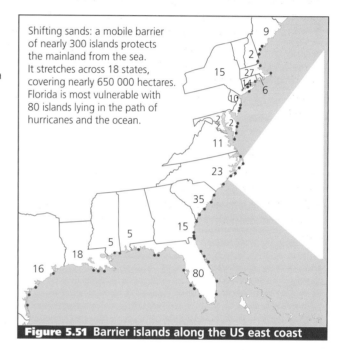

Shifting sands: a mobile barrier of nearly 300 islands protects the mainland from the sea. It stretches across 18 states, covering nearly 650 000 hectares. Florida is most vulnerable with 80 islands lying in the path of hurricanes and the ocean.

Figure 5.51 Barrier islands along the US east coast

flood prevention at Miami Beach. Erosion quickly removed 30 m of the new sand, but then the beach stabilised at 60 m wide.

Elsewhere, land use management has been introduced. Regulations vary widely. North Carolina, Maine and Massachusetts are in the forefront of restricting development. In Massachusetts, for example, there are restrictions on new developments of natural areas. Similarly, North Carolina was one of the first coastal states to legislate that land be left between the shore and new buildings to allow for erosion. Since 1979, small buildings have had to be built at least 30 times the annual rate of erosion from the shore. In 1983, the state doubled the distance from the sea for large buildings. In addition, in 1986 the state banned the construction of hard defences, such as bulkheads and groynes. Although this was a controversial decision, people and developers have adjusted, in part because the state's beaches are a major economic asset. Moreover, few people want huge seawalls and tiny beaches.

Rising sea levels and retreating coasts could pose continuing tough economic and environmental issues for Americans in the next century. A number of factors have eroded the memory of experience and history. These include:

■ competition for space

■ over-confidence in new building techniques

■ subsidised insurance

■ absence of great Atlantic storms for 25 years

■ ignorance

■ the temptation of great profits.

Some forms of protective 'hard engineering's are justified for major coastal cities, such as New York, but may not be justified for less developed areas such as Carolina Beach, North Carolina. Where to draw the border between fortification and retreat will always be the tricky issue.

Useful websites include:

Ministry of Agriculture, Fisheries and Food at
http://www.open.gov.uk.maff.maffhome.html
Sea Empress oil spill at http://www.swan.ac.uk.biosci/empress/empress.htm
World wave heights at http://www.oceanweather.com/date/data/htm

1 How are barrier islands formed?

2 Which parts of the USA are experiencing the greatest **(a)** relative rise in sea-level and **(b)** the greatest relative fall in sea level? (Use Figure 5.48 to help you. NB places that have risen will be located in areas where relative sea-level has fallen and vice versa.)

3 In what ways have human activities interfered with the natural development of barrier islands along the eastern seaboard of the USA. Use specific examples to support your answer.

4 What methods have been used to protect the eastern coastline of the USA?
How effective have they been?

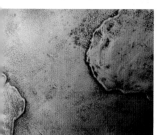

6 glaciation

From snow to glacier ice

Permanent snowfields only occur at high altitudes and/or high latitudes (Figure 6.1). The lower edge of these snowfields is the snowline and this varies in height:

- with latitude (at sea level near the poles to 5000 m in equatorial areas)
- locally with altitude
- locally with aspect (higher on south-facing slopes in the northern hemisphere).

With plentiful snowfalls and low summer temperatures (which prevent melting), snowfields enlarge and become thicker. With the increased thickness the snow changes to ice.

Snow initially falls as flakes and the accumulation is known as **alimentation**. With continued alimentation the lower snowflakes compress under the weight of more snow and gradually change into ice pellets known as **neve** or **firn**. Meltwater seeps into the gaps between the pellets and then freezes, forming a type of cement. Some places still have air, which gives the neve a white colour. With continued alimentation and increased pressure from the overlying snow, the neve pellets become tightly packed. Any remaining air is expelled and its place is filled with freezing water. Thus, the neve changes into glacier ice, by compaction and crystallisation, with a characteristic bluish colour. The change from neve to ice occurs, typically, when the neve is about 30 m thick.

Glacier ice consists of interlocking ice crystals. Between each crystal there is an extremely thin film of a solution containing chlorides and other salts. These salts lower the freezing point around the crystals and so keep the film in a liquid state. This film thus acts as a lubricant, aiding ice movement.

Types of ice mass

In 1948, H. W. Ahlmann suggested various methods of classifying ice masses, including:

- a morphological classification based on size and shape of the ice
- a dynamic classification based on the degree of activity in the ice
- a thermal classification.

Key Definitions 1

Ablation   Melting of ice.

Accumulation   Build up of snow and ice in a glacier.

Drift   Deposited material (includes till and fluvioglacial deposits).

Cirque   An armchair-shaped hollow out of which a glacier develops.

Fluvioglacial   Meltwater.

Regime   The balance between accumulation and ablation – whether a glacier grows (moves forwards) or retreats.

Periglacial   "On the edge" of a glacial area – charaterised by intense freeze-thaw activity and snow melt in summer.

Till   Material deposited by a glacier.

Figure 6.1 Snowline

Morphological classification

Niche glaciers are found on steep slopes, and are often associated with rock benches. They probably originated as snow patches, developing into triangular wedges of ice lying in shallow, funnel-shaped hollows, such as on Spitzbergen.

Cirque glaciers develop from snow patches accumulating in natural hollows which face away from the sun. Basins develop into armchair-shaped basins or cirques (by freeze-thaw weathering) allowing ice to collect (Figure 6.2).

Valley glaciers are divided into two sub-types. In the **Alpine** type, ice moves out from cirques down preglacial valleys. These types of glacier are very common. By contrast, **outlet** glaciers are found at the edges of an ice cap such as Norway and Iceland.

Transection glaciers are found in mountain areas, where they radiate out from the central point. Often the ice breaches dividing ridges to form 'cols' in the ridges; these cols may then be further eroded by meltwater, such as in Spitzbergen.

Piedmont glaciers form where a valley glacier advances out from a mountain valley, into a wider lowland region. Here ice movement slows, the ice fans out and becomes thinner. An example is the Malaspina glacier in Alaska.

Floating glacier tongues/ice shelves are restricted to polar areas, where ice reaches sea level. The most famous is the Ross Ice Shelf in Antarctica which reaches a thickness of almost 400 m. The shelf ends in sheer ice cliffs, about 50 m high above the Ross Sea. These ice shelves are worn away by marine erosion, submarine thawing of the ice and by the calving of icebergs.

Mountain ice caps are areas of ice from which outlet glaciers flow. They are commonly found on plateau surfaces such as the Vatnajokull in Iceland, consisting of ice up to 760 m thick, resting on a plateau about 1200 m above sea level.

Lowland ice caps are only found near the poles, such as the Barnes ice cap on the Baffin Island off the coast of North Canada.

Ice sheets occur in Greenland and Antarctica. They are the largest forms of ice mass, with ice over 3000 m thick; the Greenland sheet covers 1 800 000 sq km, the Antarctic sheet over 12 million sq km. The great mass of the ice has caused the land beneath to sink (isostasy) and its surface to be smoothed and rounded, scratched and polished, thus reducing the preglacial relief (in contrast to valley glaciers which increase the height differences by deepening valleys). Occasionally, high peaks project through the ice as **nunataks**.

Dynamic classification based on ice movement and ice activity

Active glaciers are normally fed by a continuous supply of ice, from regular avalanches or cirques such as the glaciers on the west side of the Southern Alps of New Zealand.

Passive/inactive glaciers exist where the supply of ice is small or is balanced by melting, such as in the lowlands of the east Scandinavian mountains.

Thermal classification based on temperature

The temperature at the base of a **temperate glacier** is above freezing. This means that there is some meltwater present. This acts as a lubricant, allowing the ice to move easily and thus erode the local rock by glacial and fluvioglacial erosion.

In **polar glaciers**, by contrast, the ice remains frozen at the base. Consequently there is little meltwater and little movement, so little erosion occurs. Where meltwater does occur it is on the surface as supraglacial streams or slush.

Valley glaciation

If an area of accumulation is sufficiently large, and if the seasonal supply of snow is sufficient, a tongue-like mass of ice may move out and down the slope as a glacier, following a line of least resistance, usually a preglacial river valley.

Size and regime

The size of a glacier depends on its regime, i.e. the balance between the rate and amount of supply of ice and the amount and rate of ice loss. The glacier will have a **positive regime** when the supply is greater than loss by ablation (melting, evaporation, calving, wind erosion, avalanche, etc.) and so the glacier will thicken and advance. A negative regime will occur when the wasting is greater than the supply (as with the Rhône glacier today) and so the glacier will thin and retreat. Any glacier though, can be divided into two sections by the equilibrium line, i.e. the area of accumulation at high altitudes generally, and an area of ablation at the snout (Figure 6.3).

Figure 6.2 A cirque

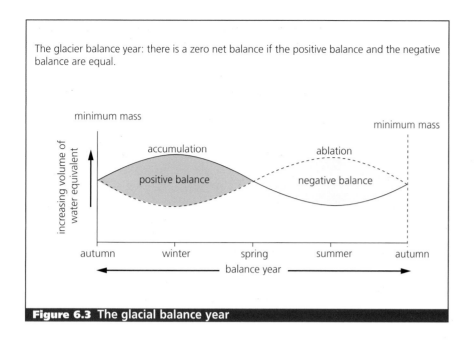

The glacier balance year: there is a zero net balance if the positive balance and the negative balance are equal.

Figure 6.3 The glacial balance year

Regime and climatic change

Glacial regimes are very closely related to climate and climatic changes, both in the short and long term over thousands of years. The last 1.5 million years have seen a number of glacial advances and retreats, resulting from climatic changes.

Figure 6.4 shows the long-term changes which have occurred during the last geological era. The events of this era were initially worked out by A. Penck and E. Bruckner in the European Alps in 1909. They identified four distinct phases in glacial advance: the **Gunz**, **Mindel**, **Riss** and **Wurm** (with the most intense glaciation during the Riss). Between these **glacial phases** the climate became relatively warmer (and perhaps wetter, more 'pluvial'); some of these **interglacial** phases may have been warmer than today's climate. The longest interglacial was that occurring between the Mindel/Riss Glaciations lasting about 190 000 years.

Many different causes of these climatic changes have been proposed and these include:

- changes in the earth's orbit
- changes in the tilt of the earth's axis
- changes in solar radiation
- changes in the position of the poles
- changes in the amount of water vapour in the atmosphere
- changes in the distribution of land and sea
- changes in the relative levels of land and sea
- changes in the nature and direction of ocean currents and a reduction of carbon dioxide in the atmosphere.

There have also been short-term glacial fluctuations, experienced during historic times such as the early Middle Ages which were very mild at least until the 12th century, causing the area of ice substantially to decrease. By the 16th century, however, conditions had become very much colder and so there was glacial advance; other glacial advances have occurred as recently as the 1850s and the 1960s. Today, however, many glaciers have negative regimes, partly as a result of global warming.

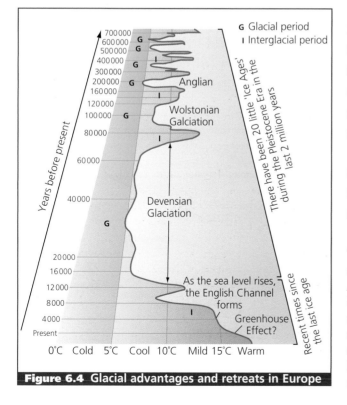

Figure 6.4 Glacial advantages and retreats in Europe

Ice movement

Ice has great rigidity and strength. Under sudden compression or tension though, it will break or shear apart; under steady pressure it will behave as a plastic body. In a typical glacier then, there are two zones of movement:

■ an upper zone of fracture (up to 60 m thick) where the ice is brittle, breaking under sudden changes in tension, to form crevasses

■ a lower zone of flow, where more steady pressure and lubricating water allow it to move as a viscous body.

The mechanisms of ice movement are very complex but include two main types of flow:

1 **gravity flow, involving a series of mechanisms, such as regelation, plastic deformation and laminar flow**

2 **extrusion flow.**

Gravity flow results from changes in the gradient over which the ice moves. During regelation – ice crystals melt under pressure. This meltwater flows downslope between existing crystals for a short distance away from the local pressure zone (Figure 6.5). This water will refreeze when no pressure is experienced. In addition, the meltwater will act as a lubricant allowing frozen ice crystals to slide bodily over each other (this is known as intergranular translation) and **laminar flow** where rigid blocks of ice are lubricated and slide along lines of weakness such fractures within the ice. Moreover, the meltwater may assist **basal sliding** – the movement of the whole glacier over the rock floor. This is common in temperate glaciers where there is much meltwater. **Plastic deformation** occurs under certain conditions that cause the ice to creep; it appears to be common in polar glaciers.

Extrusion flow is common on thick ice and appears to be the result of extreme weight and pressure (Figure 6.6). It is common in large continental ice masses, in glaciers issuing

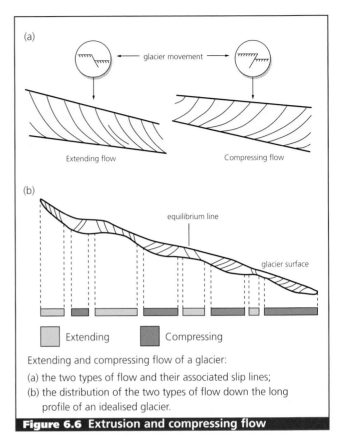

(a)

glacier movement

Extending flow

Compressing flow

(b)

equilibrium line

glacier surface

Extending   Compressing

Extending and compressing flow of a glacier:

(a) the two types of flow and their associated slip lines;

(b) the distribution of the two types of flow down the long profile of an idealised glacier.

Figure 6.6 Extrusion and compressing flow

from ice sheets, and in valley glaciers where the ice is forced to move uphill. The thickness and pressure result in a downward then outward/horizontal movement at the edges of the ice. It is a rapid and more potent type of flow than gravity flow, resulting in deep rock basins and cols where the ice flows over dividing ridges.

The velocity of flow is controlled by:

■ the gradient of the rock floor

■ the thickness of ice (controlling pressure and meltwater formation)

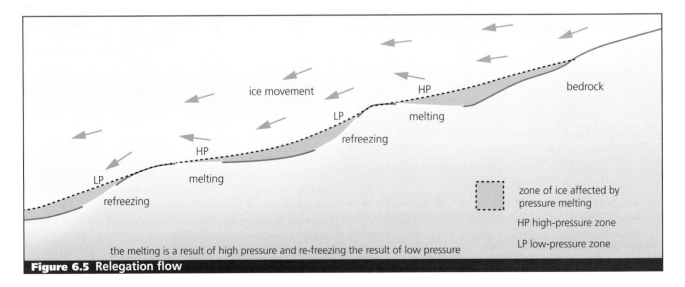

ice movement

HP

bedrock

LP

melting

refreezing

HP

melting

LP

refreezing

HP

melting

LP

refreezing

zone of ice affected by pressure melting

HP high-pressure zone

LP low-pressure zone

the melting is a result of high pressure and re-freezing the result of low pressure

Figure 6.5 Relegation flow

Figure 6.7 Crevasses and seracs on glacier

Figure 6.8 A bergschrund

- the temperatures within the ice (controlling meltwater formation).

The velocity will increase if these factors increase.

Within a body of ice there will be different rates of movement. This is known as **differential flow**. The sides and base of the ice flow will move more slowly than the ice in the middle, as a result of friction between the ice and the valley walls and bottom. Differential flow will also occur within the glacier if the gradient changes. If the gradient is small the velocity will decrease and the ice will pile up and become thicker; a steep gradient will cause the ice to accelerate, thin out and break into ice falls, with crevasses dividing the ice into blocks or seracs (Figure 6.7).

Features formed as a result of differential flow on a glacier include:

- the **bergschrund** – a gaping crack, more developed in summer, seen around the head of the ice field (Figure 6.8). It indicates the area where the ice mass is drawing away from the sides of the basin.

- **crevasses** – the cracks on the ice surface. Stresses (tension and shearing) within the moving ice result in a complicated pattern of crevasses; these may be **transverse** (across a glacier) due to the steepening of gradient or a bend in the valley; **longitudinal** or **marginal** (along the glacier) due to a local widening of the valley or thinning of the ice. Other minor types are chevron (near ice margins where ice has rotated) and splaying (which starts as longitudinal but spreads out at the ends). Major crevasses occur at an ice fall.

Glacial erosion

Glacial erosion tends to occur in upland areas, whereas deposition is more common in lowland areas. Glacial deposits are, nevertheless, also found in the hills.

There are two main types of glacial erosion.

1 **Plucking** occurs mainly at the base of the glacier but can occur at the sides. It involves downward pressure caused by the weight of the ice and then downhill drag as the ice moves, slow enough for meltwater to freeze onto obstacles. Plucking is very marked in well jointed rocks and on rocks that have been weakened by freeze-thaw weathering.

Questions

1 With reference to valley glaciers:
   (a) (i) Outline the processes of accumulation and ablation of ice.
   (ii) Describe the difference in the winter and summer balance of accumulation and ablation.
   (b) Discuss the relationship between the advance or retreat of the snout of a glacier and the annual balance between accumulation and ablation.
   (c) Provide a reasoned explanation of the relationship between the temperature of the base of a glacier and the speed of flow of the glacier.

2 **Abrasion** occurs where debris embedded in the base and sides of the ice are dragged over the local rock. The coarser material will scrape, scratch and groove the rock, leaving **striations** and **chatter marks**. The finer material will smooth and 'polish' the rock. As ice movement continues, the glacier load will itself be worn down to form a fine **rock flour**.

A glacier will not only acquire its 'load' by erosion but also from meltwater (fluvioglacial) erosion. The presence of meltwater greatly assists the freeze-thaw weathering of exposed rock, thus aiding erosion, and from rockfalls and avalanches.

The amount and rate of glacial erosion depends on:

- the local geology
- the velocity of the glacier, itself dependent on gradient
- the weight and thickness of the ice body
- the amount and character of the load carried by the ice.

Landforms created by glacial erosion

Cirques

In the British Isles **cirques** are usually found on north- and/or east-facing slopes where insolation is lowest, allowing rapid accumulation of snow. They are often found at the head of glacial valleys, carved in hard, igneous or metamorphic rocks. Cirques (also called cwms or corries) range in size from small basins in the UK to massive, armchair shaped hollows in the Himalayas.

Cirque formation occurs when a shallow, preglacial hollow accumulates large volumes of snow. The hollow is enlarged by freeze-thaw weathering at the edge of the snow patch. The freezing and thawing of the thin snow is called **nivation**. Continued nivation enlarges the hollow and so neve proper can form, and gradually, as the basin further develops, ice can accumulate. At a critical depth and weight, the ice begins to move out of the hollow by **extrusion flow** in a **rotational** manner (Figure 6.9). This rotational movement helps erode the hollow further by plucking and abrasion, and so a true cirque is formed. Meanwhile, meltwater, especially that which makes its way down the bergschrund, helps in continuing cirque growth. As the meltwater trickles down into the cirque freeze-thaw weathering occurs at the back of the basin causing this backwall to retreat and remain steep. The meltwater, once in the basin, is also involved in freeze-thaw weathering of the base, preparing the rock for erosion by the moving ice.

When the ice finally disappears, an armchair shaped hollow remains, often containing a small lake, known as a llyn or tarn in the UK, dammed back by the cirque lip left as a result of rotational movement of the ice (Figure 6.10).

If several cirques develop in a highland region, they will jointly produce other erosional features. For example, when two cirques lie back to back, cirque enlargement by rearwall recession will create a narrow, steep-sided ridge between the two, called an **arête** (Figure 6.11). If three or more cirques develop the central mass that will temporarily survive between them will become a **pyramidal peak**, usually sharpened by freeze-thaw weathering, such as the Matterhorn. These features are very much subdued today, often mantled with scree which helps to protect them from further weathering.

Glacial troughs

Glacial troughs or **U-shaped** valleys occur when a glacier occupies and modifies a pre-existing valley. A number of stages can be identified:

(i) **preglaciation** – before the actual onset of glaciation, active freeze-thaw weathering under periglacial conditions, ahead of the advancing ice, will weaken the floor and sides, so preparing it for rapid erosion.

(ii) In **interglacial** phases, periglacial periods will return, further weakening the already-eroded rock. In addition there will be **pressure release** as the weight of the ice is reduced.

(iii) During **glaciation**, the eroding power of the moving ice will cause the valley to become U-shaped in cross-section with a flat floor and steep sides. In plan, the valley will become straight (as opposed to the winding nature of the former valley) as the interlocking spurs are eroded to leave **truncated spurs**.

(iv) **Extrusion flow** in the ice can cause the ice to erode deep rock basins in the valley floor, later perhaps occupied by long, narrow **ribbon lakes**.

(v) As the ice moves along, it may also erode **rock steps** by scooping out lines of weaker rock, by opening up major lines of weakness (such as joints) or by experiencing periods of intense and potent extrusion flow (Figure 6.12).

(vi) Where a glacier did not entirely fill the preglacial valley, there will be marked change in slope on the valley sides, so leaving glacial benches, topped by steep, frost-shattered walls,

(vii) Some glacial troughs end abruptly at their 'head' in a steep wall, known as the **trough end**, above which lie a number of cirques (Figure 6.13). Probably a whole series of cirque glaciers developed and joined at one point to feed the main glacial trunk; at the point of coalescence there was a sudden increase in the amount, weight and eroding power of the ice, so forming this wall.

(viii) In the preglacial valley, any river tributary would cut down to meet the level of the main river. With **tributary glacial valleys** this is not the case. The small tributary glacier does not have the weight nor the power to cut down to the depths of the main trough; the tributary ice will only cut down far enough so that its

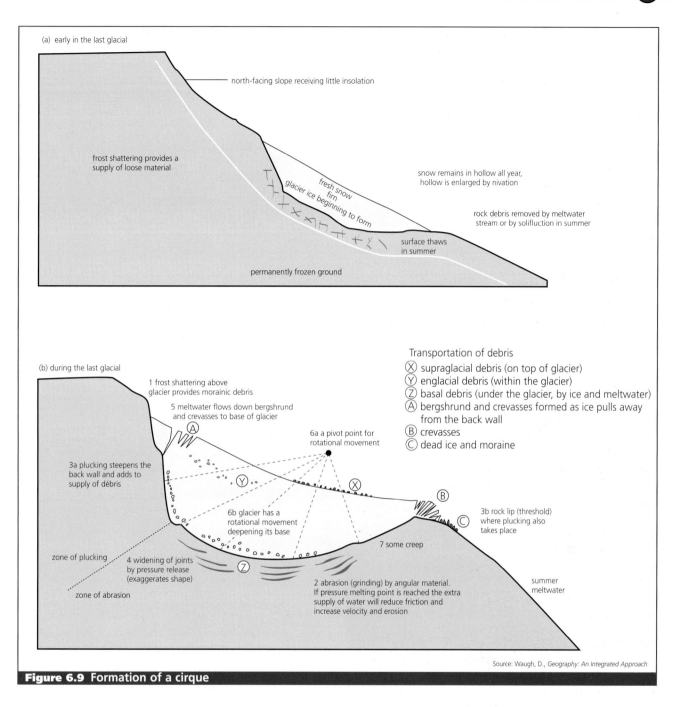

Source: Waugh, D., *Geography: An Integrated Approach*

Figure 6.9 Formation of a cirque

Text within the figure:

(a) early in the last glacial

north-facing slope receiving little insolation

frost shattering provides a supply of loose material

fresh snow firn

glacier ice beginning to form

snow remains in hollow all year, hollow is enlarged by nivation

rock debris removed by meltwater stream or by solifluction in summer

surface thaws in summer

permanently frozen ground

(b) during the last glacial

1 frost shattering above glacier provides morainic debris

5 meltwater flows down bergshrund and crevasses to base of glacier

6a a pivot point for rotational movement

3a plucking steepens the back wall and adds to supply of débris

6b glacier has a rotational movement deepening its base

zone of plucking

zone of abrasion

4 widening of joints by pressure release (exaggerates shape)

2 abrasion (grinding) by angular material. If pressure melting point is reached the extra supply of water will reduce friction and increase velocity and erosion

7 some creep

3b rock lip (threshold) where plucking also takes place

summer meltwater

Transportation of debris
X supraglacial debris (on top of glacier)
Y englacial debris (within the glacier)
Z basal debris (under the glacier, by ice and meltwater)
A bergshrund and crevasses formed as ice pulls away from the back wall
B crevasses
C dead ice and moraine

ice can 'slip' onto the top of the main glacier. When the ice disappears, these tributary glacial valleys are left high, as **hanging valleys**, today often the site of waterfalls (and potential HEP sites).

(ix) Major modifications of these smooth, straight glacial troughs have occurred in postglacial times. Frost shattering creates **scree** which now often covers the sides of the troughs, masking the former U-shape.

Moraines (once covering the sides of the ice as **lateral moraine,** the middle as **medial moraine** and the base of the glacier as **ground moraine**) now lie dumped on the trough floor, damning back the water to form more **ribbon lakes** or being reworked on the contemporary river valley floors (often forming rich soils, allowing intensive farming to take place in these sheltered valleys).

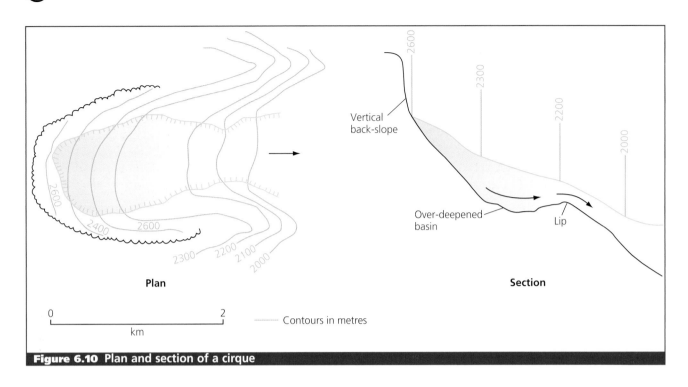

Figure 6.10 Plan and section of a cirque

Plan

Section

Vertical back-slope

Over-deepened basin

Lip

0 ——————— 2
km

Contours in metres

Figure 6.11 An arête

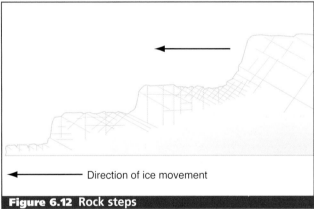

Direction of ice movement

Figure 6.12 Rock steps

Other features

A **Roche moutonée** (stoss and lee) can be large masses of ice moulded rock, projecting well above the general level of the trough floor. The upstream (stoss) side of this very resistant mass is smoothed and polished by abrasion as the ice pushes its way up and over by extrusion flow; striations can be seen. As the ice accelerates down the downstream (lee) side (prepared and weakened by freeze/thaw) plucking occurs, giving it a rough and irregular appearance (Figure 6.14).

A **crag and tail** occurs where a very large, resistant rock body (the crag) obstructs the ice flow. The ice will tend to move around the crag eroding material and solid rock in its path. Immediately in the lee of the crag, any material will be preserved and also morainic material from the nearby ice may

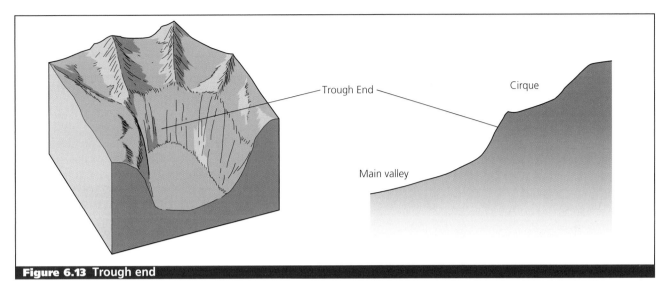

Figure 6.13 Trough end

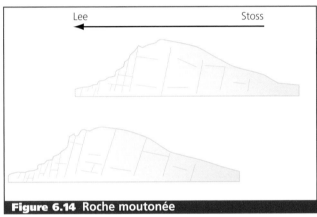

Figure 6.14 Roche moutonée

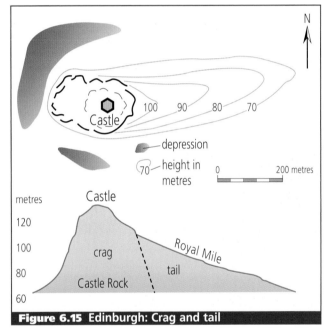

Figure 6.15 Edinburgh: Crag and tail

fall off into this area (the dead space) leaving a gently sloping tail of material. One example is the Edinburgh Castle Rock, the ancient volcanic plug which forms the crag has a tail of carboniferous limestone over lying old red sandstone, upon which the Royal Mile of Edinburgh now stands (Figure 6.15).

Fluvioglacial erosion

Meltwater coming from an area of melted ice is capable of intense erosion. Meltwater streams can carve deep channels in the land which have, in many cases, helped lead to modifications of drainage as in the North Yorkshire Moors.

Glacial lakes and overflow channels

During the last glaciation, ice sheets up to 300 m thick in some places, spread south and south-west from north England and the North Sea, surrounding the North York Moors area, leaving the Moors as an island of uncovered ice-free land. As the ice moved south it first blocked the mouth of the River Esk and stopped its waters from reaching the North Sea. At the same time, ice moving from the western edge of the Moors blocked the alternative outlet. Gradually water gathered in the low lying dales, forming lakes that found temporary outlets across the ridges between one dale and another. Here the water cut cols or spillways (Figure 6.17). Eventually the lake waters joined and the water surface stood 215 m above present sea-level. At this height the water spilt over the lowest point of the main divide and and began to flow southwards in an ever increasing torrent, eroding the trench at Newtondale. Newtondale is thus known as a overflow channel. This feature is a well defined trench 75 m deep with very steep sides cut into the surrounding plateau of the moors; there is no clear valley head.

By this time both the east and west ends of the Vale of Pickering were also blocked by ice so that another vast lake was created (Figure 6.18). The Newtondale overflow water entering this lake (Lake Pickering) deposited the Pickering Delta.

The ice had by this time moved even further south, blocking the outlet of the River Derwent into the North Sea. The river found an alternative outlet southward and so cut the deep valleys known today as Langdale and Forge Valleys. These valleys in turn led into Lake Pickering and caused the lake to rise. The enlarged lake soon cut an outlet to form the Kirkham Abbey Gorge.

Questions

1 Figures 6.16 (i), (ii) and (iii) all refer to Scotland.
  Figure 6.16 (i) shows the present day annual precipitation.
  Figure 6.16 (ii) shows altitude.
  Figure 6.16. (iii) shows the distribution of cirques.
  (a) Describe and suggest reasons for the pattern of precipitation shown.
  (b) Explain how cirques are formed.
  (c) With the help of the information provided on the maps, discuss factors that may determine the distribution of cirques.
2 Explain what is meant by
  (a) plucking
  (b) abrasion
  (c) rotational movement.
3 Why does the ice flow uphill and leave a corrie lip at the mouth?

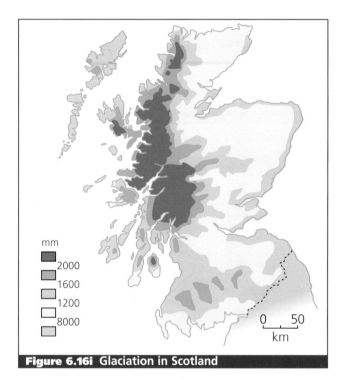

mm
2000
1600
1200
8000

0   50
km

Figure 6.16i Glaciation in Scotland

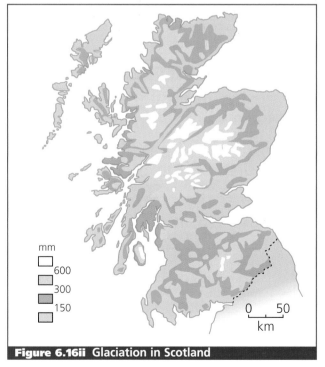

mm
600
300
150

0   50
km

Figure 6.16ii Glaciation in Scotland

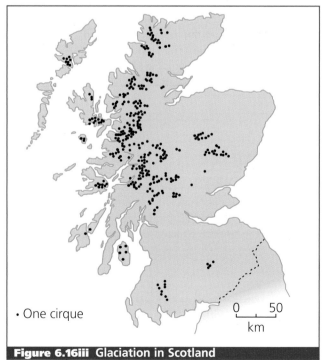

• One cirque

0   50
km

Figure 6.16iii Glaciation in Scotland

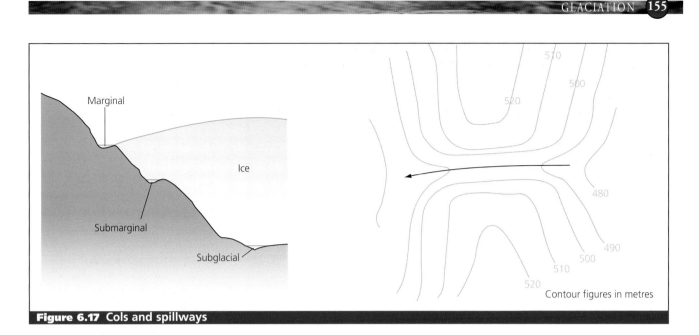

Figure 6.17 Cols and spillways

Contour figures in metres

As the ice retreated, masses of boulder clay were dumped along the east coast, south of Scarborough. These masses blocked the eastern end of the Vale of Pickering and so the Derwent had to maintain its southward-flowing, ice-diverted course, crossing the ancient bed of Lake Pickering.

To the north of the Moors, however, the River Tees was once again allowed to return more or less to its preglacial course to the North Sea, thus abandoning the gorge at Newtondale. As the ice receded various temporary links between the lakes were enlarged, either by deepening notches in the dividing ridges or cutting terraces between the highland and the ice front.

This explanation of the North York Moors was worked out by Percy Kendall in 1902. Since that time further research has has shown that some of Kendall's 'overflow channels', are in fact channels eroded by subglacial streams, i.e. subglacial channels. Other types of channel that can be attributed to meltwater are ice marginal channels (formed between the ice and the highlands) and submarginal channels (formed near the edge of the ice but under it). Thus in some cases, lakes need not have existed.

Many other examples and drainage diversion exist in Britain. For example, ice advancing southwards across the Cheshire Plain dammed back the waters of the upper Severn, which had previously been flowing north to join the River Dee. The water from the new lake (Lake Lapworth) escaped through and deepened the route now called the Ironbridge Gorge. The modern course of the Severn keeps this route (Figure 6.19).

During the penultimate glaciation, a huge lake filled much of the Midland Plain of England (Lake Harrison) (Figure 6.20). The lake finally drained towards the south-east, following the course of the contemporary River Avon which itself leads to the Severn.

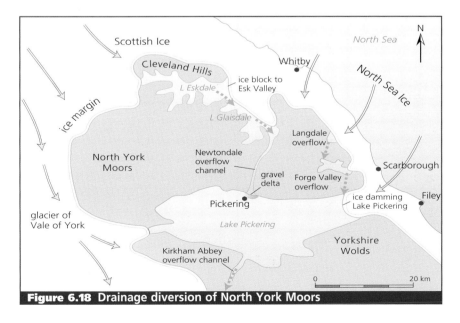

Figure 6.18 Drainage diversion of North York Moors

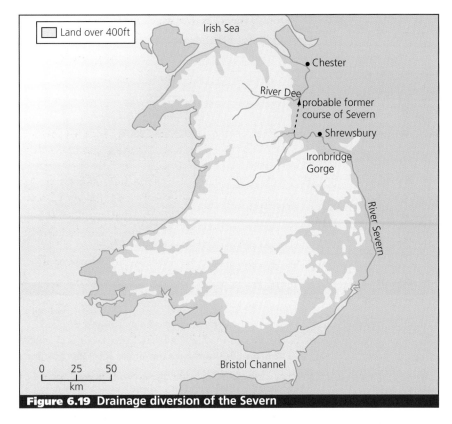

Figure 6.19 Drainage diversion of the Severn

Prior to the Pleistocene glaciations, the River Thames followed a much more northerly course across the London Basin, passing close to what is now Watford and St Albans. A southerly diversion occurred when ice filled the northern part of the London Basin (Figure 6.21)

Glacial deposition

Ice is able to transport vast amounts of material, ranging in size from fine rock flour to huge boulders. This material is carried on the ice surface, within the ice itself and dragged along the base of the ice.

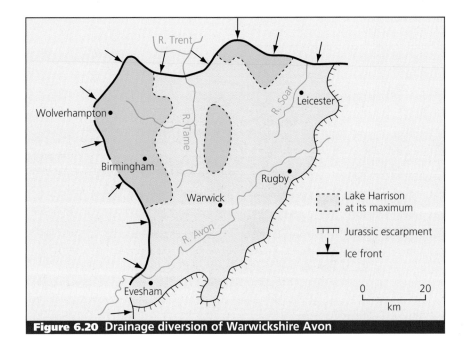

Figure 6.20 Drainage diversion of Warwickshire Avon

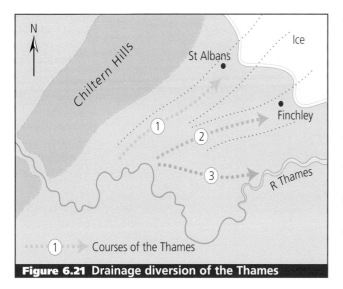

Figure 6.21 Drainage diversion of the Thames

(labels on figure: N, Chiltern Hills, St Albans, Ice, Finchley, R Thames, 1, 2, 3, Courses of the Thames)

Often, this material is deposited directly by the ice – dropped from its lowest layers or redeposited at the ice front/margin. This direct deposition usually results in unstratified material called **till** or **boulder clay**.

Meltwater, flowing from the ice, also transports and deposits material known as **fluvioglacial** deposition (Figure 6.22).

Drift

The term **drift** describes all glacial and fluvioglacial deposits left after ice retreat. In Europe, there is a distinction between the Older Drift, laid down during early glacial phases, and the Newer Drift deposited by the less extensive and younger advances. The Older Drift deposits are more extensive and being older, more altered by erosion, re-sorting and re-deposition by later ice and by post-glacial geomorphological processes (mainly weathering and rivers).

Erratics are large boulders foreign to the local geology which have been dumped by the ice, usually on flat areas. Some erratics, though, have been left stranded in precarious positions as **perched blocks**, such as the bluish micro-granite from an island off the Ayrshire coast that has been found in the Merseyside area and in Fishguard in Dyfed. If a series of erratics has been deposited in a fan-shaped pattern it is called a **boulder train**. The apex of the 'train' points towards the source area and again aids reconstruction of the flow. The location and geology of erratics can help deduce the direction of ice movement.

Till

Till or **boulder** clay is a common, widespread and unstratified glacial deposit, composed of finely grained rock flour (sands and clays) mixed together with rocks of all shapes and sizes (Figure 6.22). Its composition is variable, depending

on the nature of the rocks over which the ice moved. Till can be divided into two types:

(i) **lodgement** till dropped by actively moving glaciers

(ii) **ablation** till dropped by stagnant ice.

Moraines

Moraines are lines of loose rock fragments which have been weathered from the valley sides above the ice, and have fallen downslope onto the ice (Figure 6.23). The lines of material lying near the valley sides are lateral moraines. Where two glaciers meet, the **lateral moraines** of each will join to form a **medial** moraine down the centre of the newly formed, larger glacier. At the snout of a stationery or slowly moving glacier, much material is deposited as a crescent-shaped mound or **terminal** moraine. Similar, but much smaller, **recessional** moraines mark the site where the snout halted for brief periods during the retreat of the ice front.

The character of the terminal moraine will depend upon:

■ the amount of material carried by the ice

■ the rate of ice movement

■ the rate of ablation and thus the amount of meltwater.

The angle of slope of the moraine facing the ice (ice contact slope) is always steeper than the lee slope. The largest terminal moraine in Britain is the Cromer Ridge in Norfolk: it forms a belt of hummocky hills, composed of sands and gravels, 8 km wide and 90 m high.

Debris is also carried within the ice (**englacial moraine**) having made its way down crevasses and moulins within the ice; the debris found at the base of the ice is called **subglacial moraine**; this is left after a steady retreat of the ice.

Morainic material is usually ungraded, ranging in size from clay to boulders. It is usually modified by subsequent glacial action such as **push moraines** where ice pressure has bulldozed other older moraines, causing folding of the material, or by meltwater.

Drumlins

In some low-lying, undulating regions, particularly where a valley glacier flows into a lowland plain (causing a thinning of the ice), the till has been deposited as small, oval mounds a few metres long and high, to considerable hillocks 1.5 km long and 90 m high such as in Cavan, in Ireland, the Central Lowlands of Scotland and the Solway Plain. These **drumlins** (Figure 6.24) usually consist of sandy rather than clayey till. The long axis of the drumlin is aligned with ice flow with the steeper, thicker end of the feature facing the direction whence the ice came.

It seems that the ice deposited each mass of till (possibly from an area of the ice where the base was more heavily loaded with material), where friction between the ice and the rock floor was stronger than the adhesion between the till and the ice. The mass of till was then streamlined by the moving ice. It is also possible that:

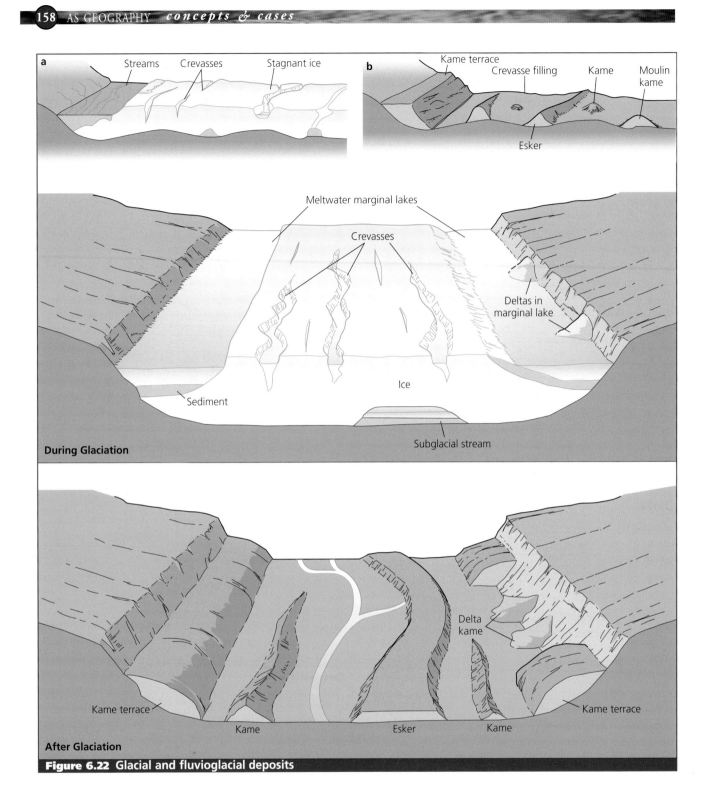

a Streams Crevasses Stagnant ice

b Kame terrace Crevasse filling Kame Moulin kame

Esker

Meltwater marginal lakes

Crevasses

Deltas in marginal lake

Ice

Sediment

Subglacial stream

During Glaciation

Delta kame

Kame terrace Kame Esker Kame Kame terrace

After Glaciation

Figure 6.22 Glacial and fluvioglacial deposits

- some drumlins were produced by the pressure of the active ice moulding sheets of older drift

- other drumlins were formed where a rock obstacle has been coated with till. This coating is often very thin. The drumlins so formed are called **false drumlins**.

Fluvioglacial deposition

The wasting of ice results in the formation of meltwater streams on the ice surface (supraglacial streams), within the ice (englacial streams) and beneath the ice (subglacial streams). These meltwater streams carry much debris and

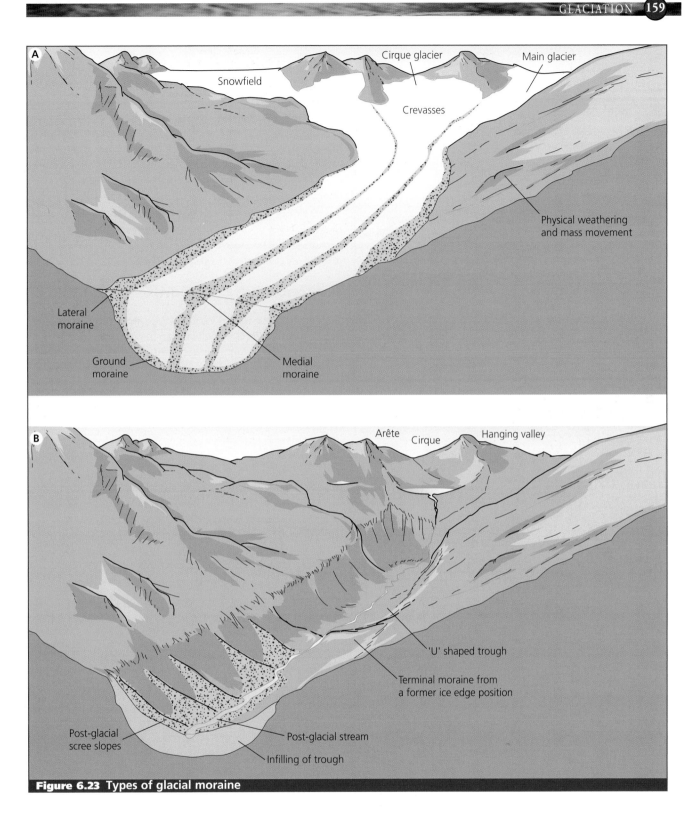

A
Snowfield
Cirque glacier
Main glacier
Crevasses
Physical weathering and mass movement
Lateral moraine
Ground moraine
Medial moraine

B
Arête
Cirque
Hanging valley
'U' shaped trough
Terminal moraine from a former ice edge position
Post-glacial scree slopes
Post-glacial stream
Infilling of trough

Figure 6.23 Types of glacial moraine

deposit this during active ablation or deglaciation, or during a prolonged period of ice stagnation, in stratified drift formations.

There are two major forms of stratified fluvioglacial drift:

- **prolonged drift**
- **ice contact stratified drift.**

Prolonged drift

There are two forms of prolonged drift. The first consists of material dropped in glacial lakes (such as **varved clays**), along lake shorelines and as lake deltas such as Vale of Pickering. With the recession of continental ice sheets and large ice bodies, lakes frequently formed as temporary features.

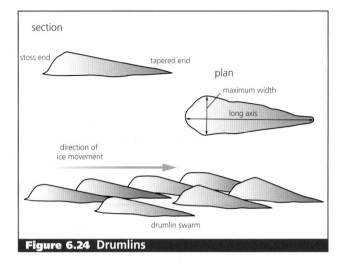

section

stoss end

tapered end

plan

maximum width

long axis

direction of
ice movement

drumlin swarm

Figure 6.24 Drumlins

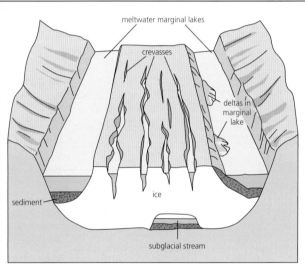

meltwater marginal lakes

crevasses

deltas in
marginal
lake

sediment

ice

subglacial stream

During glaciation

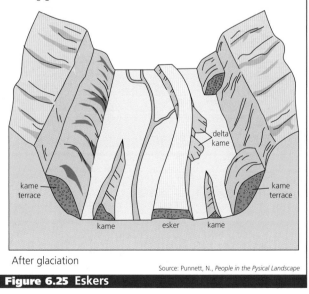

delta
kame

kame
terrace

kame
terrace

kame   esker   kame

After glaciation

Source: Punnett, N., *People in the Pysical Landscape*

Figure 6.25 Eskers

Material was deposited into these lakes, brought down by seasonal meltwaters. This seasonal transport produces layered deposits known as **varves**. One varve represents one season's melting. Summer deposition is indicated by a layer of coarse deposits (more vigorous meltwater is able to carry coarser, heavier material); winter deposition is indicated by a thinner layer of finer material (less meltwater, more sluggishly flowing water cannot carry coarse material).

The second form is outwash plains (also called **sandur**) that contain material that is well sorted and stratified by the meltwater streams. The coarsest materials (gravels) are deposited first near the ice margin, while the finer materials (sands) are carried further down the valley. The seasonal flow of meltwater also causes vertical layering.

Ice contact stratified drift

Ice contact stratified drift consists of a more varied class of depositional features, which change in character over short distances. In addition, they are often modified as a result of ice retreat.

Eskers

Eskers are elongated ridges of coarse, stratified fluvioglacial material (sands and gravels) (Figure 6.25). The ridges are usually meandering such as in Scandinavia where they wind for over 100 km between lakes and marshes. The popular theory to explain these is that material was deposited in subglacial meltwater tunnels during a period of lengthy ice stagnation. Another theory is that some eskers could be elongated deltas, deposited by streams flowing out from tunnels at the front of a continuously and rapidly retreating ice front. This theory is supported by the occurrence of some 'beaded eskers' – the 'bead' or 'bump' in the esker represents summer deposition when rapid ice wasting produced increased meltwater, able to transport and deposit much more material; in winter, deposition would be less but more regular.

Kames

Kames are irregular mounds of bedded sands and gravels, arranged in a chaotic manner. There are several types of kame:

(1) The true **kame** (or kame delta) which probably represents a small delta, formed where a meltwater stream flowed out beneath an area of stagnant or slowly decaying ice, into a lake dammed between the ice front and drift material (probably the terminal moraine) (Figure 6.26). Many kames are found along the northern sides of the Southern Uplands. One of the most characteristic features of a kame landscape is the small, shallow hollow (called a **kettle hole**) (Figure 6.27)) amid the kame mounds, hence the term 'kame and kettle country'. The holes are probably due to deposition of material around blocks of

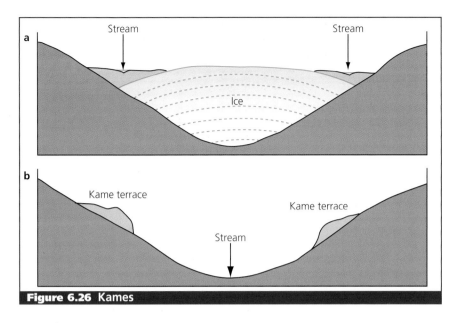

Figure 6.26 Kames

1 Explain how you would tell the difference between each of the following pairs:
   (a) terminal moraine and a kame;
   (b) lateral moraine and kame terraces,
   (c) medial moraine and eskers.
2 Explain how the following were formed: drumlins, erratics, terminal moraine, kettle holes, eskers and kame terraces.
3 Study the map extract (Figure 6.28) which shows Wastwater in the Lake District.
   (a) Describe the shape of Wastwater. How long is it? How wide is it? What type of lake is Wastwater?
   (b) How does it differ from Blea Tarn and Burnmoor Tarn?
   (c) What is the difference in height between Illgil Head and Wastwater?
   (d) What is the average gradient between Illgil Head at the nearest part of Wastwater?
   (e) Describe the likely nature of The Screes. (You may need to refer to pages 111–113 to help you.)
   (f) Suggest how The Screes were most likely formed.
   (g) What is the map evidence for glacial activity in this area?
   (h) Suggest why it is not always easy to find clear map evidence for glacial activity on a map.

ice broken off from the front of the stagnant ice body. Each block of ice would finally melt leaving a hole. The holes are often filled with water as numerous small lakes.

(2) **Kame terraces** are formed along the ice edge, laid down by streams occupying the trough between the ice and valley side. They appear as narrow flat-topped, terrace-like ridges such as along the edges of valleys in the Lammermuir Hills in eastern Scotland.

Useful websites include:

The British Antarctic Survey on http://www.nerc-bas.ac.uk/

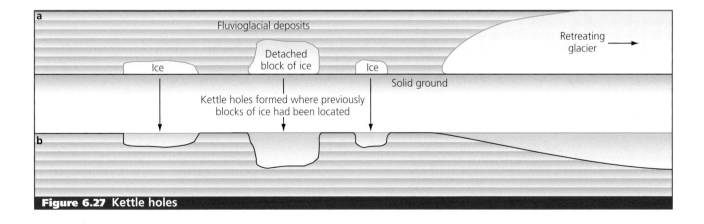

Figure 6.27 Kettle holes

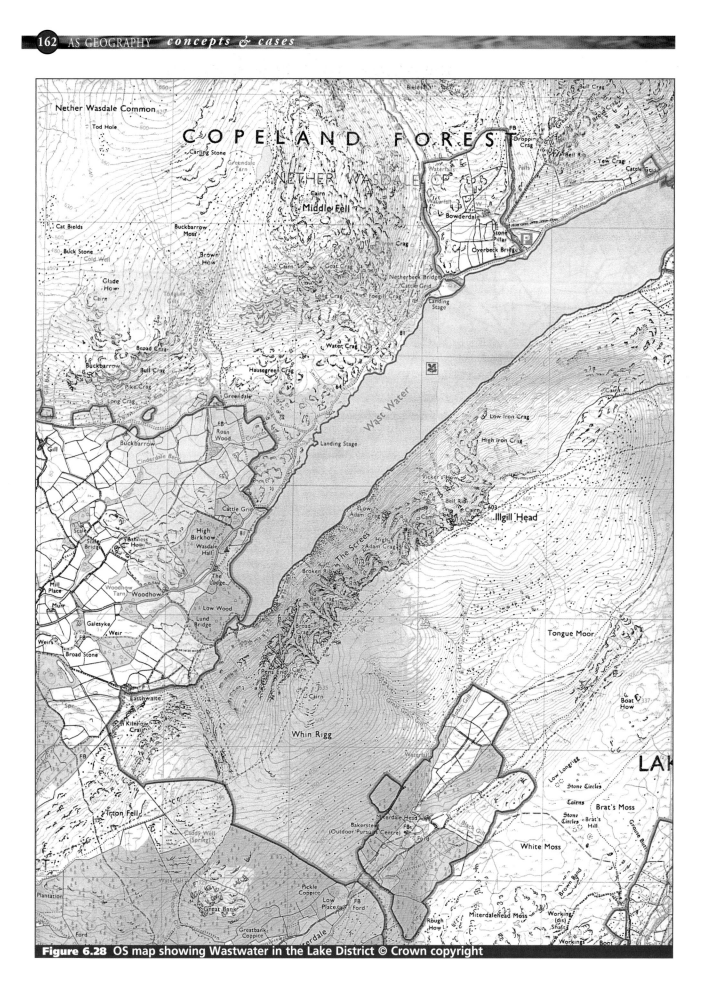

Figure 6.28 OS map showing Wastwater in the Lake District © Crown copyright

7 population dynamics and *resource relationships*

Early humankind

According to two groups of geneticists working independently at the University of Arizona and at Stanford University in California all men and women living today can trace their ancestry back to one man who lived 150 000 years ago and whose closest living relatives are a small tribe in South Africa (Figure 7.1). Human evolution, however, goes back much further than this.

The first hominids appeared in Africa around 5 million years ago, on a planet which is generally accepted to be 4600 million years old. They differed from their predecessors, the apes, in the fact that they walked on two legs and did not use their hands for weight-bearing. Other uses were soon found for these now liberated hands, with the acquisition of new skills charted in the evolutionary record as an increase in the size of the brain. After 2 million years cranial capacity had increased by 50% from the 600 cc of the earliest hominid, *Australopithecus*, to the 900 cc of the primitive man named *Homo erectus*. The final increase to *Homo sapiens'* current average of 1450 cc took place about 100 000 years ago.

The evolution of humankind was matched by its geographical diffusion. Whereas the locational evidence for *Australopithicus* is confined to Africa, remains of *Homo erectus* have been found stretching from Europe to South East Asia. *Homo sapiens* roamed even further, making the first incursions into the cold environments of high latitudes.

During most of the period since *Homo sapiens* first appeared, global population was very small, reaching perhaps some 125 000 people a million years ago, although there is insufficient evidence to be very precise about population in the distant past. It has been estimated that 10 000 years ago, when people first began to domesticate animals and cultivate crops, world population was no more than 5 million. This period of economic change, known as the Neolithic Revolution, significantly altered the relationship between people and their environments. But even then the average annual growth rate was less than 0.1% per year, extremely low compared with contemporary trends.

Technological advance improved the carrying capacity of the land and population increased so that by 3500 BC global population had reached 30 million and by the time of Christ it had risen to about 250 million.

Demographers estimate that world population reached 500 million by about 1650. From this time on population grew at an increasing rate. By 1830, global population had doubled to reach one billion (Figure 7.3). Figure 7.4 shows the time taken for each subsequent billion to be reached. It had taken the entire evolution of humankind until 1960 to reach a global population of 3 billion but it

Key Definitions

Demography The scientific study of human populations.

Crude birth rate (generally referred to as the 'birth rate') The number of births per thousand population in a given year. It is only a very broad indicator as it does not take into account the age and sex distribution of the population.

Crude death rate (generally referred to as the 'death rate') The number of deaths per thousand population in a given year. Again only a broad indicator as it is heavily influenced by the age structure of the population.

Rate of natural change The difference between the birth rate and the death rate.

Infant mortality rate The number of deaths of infants under one year of age per thousand live births in a given year.

Life expectancy (at birth) The average number of years a person may expect to live when born, assuming past trends continue.

Demographic transition The historical shift of birth and death rates from high to low levels in a population.

Census An official periodic count of a population including such information as age, sex, occupation and ethnic origin.

Carrying capacity The largest population that the resources of a given environment can support.

Population structure The composition of a population, the most important elements of which are age and sex.

DNA tests trace Adam to Africa

by Steve Connor
Science Correspondent

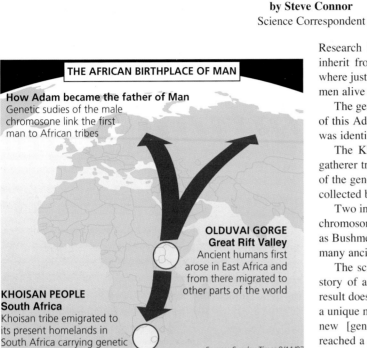

THE AFRICAN BIRTHPLACE OF MAN

How Adam became the father of Man
Genetic sudies of the male chromosone link the first man to African tribes

OLDUVAI GORGE Great Rift Valley
Ancient humans first arose in East Africa and from there migrated to other parts of the world

KHOISAN PEOPLE South Africa
Khoisan tribe emigrated to its present homelands in South Africa carrying genetic remnants of Adam with them

Source: *Sunday Times* 9/11/97

Research into the human Y chromosome – which sons only inherit from their fathers – has pinpointed the time and place where just one man gave rise to the male genetic ingredients of all men alive today.

The geneticists have also located the oldest direct descendants of this Adam, who they say lived alongside an African Eve who was identified in similar studies 10 years ago.

The Khoisan people of South Africa, some with a hunter-gatherer tradition stretching back thousands of years, share most of the genetic traits that first arose when Adam hunted game and collected berries in his African Garden of Eden.

Two independent investigations of minute mutations on the Y chromosome pinpointed the Khoisan people, who are also known as Bushmen or Hottentots, as the only ethnic group to possess so many ancient remnants of the original Adam.

The scientists said the research does not support the biblical story of a single man and woman in a Garden of Eden. "This result does not mean there was ever only one male but rather that a unique mutation occurred, resulting in one son who defined the new [genetic] line and whose male descendants eventually reached a majority in Africa. Some offspring of this lineage left Africa to populate the entire globe," Professor Oefner said.

Source: *Sunday Times*, November 1997

Figure 7.1 Early humankind

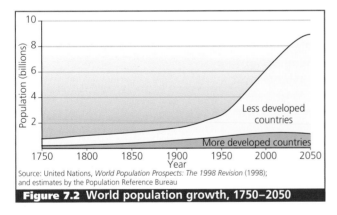

Source: United Nations, *World Population Prospects: The 1998 Revision* (1998); and estimates by the Population Reference Bureau

Figure 7.2 World population growth, 1750–2050

Figure 7.3 Increase in the world's population and the time taken to add another billion

1830	1 billion	
1930	2 billion	(100 years)
1960	3 billion	(30 years)
1975	4 billion	(15 years)
1987	5 billion	(12 years)
1999	6 billion	(12 years)

Figure 7.4 Ruins of the original city of Carthage

would take less than another 40 years for this population to double. In 1998, the population of the world was increasing at a rate of 78.7 million a year, equivalent to 215 000 people a day or 150 every minute. The United Nations estimated that world population reached 6 billion on 12 October 1999 with the birth of a baby in the Bosnian capital, Sarajevo.

Demographic transition

Although the populations of no two countries have changed in exactly the same way, some generalisations can be made about population growth since the middle of the 18th century. These are illustrated by the model of demographic transition (Figure 7.6) which is based on the experience of north-west Europe, the first part of the world to undergo such changes as a result of the significant industrial and agrarian advances that occurred there during the 18th and 19th centuries.

No country as a whole retains the characteristics of stage 1, which applies only to the most remote societies on earth such as isolated tribes in New Guinea and the Amazon, which have little or no contact at all with the outside world. All the developed countries of the world are now in stage 4, most

Figure 7.5 World's six billionth citizen

having experienced all of the previous stages at different times. The poorest of the developing countries (such as Bangladesh, Niger and Bolivia) are in stage 2 but are joined in this stage by the oil-rich nations of the Middle East where increasing affluence has not been accompanied by a significant fall in fertility. Most developing countries that have registered significant social and economic advances are in stage 3 (such as Brazil, China and Turkey), while some of

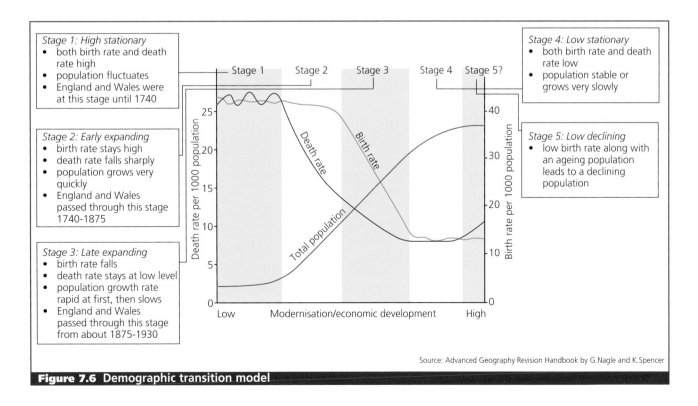

Stage 1: High stationary
- both birth rate and death rate high
- population fluctuates
- England and Wales were at this stage until 1740

Stage 2: Early expanding
- birth rate stays high
- death rate falls sharply
- population grows very quickly
- England and Wales passed through this stage 1740-1875

Stage 3: Late expanding
- birth rate falls
- death rate stays at low level
- population growth rate rapid at first, then slows
- England and Wales passed through this stage from about 1875-1930

Stage 4: Low stationary
- both birth rate and death rate low
- population stable or grows very slowly

Stage 5: Low declining
- low birth rate along with an ageing population leads to a declining population

Source: Advanced Geography Revision Handbook by G.Nagle and K.Spencer

Figure 7.6 Demographic transition model

the newly industrialised countries such as South Korea and Taiwan have just entered stage 4. With the passage of time more countries will reach the final stage of the model. The basic characteristics of each stage are:

The high fluctuating stage (stage 1) The birth rate is high and stable while the death rate is high and fluctuating due to the sporadic incidence of famine, disease and war. Population growth is very slow and there may be periods of considerable decline. Infant mortality is high and life expectancy low. Society is pre-industrial with most people living in rural areas, dependent on subsistence agriculture.

The early expanding stage (stage 2) The death rate declines to levels never before witnessed. The birth rate remains at its previous level as the social norms governing fertility take time to change. As the gap between the two vital rates widens, the rate of population growth increases to a peak at the end of this stage. Infant mortality falls and life expectancy increases. The proportion of the population under 15 increases. Although the reasons for the decline in mortality vary in intensity and sequence from one country to another, the essential causal factors are: better nutrition; improved public health particularly in terms of clean water supply and efficient sewage systems; and medical advance. Considerable rural to urban migration occurs. However, in recent decades urbanisation in developing countries has often not been accompanied by the industrialisation that was characteristic of the developed nations during the 19th century.

The late expanding stage (stage 3) After a period of time social norms adjust to lower mortality and the birth rate begins to decline. Life expectancy continues to increase and infant mortality to decrease. Countries in this stage usually experience lower crude death rates than nations in the final stage due to their relatively young population structures. The rate of urbanisation generally slows.

The low fluctuating stage (stage 4) Both birth and death rates are low. The former is generally slightly higher, fluctuating somewhat due to changing economic conditions. Population growth is slow and may even be negative if the birth rate dips below the death rate. Crude death rates rise slightly as the average age of the population increases. However, life expectancy still improves as age specific mortality rates continue to fall. This stage completes the population cycle according to the model but will a further stage (stage 5) of prolonged natural decrease, be evident in a few decades time?

Geographical models

A model is a simplification of reality. Concentrating only on major characteristics and omitting the detail, it makes understanding easier at the start of the learning process. However, once clear about the framework of a situation or process, it is then not too difficult to unravel the detail. A good model will be helpful in this respect from the start. If, however, the model has clear limitations, then it needs to be improved (refined). In geography, models are useful in most areas of the subject. The model of demographic transition is the first to be examined in this chapter.

Demographic transition in England and Wales

In England and Wales in medieval times (stage 1) both the birth rate and the death rate hovered around 35 per 1000. The birth rate was generally a little higher, resulting in a slow rate of natural increase. While the birth rate tended to remain at a relatively stable level the death rate varied considerably at times. For example, the 1348–49 epidemic of bubonic plague known as the Black Death killed something like a third of the population. Also of great demographic consequence were the bubonic plagues of 1603, 1625 and 1665, the latter referred to as the Great Plague. The increase in mortality between 1720 and 1740 (Figure 7.7) has been attributed to the availability of cheap gin during this period which took a considerable toll on the working class. These conditions of high fertility and high mortality persisted until about 1740.

Stage 2, which lasted until about 1875, witnessed a period of rapid urbanisation which alerted both public officials and factory owners to the urgent need for improvements in public health. Factory owners soon realised that an unhealthy workforce had a huge impact on efficiency. The provision of clean, piped water and the installation of sewage systems, allied to better personal and domestic cleanliness, saw the incidence of the diarrhoeal diseases and typhus fall rapidly.

Although in many ways life in the expanding towns was little better than in the countryside, there was a greater opportunity for employment and a larger disposable income so that more

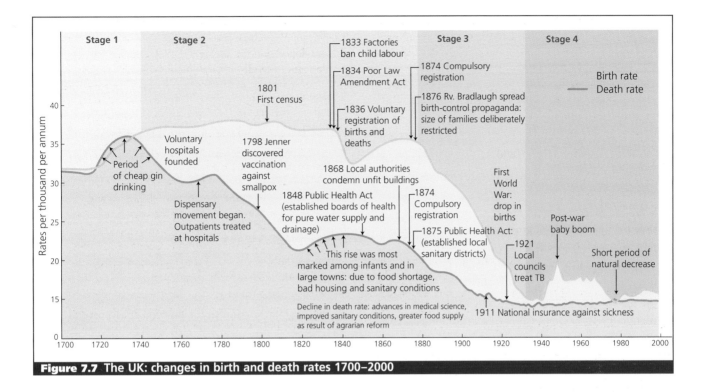

Figure 7.7 The UK: changes in birth and death rates 1700–2000

food and a wider range of food products could be purchased. Contemporary studies in developing countries show a very strong relationship between infant nutrition and infant mortality. Infant mortality in England fell from 200 per 1000 in 1770 to just over 100 per 1000 in 1870.

The virulence of the common infectious diseases diminished markedly. For example, scarlet fever, which caused many deaths in the 18th century, had a much reduced impact in the following century. From about 1850 mortality from tuberculosis also began to fall. A combination of better nutrition and the general improvements in health brought about by legislation such as the Public Health Acts of 1848 and 1869 were the most likely causal factors.

The final factor to be considered in stage 2 is the role of medicine. Although some important milestones were reached, such as Jenner's discovery of a vaccination against smallpox, there was no widespread diffusion of medical benefits at this time. Of all the drugs available in 1850, fewer than ten had a specific action, so their impact on mortality was negligible. Surgery was no more advanced and anaesthesia was unavailable until the last years of the century.

We can be much surer about the accuracy of demographic data during this period. The first census of England and Wales was taken in 1801 and every ten years thereafter, and from 1836 the registration of births and deaths was introduced on a voluntary basis. The latter became compulsory in 1874.

After 1875 the continued decline of the death rate was accompanied by a marked downturn in the birth rate (stage 3). Medical science began to play an important role in controlling mortality and doctors were now able to offer potent, specifically effective drugs. From about 1906 increasing attention was paid

to maternity and child welfare, and to school health. More measures to improve public health were introduced while there were further gains in nutrition. From the late 1870s onwards, cheap North American wheat began to arrive in Britain in large quantities, along with refrigerated meat and fruit from Australia and New Zealand.

The beginning of the decline in fertility coincided with the widespread publicity that surrounded the trial of Charles Bradlaugh and Annie Besant. These two social reformers were prosecuted (and later acquitted) for publishing a book that gave contraceptive advice. However, perhaps the most important factor was the desire for smaller families now that people could be sure that the decline in mortality was permanent and because the monetary cost of children was higher in urban compared to rural areas.

Family size varied by social group, with the upper and professional middle classes leading the way in contraception. The birth rate, which had been 30.5 per 1000 in 1890, fell to 25 per 1000 in 1910 and was down to 17 per 1000 by 1930, at which time it is reasonable to assert that England and Wales was entering the final stage of demographic transition.

By 1940 the birth rate had fallen further to 14.5 per 1000 but this was undoubtedly influenced by the outbreak of war the previous year. The higher figures in the three decades following the end of the war are generally accounted for by the phenomenon known as the 'post-war baby boom'. By 1980, however, the birth rate was down again to 14 per 1000, and has remained very close to that figure ever since. The introduction of the oral birth pill in 1960 and improvements in other forms of contraceptives meant that the relationship between desired family size and achieved family size had never been stronger.

Demographic transition in the developing world

There are a number of important differences in the way that developing countries have undergone population change compared to the experiences of most developed nations before them. In the developing world:

■ birth rates in stages one and two were generally higher, in some cases over 50 per 1000

■ the death rate fell much more steeply and for different reasons

■ some countries had much larger base populations and thus the impact of high growth in stage 2 and the early part of stage 3 has been far greater

■ for those countries in stage 3 the fall in fertility has also been steeper

■ the relationship between population change and economic development has been much more tenuous.

The components of population change

The relationship between births and deaths (natural change) is not the only factor in population change. The balance between immigration and emigration (net migration) must also be taken into account (Figure 7.8). The corrugated divide indicates that the relative contributions of natural change and net migration can vary over time. The model is a simple graphical alternative to the population equation P = (B − D) ± M, the letters standing for population, births, deaths and migration respectively.

The data in Figure 7.8 gives a natural increase of 94 000 for the UK in 1996 (birth rate 12.5 per 1000, death rate 10.9 per 1000) with a net immigration of

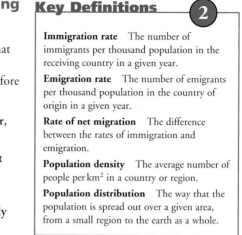

Key Definitions ②

Immigration rate The number of immigrants per thousand population in the receiving country in a given year.

Emigration rate The number of emigrants per thousand population in the country of origin in a given year.

Rate of net migration The difference between the rates of immigration and emigration.

Population density The average number of people per km² in a country or region.

Population distribution The way that the population is spread out over a given area, from a small region to the earth as a whole.

56 000. London accounted for almost half this natural increase, with nearly half of the ethnic minority population of the country living in the capital city. Since the mid-1980s there has been net migration into the UK. It should be noted that in recent years government publications have used the terms 'inflow' and 'outflow' rather than 'immigration' and 'emigration'. In 1996, the population of the UK was 58.8 million, an increase of 11% since 1961.

The UK census

A full census in the UK takes place every ten years. Preparation for the census in 2001 began as early as 1993 with a development programme taken forward by the Office for National Statistics, the General Register Office for Scotland and the Northern Ireland Statistics and Research Agency. Various tests were conducted particularly with regard to new and revised questions. It is important that all questions are clear and allow people to respond with accuracy. The census is the most valuable source of information about the characteristics of people in the UK, particularly for small areas.

The census provides a benchmark for official population estimates. Between censuses the population figures are rolled forward using annual estimates of the components of population change. As the decade proceeds, problems with estimating migration, in particular, progressively affect these rolled-forward figures. A new census is used for both revising previous years' data and for providing a base for the estimates in the following decade.

Distribution and density

Figures 7.11 and 7.12 show the distribution and density of population by world region. The huge contrast between

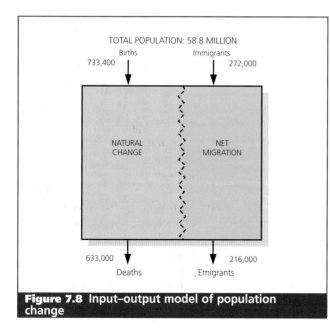

Figure 7.8 Input–output model of population change

TOTAL POPULATION: 58.8 MILLION

Births 733,400

Immigrants 272,000

NATURAL CHANGE

NET MIGRATION

Deaths 633,000

Emigrants 216,000

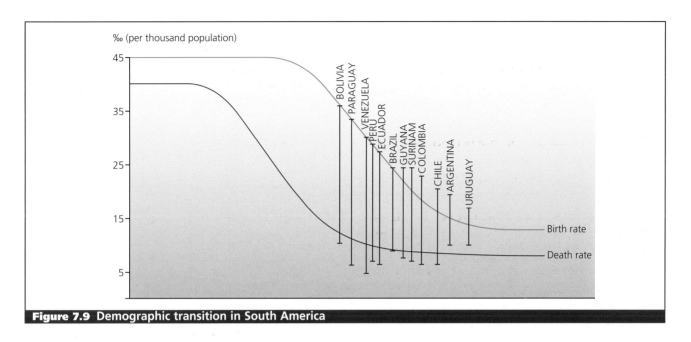

Figure 7.9 Demographic transition in South America

the developed and developing worlds is readily apparent. The developed world has only about 20% of world population and its population is growing at only one quarter of the global rate. However, within the developed world itself there is a considerable difference in natural change between Europe

on the one hand (–0.1%), and North America and Oceania (0.6% and 1.1% respectively) on the other. There are also significant differences in the developing world, with Africa having by far the highest natural increase at 2.5%.

The global average for population density also covers a wide regional range. The overall difference between the developed and developing worlds is largely accounted for by the extremely high figure for Asia. As for the other developing regions, Europe is about 50% higher and North America 50% lower in population density.

The average density figure for each region masks considerable disparity. The most uniform distributions of population occur where there is little variation in the physical and human environments. Sharp contrasts in these environments are strongly reflected in settlement patterns. People have always avoided hostile environments if a reasonable choice has been available. Visual correlation between an atlas map of the world showing population density with maps of relief, temperature, precipitation and vegetation illustrates that low densities are associated with high altitudes, polar regions, deserts and rain forests. More detailed maps can show the influence of other physical factors such as soil fertility, natural water supply and mineral resources. At the physical/human interface the spatial incidence of disease and pests, particularly in developing countries, can seriously limit human settlement.

The more advanced a country the more important the elements of human infrastructure become in influencing population density and distribution. While a combination of physical factors will have decided the initial location of the major urban areas, once such entities reach a certain size, economies of scale and the process of cumulative causation (see Chapter 9) ensure further growth. As the importance of agriculture decreases, employment relies more and more on the secondary and tertiary sectors which are largely urban

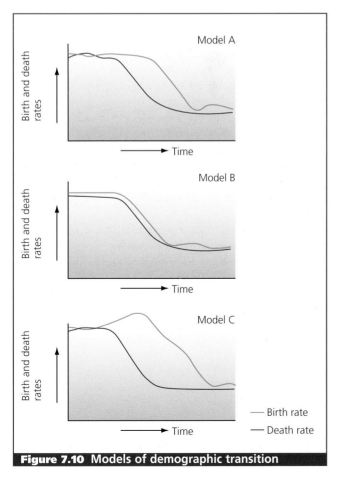

Figure 7.10 Models of demographic transition

Figure 7.11 Distribution of world population

	Population mid-1999 (millions)	Population density km²	Natural change (%)	Doubling time* (years)	Projected Population 2010	2025
World	5982	42	1.4	49	6883	8054
More developed	1181	22	0.1	533	1216	1241
Less developed	4800	54	1.7	40	5667	6813
Africa	771	23	2.5	28	979	1290
North America	303	13	0.6	119	333	374
Latin America and Caribbean	512	23	1.8	38	600	709
Asia	3637	107	1.5	46	4206	4923
Europe	728	32	−0.1	–	731	718
Oceania	30	3	1.1	64	34	41

*At current rate of growth

Source: 1999 World Population Data Sheet published by the Population Reference Bureau, Washington, DC

based. The lines of communication and infrastructure between major urban centres provide opportunities for further urban and industrial location.

Scale

Patterns and processes in human and physical geography can be examined at a range of scales from the global or macro-scale, through the intermediate or meso-scale, to the local or micro-scale. When analysing data it is important to be clear about the scale to which it refers.

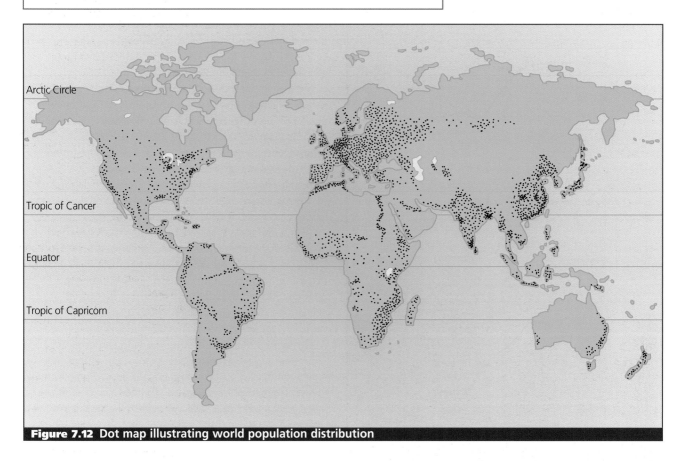

Figure 7.12 Dot map illustrating world population distribution

Population change in Brazil

The most rapid changes in population density in human history have occurred in the developing world in the latter half of the 20th century. The first Brazilian population census, collected in 1872, recorded a total of 9.9 million people (Figure 7.14). By the turn of the century this had increased to 17.4 million. At this time the birth rate was approximately 46 per 1000 and the death rate 30 per thousand. However, while the birth rate continued at a high level until 1960, the death rate declined steadily resulting in an increasing rate of population growth. The total population reached 50 million in the late 1940s and 100 million in the early 1970s. Since 1940 Brazil's population has nearly quadrupled.

In recent decades, however, the rate of population growth has slowed significantly. From an annual growth rate of almost 3% in the 1960s, the rate dropped to 2% in the 1980s, and 1.5% for the 1990s. The cause has been a rapidly falling fertility rate.

In Brazil (Figure 7.16) all regions have been affected by population growth, creating pressures on both the physical environment and on human infrastructure. The sprawling cities of the Southeast and Northeast (Figure 7.17) are an obvious indicator of increasing density and urban pressure. In the Southeast the percentage of the population classed as urban increased from 57% in 1960 to 89.3% in 1996. During the same period the level of urbanisation in the Northeast increased from 33.4% to 65.2%.

The changing landscapes of the Centre-west and the North are another consequence of population growth. These latter regions were largely undeveloped prior to the 1950s, but significant changes since that time have brought them firmly into the national economy. The construction of Brasília as the new capital city was the fundamental catalyst for significantly populating the Centrewest. In the North, however, the resources of the Amazon rain forest have been exploited more to satisfy the demands of population pressures in other parts of Brazil and in the developed world than in the region itself.

At the dawn of the 19th century, overall population density in Brazil was 2.0 per km²: by 1940 it was 4.8 and by 1970 10.9. Even in 1970 the pattern of settlement still demonstrated the predominantly coastal distribution typical of colonial times. In spite of the considerable changes in density and distribution in the latter part of the 20th century the average population density for the country as a whole was only 19 per km² in 1996, compared with a global average of 39 per km².

Figure 7.13 The sparsely populated Paraná region

1 Study Figure 7.9.
  (a) Define the terms (i) birth rate, (ii) death rate, (iii) rate of natural change.
  (b) On a copy of Figure 7.9 insert vertical dotted lines to show the four stages of demographic transition.
  (c) Indicate on your diagram where the rate of natural change is greatest.
  (d) Which of the countries illustrated has (i) the highest birth rate, (ii) the lowest death rate, (iii) the greatest rate of natural change?
  (e) Suggest why the demographic situation in South America does not fit the model perfectly.
  (f) Discuss the possible reasons for the demographic variations in South America illustrated by the model.

2 **(a)** Describe the differences between the three models in Figure 7.10.
  (b) In model A, the experience of most countries, explain why the birth rate declined much later than the death rate.
  (c) Suggest possible reasons for the absence of a period of rapid growth in model B.
  (d) Why did the birth rate rise as the death rate fell in some countries such as Japan and Mexico?
  (e) Discuss the main differences in the way that developing countries are undergoing population change compared to the experiences of most developed nations before them.

3 **(a)** Using the information in Figure 7.8, calculate the following average annual rates per thousand for the UK in 1996: birth, death, natural change, immigration, emigration, net migration.
  (b) By how much did the total population change over the year?
  (c) In the late 1970s the UK experienced a brief period of natural decrease. Suggest the likely cause of the fall in the birth rate at this time.
  (d) Suggest reasons for the relatively high birth rate between the late 1940s and late 1960s compared to the periods immediately before and after this time.
  (e) Why has there been such little variation in the crude death rate in the UK since the 1920s?

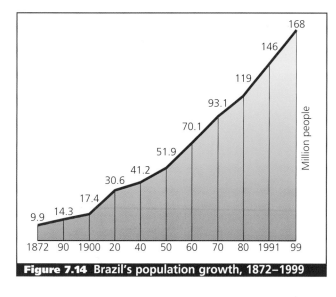

Figure 7.14 Brazil's population growth, 1872–1999

Fertility Rate*

Population Growth
Average annual % increase

Sources: World Bank; UNDP
*Average number of babies born per woman of child-bearing age

Figure 7.15 Brazil: fertility and population growth

Figure 7.16 Brazil: population growth by region

Grandes Regiões	Resident population						
	1940	**1950**	**1960**	**1970**	**1980**	**1991**	**1996**
Brazil	**41 165 289**	**51 941 767**	**70 070 457**	**93 139 037**	**119 002 706**	**146 825 475**	**157 079 573**
North	1 467 940	1 834 185	2 561 782	3 603 860	5 880 268	10 030 556	11 290 093
Northeast	14 426 185	17 992 094	22 181 880	28 111 927	34 812 356	42 497 540	44 768 201
Southeast	18 304 317	22 549 386	30 630 728	39 853 498	51 734 125	62 740 401	67 003 069
South	5 722 018	7 835 418	11 753 075	16 496 493	19 031 162	22 129 377	23 516 730
Centre–West	1 244 829	1 730 684	2 942 992	5 073 259	7 544 795	9 427 601	10 501 480

Source: IBGE, Censos Demográficos de 1940 a 1991 e Contagem da População de 1996.

Figure 7.17 Population density in Brazil

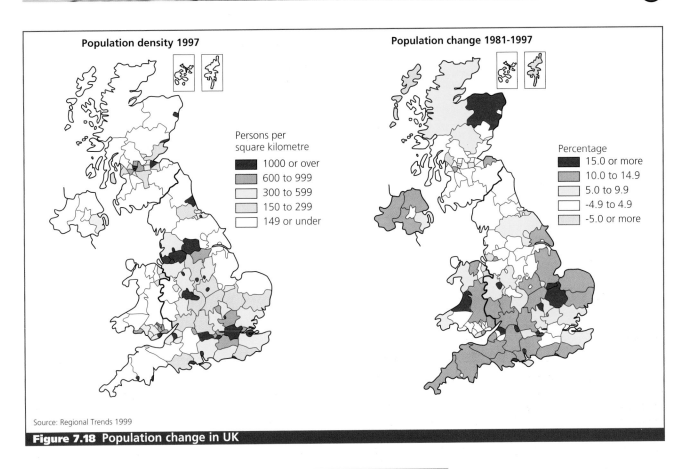

Population density 1997

Persons per square kilometre

- ■ 1000 or over
- ▨ 600 to 999
- ▨ 300 to 599
- ▨ 150 to 299
- ▨ 149 or under

Population change 1981-1997

Percentage

- ■ 15.0 or more
- ▨ 10.0 to 14.9
- ▨ 5.0 to 9.9
- □ -4.9 to 4.9
- ▨ -5.0 or more

Source: Regional Trends 1999

Figure 7.18 Population change in UK

Figure 7.19 Population change in the South–East

	Population 1961 (millions)	Population 1991 (millions)	Population density 1991 per km²
The whole of the South–East	16.1	17.0	621
Areas of the South–East:			
Greater London	8.0	6.9	4338
Outer Metropolitan area	4.4	5.4	593
Rest of South–East	3.7	4.7	289

The highest densities are along the coast, stretching from the far north to the extreme south. In fact, 80% of Brazilians live within 320 km of the Atlantic coast. Density in the North is less than 3 people per km², making it one of the world's great areas of low density. In contrast the state of São Paulo, the hub of the national economy, has over twice as many people as the next state in demographic order (Minas Gerais). In essence, the concentration of population has resulted from that of another concentration, economic activity. Brazil is highly urbanised. At 76% the level of urbanisation is similar to the developed world.

Figure 7.20 São Paulo – the greatest concentration of population in South America

Fertility

The factors influencing fertility

In most parts of the world fertility exceeds both mortality and migration and is thus the main determinant of population growth (Figure 7.21). Its importance has increased over time with the worldwide fall in mortality. According to the 1999 World Population Data Sheet produced by the Population Reference Bureau based in Washington, Niger has the highest crude birth rate at 54 per 1000 followed by Chad (50 per 1000) and Gaza (49 per 1000). At the other end of the scale, Latvia and Bulgaria had birth rates of only 8 per 1000.

Figure 7.21 Fertility and contraception

	Birth rate	Total fertility rate	% of married* women using contraception (all methods)
World	23	2.9	58
More developed	11	1.5	72
Less developed	26	3.2	55
Africa	39	5.4	24
North America	14	2.0	77
Latin America and Caribbean	24	2.9	68
Asia	23	2.8	60
Europe	10	1.4	71
Oceania	18	2.4	59

Source: *1999 World Population Data Sheet* (For developed countries nearly all data refers to 1997 or 1998 and for less developed countries to some point in the mid to late 1990s.)

*married or 'in union'

Questions

1 **(a)** Define the terms (i) population density, (ii) population distribution, (iii) doubling time.
  (b) Describe and explain the differences in natural change by world region [Figure 7.11].
  (c) How will the distribution of world population change by 2025?
  (d) To what extent does Figure 7.12 add to the information provided by Figure 7.11?
  (e) For a country you have studied, explain the distribution of its population.

2 **(a)** Discuss three reasons for the differences in population density shown in Figure 7.18.
  (b) Identify the main advantages and disadvantages of living in (i) an area of high population density, (ii) an area of low population density.
  (c) Describe and explain the regional population changes illustrated.
  (d) Suggest reasons for the variations in population density within the Southeast region [Figure 7.19].

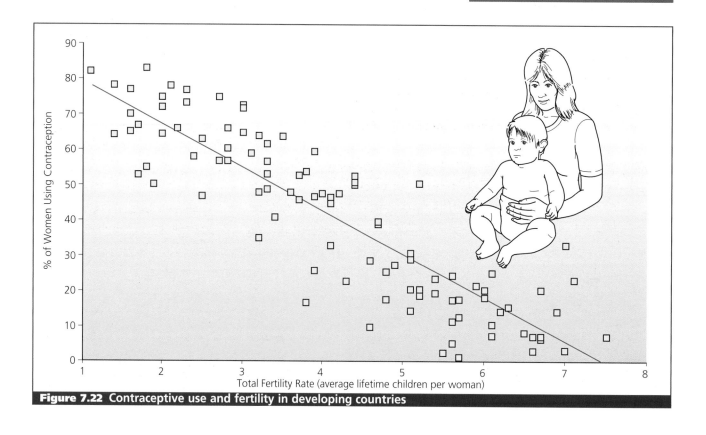

Figure 7.22 Contraceptive use and fertility in developing countries

Any meaningful analysis of fertility, however, must look beyond the birth rate, with the most detailed studies examining age-specific rates for each year of age. In terms of the total fertility rate (TFR) Niger again headed the list at 7.5, followed by Qatar (7.1), Gaza (7.0) and Uganda (6.9). The lowest TFR, which was 1.1, was recorded by Latvia and Bulgaria.

The factors affecting fertility levels can be grouped into four categories:

■ **Demographic** – other demographic factors, particularly mortality rates influence the social norms regarding fertility. One study of sub-Saharan Africa, where the average infant mortality is over 100 per 1000, calculated that a woman must have an average of ten children to be 95% certain of a surviving adult son.

■ **Social/Cultural** – in some societies, particularly in Africa, tradition demands high rates of reproduction. Here the opinion of women in the reproductive years will have little influence weighed against intense cultural expectations. Education, especially female literacy, is the key to lower fertility (Figure 7.23). With education comes a knowledge of birth control, greater social awareness, more opportunity for employment and a wider choice of action generally. The more educated women are, the later they marry; this widens the age gap between successive generations. Most countries exhibit different fertility levels according to social class with fertility decline occurring in the highest social classes first. In some countries religion is an important factor. For example, the Muslim and Roman Catholic religions oppose artificial birth control. However, the degree of adherence to religious doctrine tends to lessen with economic development.

■ **Economic** – in many of the least developed countries children are seen as an economic asset. They are viewed as producers rather than consumers and as security when parents reach old age. In the developed world the general perception is reversed and the cost of the child dependency years is a major factor in the decision to begin or extend a family.

■ **Political** – there are many examples in the twentieth century of governments using direct policies to change the rate of population growth for economic and strategic reasons. During the late 1930s Germany, Italy and Japan all offered inducements and concessions to those with large families. In recent years Malaysia has adopted a similar policy. Today, however, most governments that are interventionist in terms of fertility want to reduce population growth. In many countries the government has no official policy with regard to population but indirectly a range of decisions made by government may influence population trends. Such factors include the level of family allowances, investment in sex education and family planning, and care of the elderly.

Figure 7.24 summarises the behavioural and biological factors influencing fertility while Figure 7.25 examines the possible economic effects of reducing fertility.

The fertility issue: population explosion followed by implosion?

In the UN's latest long-range population projections, the 'most probable' medium-fertility scenario assumes that fertility will stabilise at slightly above two children per woman. The result would be a global population of 10.4 billion by the end of the 21st century (Figure 7.26). This is in sharp contrast to warnings in earlier decades of population 'explosion'. The main reason for the slowdown in population growth is that fertility levels in most parts of the world are falling faster than previously expected.

Figure 7.23 Secondary school in Tunisia

Key Definitions 3

Fertility rate The number of live births per 1000 women aged 15–49 years in a given year.

Age-specific fertility rate The fertility rate conventionally divided into seven age groups (15–19, 20–24 etc.) or for even more detail, into each year of age from 15–49 years.

Total fertility rate The average number of children that would be born alive to a woman (or group of women) during her lifetime, if she were to pass through her child-bearing years conforming to the age-specific fertility rates of a given year.

Replacement level fertility The level at which each generation of women has just enough daughters to replace themselves in the population. Although the level varies for different populations a total fertility rate of 2.12 children is usually considered as replacement level.

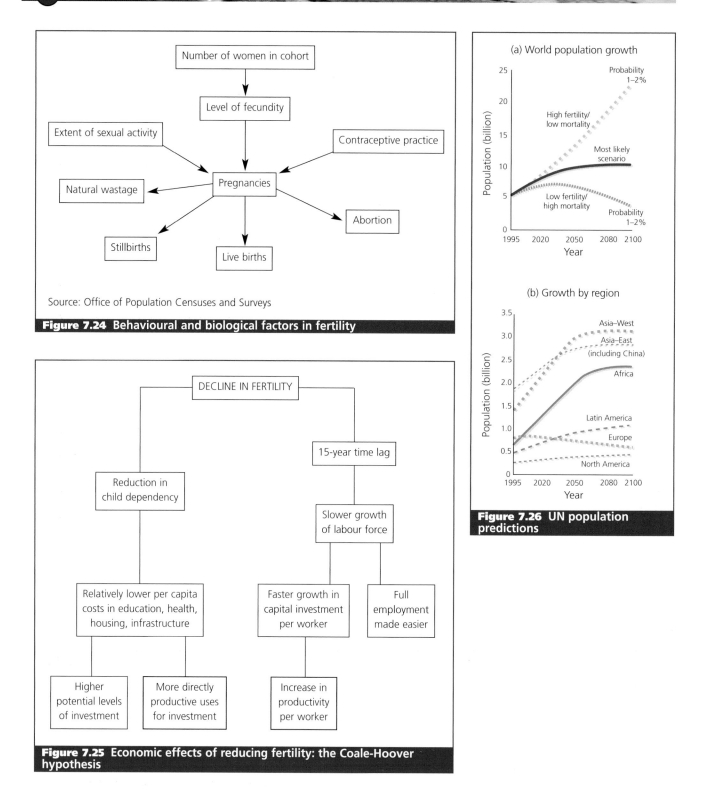

Source: Office of Population Censuses and Surveys

Figure 7.24 Behavioural and biological factors in fertility

Figure 7.25 Economic effects of reducing fertility: the Coale-Hoover hypothesis

(a) World population growth

(b) Growth by region

Figure 7.26 UN population predictions

In the second half of the 1960s, after a quarter century of increasing growth, the rate of world population growth began to slow down (Figure 7.27). Since then some developing countries have seen the sharpest falls in fertility ever known and thus earlier population projections did not materialise. The UN predicted a global growth rate of 2.0–2.1% for the 1970s but by 1975–80 growth was down to only 1.7%.

Only since the Second World War has population growth in the poor countries overtaken that in the rich. However, the 1960s saw population growth in the developing world peak at 2.4% a year and by the late 1990s it was down to 1.8%. But, even though the rate of growth has been falling for three decades, demographic momentum meant that the numbers being added each year did not peak until the late 1980s.

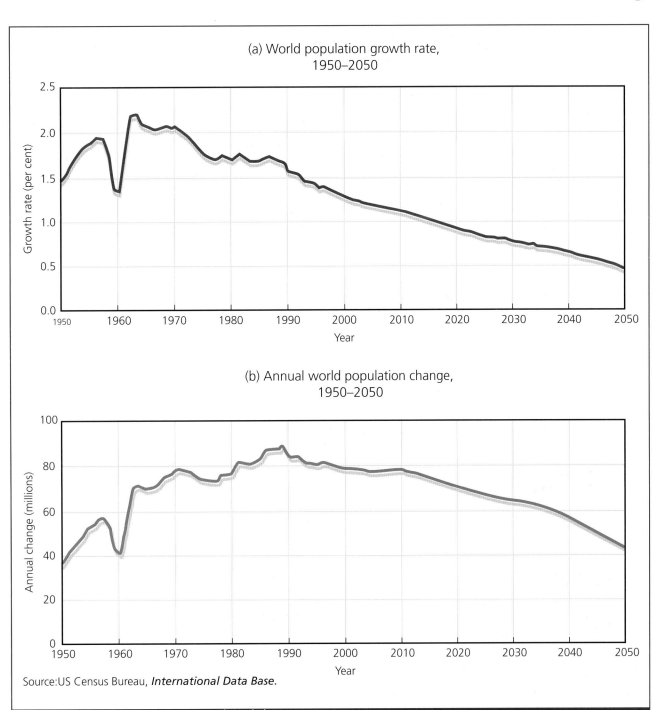

Figure 7.27 World population growth and change

The demographic transformation, which took a century to complete in the developed world, has occurred in a generation in some developing countries. Fertility has dropped further and faster than most demographers foresaw 20 or 30 years ago. Except in Africa and the Middle East, where in almost 50 countries families of at least six children are the norm and population growth is still over 2.5% per year, birth rates are now declining in virtually every country. In the poorest nations there is often a large gap between the private gains from having many children and the social gains from reducing population growth.

Modern contraceptive methods have played a key role in lowering global fertility. Among women of reproductive age who are married (or in non-marital unions), half now depend on such methods as female sterilization (the most popular), male sterilization, hormonal implants, injectibles, intrauterine devices, birth control pills, condoms and diaphragms. The first four methods are almost totally effective. Next are IUDs, followed by the pill and the male condom. Worldwide, 7% of all women depend on traditional methods of birth control which are far less reliable than the methods listed above.

Fertility in India

India is on the way to catching China as the most populous country on earth. At the first census after independence, in 1951, India had a population of 361 million. It is now close to 1 billion and by 2050 it should overtake China. This assumes annual population growth of around 0.9% for India, compared with only 0.4% a year for China. In southern states such as Tamil Nadu and Kerala where literacy rates are high, fertility rates have fallen sharply. However, in the impoverished 'Hindi belt' in the north traditional attitudes dominate, ensuring large numbers of children. Nevertheless, in India as a whole fertility has dropped more than 40% in the past 25 years.

In 1952, India became the first developing country to introduce a population policy designed to reduce fertility and to aid development with a government-backed family planning programme. Rural and urban birth control clinics rapidly increased in number. Financial and other incentives were offered in some states for those participating in programmes, especially sterilisation. In the mid-1970s the sterilisation campaign became increasingly coercive, reaching a peak of 8.3 million operations in 1977. Abortion was legalised in 1972 and in 1978 the minimum age of marriage was increased to 18 years for females and 21 years for males. The birth rate fell from 45 per 1000 in 1951–61 to 41 per 1000 in 1961–71. By 1987 it was down to 33 per 1000, falling further to 29 per 1000 in 1995.

In 1995, India's total fertility rate was 3.5. The maternal mortality rate, at 570 per 100 000, is one of the highest in the world. Every year more than 75 000 women die due to pregnancy and childbirth-related reasons. Only a third of all births are attended by trained medical staff. Most births occur at home, assisted by a traditional birth attendant.

Although the infant mortality rate has fallen in recent decades, in 1995 it was 75 per thousand for the country as a whole. However, infant mortality varies considerably between the richest and poorest states. It is an important factor influencing family decisions. In the early 1990s, 45% of Indians in the reproductive age range used contraception. 72% of all contraceptive users

Figure 7.28 Family planning poster in India

relied on sterilisation as a form of birth control, 11.8% on condoms, 4.2% on inter-uterine devices, and 3.1% on oral pills. In 1993, 96% of all sterilisations were performed on women. Of the 6.7 million induced abortions that year, only 600 000 were legally performed.

Time

Patterns and processes can be examined at different *time scales*. In this chapter reference has already been made to geological time, historical time and recent time. Human geography refers to the latter two and particularly to recent time. Each of the three time scales already mentioned can of course be sub-divided. Trends in recent time may be examined on a daily, weekly, monthly, annual, or decade-by-decade basis. The important thing is to choose the most appropriate time scale for your analysis.

The concept of *time lag* is important in various types of geographical analysis. For example in this chapter the time lag between fertility falling to population replacement level in a country and the total population actually beginning to decline is a significant factor in population change.

The 'under 2.1 club'

A fertility rate of 2.1 children per woman is the replacement level below which populations eventually start falling. There are already 51 nations with total fertility rates at or below 2.1. By 2016 the UN predicts there to be 88 nations in this category. China is in the 'under 2.1 club' but its population will not begin to decline until after 2045. The UN's Population Division forecasts that by 2016 India and Indonesia, among many other countries, will have joined the club. For the world as a whole, because of the time lag of up to 40 years between reaching replacement level fertility (as a result of falling fertility) and actual population decline, population will increase by almost as much in the twenty-first century as it has in the twentieth. This is because there are larger numbers of people about to enter their child-bearing years (the legacy of years of high fertility). An estimate from the British Government Actuary for UK population, published in December 1997, forecasts a slow increase from the current 59 million to a peak of just under 63 million in 2031, before starting a slow fall. In the past 20 years the number of British couples who have decided to have only one son or daughter has risen by 40% to 2.9 million. The trend is particularly marked among professional and older parents.

In Italy, where the fertility rate is just 1.2 children per woman, population decline has already begun and the UN forecasts a fall in population of 4 million by 2020. Decline is also in progress in Russia, Ukraine, Georgia, Romania, Bulgaria, Belarus, Lithuania, Latvia, Estonia, Hungary, the Czech Republic, Slovenia, Croatia and Portugal. This list is dominated by former Warsaw Pact countries where economic collapse and uncertainty following the end of communism has made many women postpone or abandon having children.

Population policies with regard to fertility

Figure 7.29 shows the officially stated position of governments on the level of the national birth rate. What is perhaps surprising is the number of countries that perceive their birth rate to be too low. However, there can be little doubt that more nations will come into this category in the future. Forming an opinion on demographic issues is one thing but establishing a policy to do something about it is much further along the line. Thus not all nations stating an opinion on population have gone as far as establishing a formal policy.

The 1930s was a period of pronatalist policies in several European countries which saw population size as an important aspect of power. Even today there are nations conscious about population size in relation to their neighbours. However, in the post-war period most countries which have tried to control fertility have sought to curtail it.

Figure 7.29 Government views on birthrate*				
	Too high	**Satisfactory**	**Too low**	**No statement**
Africa	41	11	1	2
North America	–	2	–	–
Latin America and Caribbean	17	14	2	5
Asia	19	21	6	5
Europe	1	25	16	–
Oceania	6	3	–	4
Total	84	76	25	16

Source: *1999 World Population Data Sheet*

*The officially stated position of country governments on the level of the national birth rate. Most indicators are from the UN Population Division. 'Global Population Policy Data Base, 1997'.

Mortality and life expectancy

From what has already been said about the influence of age structure it is not surprising that the lowest crude death rates are in the developing nations (Figure 7.30). The 1999 Population Data Sheet recorded a rate of only 2 per 1000 for Kuwait, Qatar and the United Arab Emirates followed by Brunei, Macao and Andorra on 3 per 1000. However, the highest death rates were also in the developing world – in the poverty belt of sub-Saharan Africa. Malawi and Niger head the mortality list of 24 per 1000 with the following countries also recording death rates of 20 per 1000 or over: Guinea-Bissau, Ethiopia, Zimbabwe, Botswana, Namibia, Uganda and Zambia. Even the impact of very high fertility cannot mask high age-specific mortality resulting in an average life expectancy in Africa as a whole of 52 years, well below any other world region. Indeed, life expectancy is the most common 'true' measure of mortality.

The infant mortality rate is generally regarded as a prime indicator of socio-economic progress. It is the most sensitive of the age-specific rates. Although it has fallen rapidly in recent years in most countries, wide spatial variations remain at the global scale. The gap between North America and Africa is very telling indeed. The highest rates recorded by the 1999 Population Data Sheet were as follows: Afghanistan and Western Sahara 150 per 1000; Malawi 137 per 1000; Guinea-Bissau and Sierra Leone 136 per 1000.

Over the world as a whole infant mortality has declined sharply during the last half century. Between 1950 and 1955 the global average was 138 per 1000, but by 1975–80 it was down to 88 per 1000 and recently it has dipped below 60 per 1000.

Mortality can of course also vary significantly within individual countries. This holds true for both developed and developing nations. Regions benefiting from a higher level of medical infrastructure and a better quality of life generally will control mortality to a greater extent than worse-off regions. Figure 7.31 shows how infant mortality has changed in Brazil from 1940 to 1995, illustrating also the differences between the richest and poorest regions of the country.

The decline in levels of mortality and the increase in life expectancy have been the most tangible rewards of development. On a global scale, 75% of the total improvement in longevity has been achieved in the 20th century. In 1900, the world average for life expectancy is estimated to have been about 30 years but by 1950–55 it had risen to 46 years. By 1980–85 it had reached a fraction under 60 years and is presently 66 years.

Figure 7.30 Mortality and life expectancy

	Death rate	Infant mortality rate	Maternal deaths per 100 000 live births	Life expectancy at birth
World	9	57	460	66
More developed	10	8	10	75
Less developed	9	62	500	64
Africa	14	88	880	52
North America	8	7	8	77
Latin America and Caribbean	6	35	180	69
Asia	8	56	410	66
Europe	11	9	10	73
Oceania	7	29	290	74

Source: *1999 World Population Data Sheet* (For developed countries nearly all data refers to 1997 or 1998 and for less developed countries to some point in the mid to late 1990s.) The data for maternal deaths is from the *1997 World Population Data Sheet*.

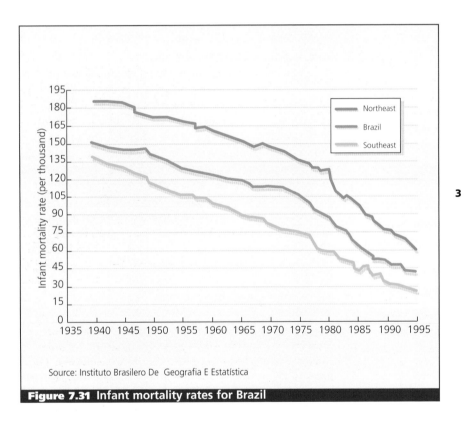

Source: Instituto Brasilero De Geografia E Estatística

Figure 7.31 Infant mortality rates for Brazil

1 (a) Define the terms (i) birth rate, (ii) total fertility rate.
(b) Why is the total fertility rate a better measure of fertility than the birth rate?
(c) Describe the variations in total fertility rate by world region.
(d) Comment on the relationship between the use of contraception and the total fertility rate.
(e) Discuss the other factors that affect the level of fertility.
2 (a) In terms of Figure 7.24, what is meant by 'number of women in cohort'?
(b) Discuss the factors that the diagram shows as influencing the number of pregnancies in a population.
(c) Comment on the factors that govern the relationship between pregnancies and live births.
(d) Examine the chains of causation illustrated by Figure 7.25.
(e) Suggest reasons why such chains of causation might not materialise in reality.
3 (a) Why does the United Nations produce different projections (scenarios) for population growth?
(b) Explain the wide variation in projected growth by world region.
(c) Describe the trends shown in Figures 7.26(a) and 7.26(b).
(d) Explain the relationship between the two graphs.
(e) Suggest why 84 governments (Figure 7.29) view their countries birth rates as being too high.
(f) Why are more and more governments concerned that the birth rates of their countries are too low?

The 20th-century fall in mortality was particularly marked after the Second World War which had provided a tremendous boost for research into tropical diseases. It is thus not surprising that the pace of mortality reduction was especially rapid in the 1950s and 1960s. Mortality reduction slowed in the 1970s as large-scale disease eradication programmes reached their limits. Thereafter, the most obvious aspects of poverty, poor nutrition and lack of clean water and sanitation have slowed improvement in much of the developing world.

Causes of death

The causes of death vary significantly between the developed and developing worlds (Figure 7.32). The cause making the largest contribution to the general reduction in mortality is respiratory disease (influenza, pneumonia, bronchitis). Such diseases on average account for about 25% of mortality change although, not surprisingly, the decline of these diseases is greater at higher mortality levels. Other infectious and parasitic diseases account for about 15% of mortality change. Respiratory TB and diarrhoeal diseases are responsible for about 10% of change each. Clearly, causes of death at any particular mortality level vary. Cancer and heart disease, which have a significant impact in the most developed nations, are much less influential in terms of mortality in poorer countries.

Warfare has had a major impact on mortality in some countries at particular times in their history. For example, in the Second World War Russian losses totalled over 20 million (Figure 7.33).

In the 1998 World Health Report, which included deaths below 50 years as a separate category for the first time, Britain came joint first in the world with Sweden. The figures show that 19 out of 20 people in Britain live to celebrate their half-century, a combination of good health-care and a low accident rate. In Britain, average life expectancies for men and women are now 74 and 82 respectively.

By the late 1990s road accidents were killing half a million people worldwide a year, many more than are killed in wars and natural disasters. During the 20th century (the first recorded pedestrian death was in 1896), motor vehicles claimed over 30 million lives. The annual death toll will rise with increasing car ownership in the developing world. The Federation of Red Cross and Red Crescent Societies calculate that road accidents cost developing countries about £32 billion a year, almost as much as all the aid they receive.

Figure 7.33 War memorial in St Petersburg

The AIDS epidemic

The continual reduction in mortality cannot be taken for granted. For example, AIDS is taking a deadly toll in some countries as it tightens its grip on the developing world. The number of people infected has risen dramatically (Figure 7.34): 63% of people worldwide currently infected live in sub-Saharan Africa. The developing world will not be able to afford the expensive drugs which appear to be cause for some optimism in the developed world. By 1997 there had been 11.7 million AIDS deaths since the beginning of the epidemic. In an alarming report published in November 1997, the UN admitted that it had 'grossly underestimated' the scale of the global AIDS epidemic. The recent increase illustrated by Figure 7.34 partly reflects a more accurate method of collecting data. The epidemic is beginning to affect life expectancy in some southern African countries. In Zambia and Zimbabwe, the infant mortality rate is up 25% and in Botswana, life expectancy has fallen to levels last seen in the 1960s.

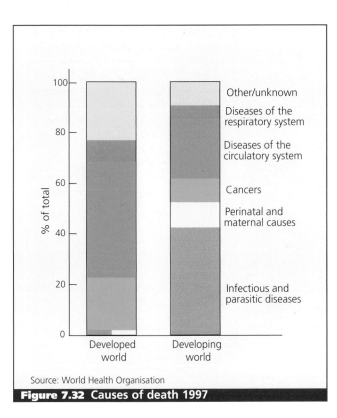

Source: World Health Organisation

Figure 7.32 Causes of death 1997

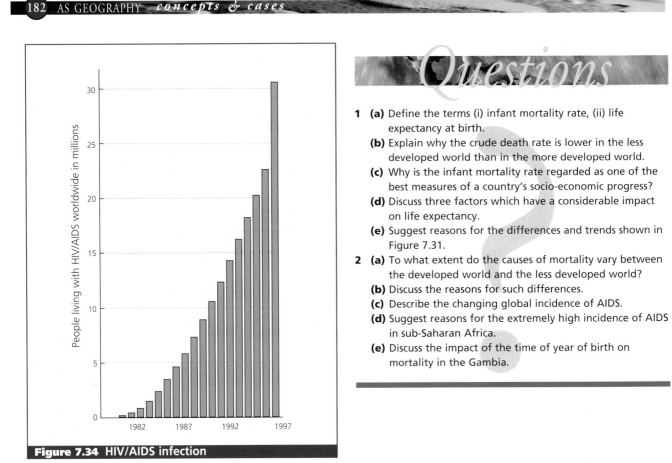

Figure 7.34 HIV/AIDS infection

1 **(a)** Define the terms (i) infant mortality rate, (ii) life expectancy at birth.
(b) Explain why the crude death rate is lower in the less developed world than in the more developed world.
(c) Why is the infant mortality rate regarded as one of the best measures of a country's socio-economic progress?
(d) Discuss three factors which have a considerable impact on life expectancy.
(e) Suggest reasons for the differences and trends shown in Figure 7.31.
2 **(a)** To what extent do the causes of mortality vary between the developed world and the less developed world?
(b) Discuss the reasons for such differences.
(c) Describe the changing global incidence of AIDS.
(d) Suggest reasons for the extremely high incidence of AIDS in sub-Saharan Africa.
(e) Discuss the impact of the time of year of birth on mortality in the Gambia.

Mortality in the Gambia

It is no surprise that mortality varies by social class. However, in the Gambia another factor is at work: the time of year of birth (Figure 7.35). By the time they reach 15, only two thirds of Leos, Virgos and Librans can expect still to be alive at 45. In contrast 93% of 15-year olds born under Pisces and Aries can expect to be alive in middle age. The precise reasons for such a contrast are not absolutely clear but seem to lie in the fact that the Gambia has two principal seasons – wet and dry. The wet season, which lasts from July to October, is the season of hunger and pregnant women may be going hungry during a crucial period of their unborn children's development. The hard work that women do in the fields can compound the problem. To make matters even worse, malaria is rife during the wet season. The researchers studying mortality in this poor African nation think that the low weight of babies born in and immediately after the wet season may be the key. It is likely that there are not enough nutrients lapping the foetus at a certain crucial period in the womb, hindering normal development of the child's immune system. This theory is tempting because many of the adulthood deaths are from infectious diseases. Thus, although this research is far from complete it does seem that poor maternal nutrition increases people's chances of developing diseases later in life.

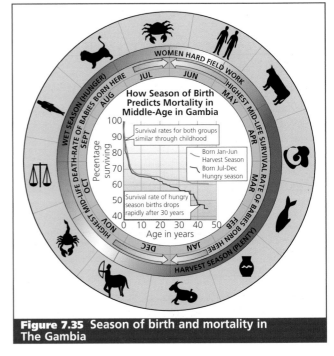

Figure 7.35 Season of birth and mortality in The Gambia

Population composition

The composition or structure of a population is the product of the processes of demographic change (fertility, mortality, migration). It is not precisely defined but generally taken to include those characteristics for which data, particularly census data, are available. The most studied aspects of population composition are age and sex. The age, sex and life expectancy of a population has implications for a country's future economic and social development. Other structures that can be studied include race, language, religion, and social/occupational groups.

Age structure

Age structure is conventionally illustrated by the use of population pyramids, as the diagrams in Figure 7.36 are known. Pyramids can be used to portray either absolute or relative data. The latter is most frequently used as it allows for easier comparison of countries of different population sizes. Each bar represents a five-year age group, apart from the uppermost bar which usually illustrates the population 85 years old and over. The male population is represented to the left of the vertical axis with females to the right.

Population pyramids change significantly in shape as a country progresses through demographic transition. The wide base in the Congo's pyramid reflects extremely high fertility. The marked decrease in width of each successive bar indicates high mortality and limited life expectancy. The base of the second pyramid (South Africa) is narrower, reflecting a considerable fall in fertility in recent decades. Falling mortality and lengthening life expectancy is reflected in the slower decrease (compared with the Congo) from bar to bar with movement up the age groups.

In the pyramid for Argentina lower fertility still is illustrated by narrowing of the base. The reduced narrowing of each successive bar indicates a further decline in mortality and greater life expectancy. The final pyramid (Japan) has a distinctly inverted base reflecting the lowest fertility yet. The width of the rest of the pyramid is a consequence of the lowest mortality and highest life expectancy of all four countries.

The problem of demographic ageing in the developed world

The populations of the developed nations are ageing at a rapid rate and, in some countries, the total population will shrink as well. These developments will put healthcare systems and public pensions, and indeed government budgets in general, under increasing pressure. Some 4% of the USA's population was 65 years of age and older in 1900. By 1995 this had risen to 12.8% and by 2030 it is likely that one in five Americans will be senior citizens. The fastest-growing segment of the population is the so-called 'oldest-old': those who are 85 years or more. It is this age group which is most likely to need expensive residential care. The situation is similar in other developed countries.

The OECD has constructed long-term fiscal projections to examine the magnitude of this problem. Some countries have made relatively good pension provision by investing wisely over a long period of time. Others, however, have more or less adopted a pay-as-you-go system, as the elderly dependent population rises. It is this latter group who will be faced with the biggest problems in the future. Very few countries are generous in looking after their elderly at present. Poverty among the elderly is a considerable problem but technological advance might provide a solution by improving living standards for everyone. If not, other less popular solutions, such as increased taxation will have to be examined.

In Britain, the ratio of pensioners to working-age people is actually due to fall between 1994 and 2021, partly because the retirement age for women is to rise from 60 to 65 years between 2010 and 2020. After that, elderly dependency will rise sharply to 2061. However, over the whole period increased elderly dependency will be partly offset by fewer young dependents (Figure 7.38). Figures 7.37 and 7.38 provide the necessary information to explain these trends. It is also of course possible that older people, who should be in better health in the future, will want to stay at work longer. Another way of reducing dependency is to allow more immigrants into the country. It will be interesting to see how far this option is exercised in the future.

Sex composition

This is usually expressed as the number of males per 100 females in a population. Male births consistently exceed female births due to a combination of biological and social reasons. For example, in terms of the latter, more couples decide to complete their family on the birth of a boy than on the birth of a girl. In the UK in 1996, 105 boys were born for every 100 girls. However, after birth the gap generally begins to narrow until eventually females outnumber males, as at every age male mortality is the higher of the two. This happens most rapidly in the poorest countries where infant mortality is markedly higher among males than females. Here the gap may be closed in less than a year. In the UK it is not until the 45–59 age group that females outnumber males. In the age group 85 and over females make up 74% of the population. Figures 7.40 and 7.41 summarise the age and gender data for the UK.

There are anomalies to the picture just presented. In countries where the position of women is markedly subordinate and deprived, the overall sex ratio may show an excess of males. Such countries often exhibit high mortality rates in childbirth. For example, in India in 1985 there were 107 males per 100 females.

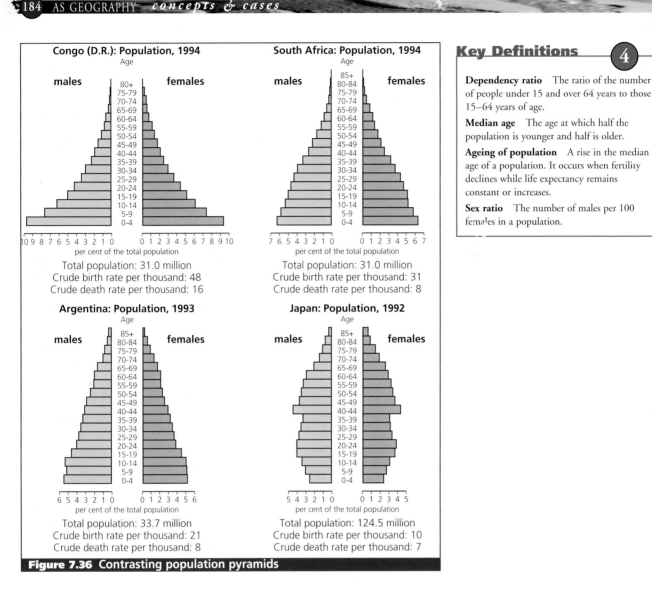

Congo (D.R.): Population, 1994
Total population: 31.0 million
Crude birth rate per thousand: 48
Crude death rate per thousand: 16

South Africa: Population, 1994
Total population: 31.0 million
Crude birth rate per thousand: 31
Crude death rate per thousand: 8

Argentina: Population, 1993
Total population: 33.7 million
Crude birth rate per thousand: 21
Crude death rate per thousand: 8

Japan: Population, 1992
Total population: 124.5 million
Crude birth rate per thousand: 10
Crude death rate per thousand: 7

Figure 7.36 Contrasting population pyramids

Key Definitions 4

Dependency ratio The ratio of the number of people under 15 and over 64 years to those 15–64 years of age.

Median age The age at which half the population is younger and half is older.

Ageing of population A rise in the median age of a population. It occurs when fertility declines while life expectancy remains constant or increases.

Sex ratio The number of males per 100 females in a population.

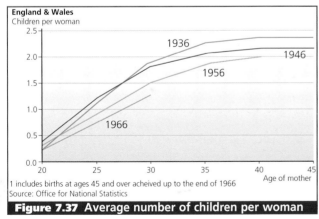

England & Wales
Children per woman
1 includes births at ages 45 and over acheived up to the end of 1966
Source: Office for National Statistics

Figure 7.37 Average number of children per woman

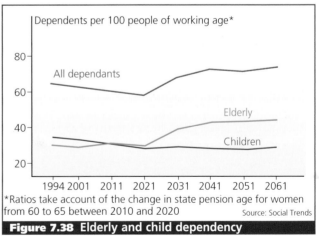

Dependents per 100 people of working age*

All dependants

Elderly

Children

*Ratios take account of the change in state pension age for women from 60 to 65 between 2010 and 2020
Source: Social Trends

Figure 7.38 Elderly and child dependency

Migration can have a substantial impact on sex composition. Historically, males have dominated the migration process causing imbalance in both the donor and receiving nations. In Australia there were 10% more males than females at the beginning of the 20th century.

The nation facing the most serious demographic problem in the world is China. While its population policies have worked in one sense a range of other problems have been created.

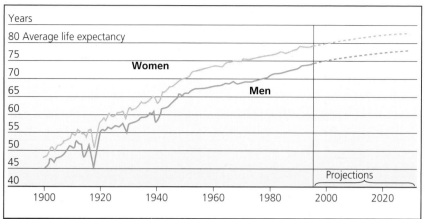

Figure 7.39 How we are living longer

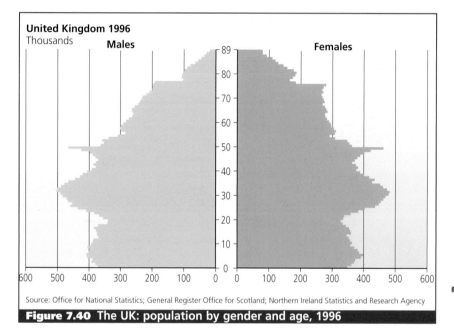

Source: Office for National Statistics; General Register Office for Scotland; Northern Ireland Statistics and Research Agency

Figure 7.40 The UK: population by gender and age, 1996

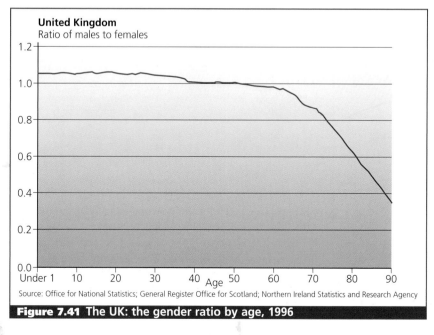

Source: Office for National Statistics; General Register Office for Scotland; Northern Ireland Statistics and Research Agency

Figure 7.41 The UK: the gender ratio by age, 1996

1 **(a)** Define the term 'dependency ratio'.
 (b) Describe the trends illustrated by Figure 7.38.
 (c) To what extent do Figures 7.37 and 7.39 explain the trends in Figure 7.38?
 (d) Discuss the implications for the UK of (i) increasing aged dependency, and (ii) falling child dependency.

2 Study the four population pyramids for China from 1953 to 1995 (Figure 7.43).
 (a) Describe the changes in the population aged under 15 from 1953 to 1995.
 (b) Outline two possible alternative explanations for these changes.
 ( ) How can such changes in youth dependency affect the economy of a country?
 (d) Describe the changes to the population aged 15 and over between 1953 and 1995.
 (e) Discuss the likely impact of such changes on the economy.
 (f) In which stages of demographic transition would you place China in 1953, 1964, 1982 and 1995? Justify your opinions.

China, a demographic time bomb

China represents an extreme example of the problems countries experiencing rapid demographic transition will have to face in the coming decades. A recent study estimates that China's population will peak at 1.6 billion in 2040, compared with 1.2 billion in 1995. It is expected to fall below 1.4 billion by 2100. These are massive demographic fluctuations, affecting 20% of the world's population, and they raise a number of serious issues concerning food, jobs, urbanisation and ageing. Take grain production. It will need to increase by 4.47 billion kg per year to keep pace with China's population growth between now and 2020. This will require significant improvements in agricultural technology and land resource management. China may have to increase its annual grain imports to 40–50 million tons. But if China lags in its agricultural development and research, it may find itself importing a lot more than that, perhaps as much as 300 million tons. That would be good news for its larger suppliers, such as the United States and Australia, but could be a disaster for poor people if prices rise too. As for employment, China's working-age population will peak at 955 million in 2020 (732 million in 1995). The massive increase in the supply of labour will be directed to the urban market. This will pose severe social and environmental problems, although it will initially provide an opportunity for investors seeking cheap labour. The working age population will decline after 2020, to about 800 million towards the end of the century. This will slow the improvement in education and skill levels among the working-age population as the rate of new

entrants declines, and that will bear down on labour productivity. The 21st century will be a period of rapid urbanisation for China. Some 90% of the population will live in towns and cities by the end of the century, compared with 37% in 1995. In absolute terms, the urban population is expected to peak at 1.2 billion in 2060 – which is broadly the same as today's total population for the entire country – compared with 450 million in 1995. This near-tripling of the urban population will have clear implications for construction and resource management.

Another demographic trend to watch out for is ageing, as China experiences a dramatic fall in fertility rates to below replacement level. In Beijing, births may already be down to 1.4–1.5 per woman. In Shanghai the ratio appears to be 0.96 births per female; in other words, more and more women are not having children at all. The upshot of all this will be a rapid ageing of the population. By 2025 the average age in China will be 40. In 1995 it was 27. Care of the elderly is clearly going to become a massive problem for the Chinese authorities, since the only social security system for most of the country's poor is their family, and in 2025 parents will have few offspring on which to depend. More and more Chinese parents only have one child, and they mostly want that child to be a boy. In fact, there is an intense social, almost peer, pressure for families to make sure their child is male. Selective abortion and female infanticide are common.

Bridal angst

China's gender ratio is unbalanced as a result. The trouble is that the men which the custom of selectivity sought to 'produce' will have to pay for their privilege by suffering a shortage of brides. And in future that means fewer children.

Indeed, by 2020, the surplus of Chinese males in their 20s will exceed the entire female population of Taiwan. The gender imbalance will get worse before it gets better; selective abortion is set to increase, not just for second and third births, but also for first births. This will cause social strains. Many young men will have to accept bachelorhood, a condition which often drives men to crime, even suicide and depression. Women on the other hand, will be scarcer, though whether that will lead to an improvement in their current low and often abused status is questionable.

To ensure that it does, anti-discrimination laws and rules on equality and female rights will have to be strengthened. The Chinese authorities could also abandon their policy of one child per family and allow family size to grow. However, without family planning and a proper revolution in recognising the rights of women, that may simply lead back to rapid population growth again. ■ M.A.

Source: OECD Observer No 217/218, Summer 1999

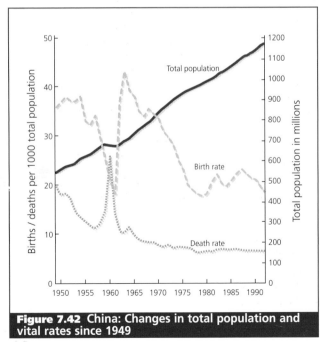

Figure 7.42 China: Changes in total population and vital rates since 1949

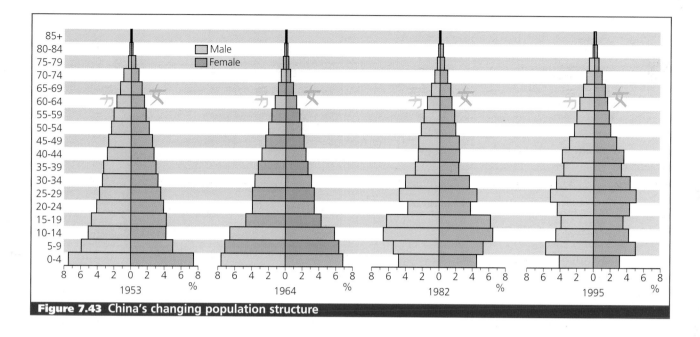

Figure 7.43 China's changing population structure

Migration

Migration is more volatile than fertility and mortality, the other two basic demographic variables. It can react very quickly indeed to changing economic, political and social circumstances. However, the desire to migrate may not be achieved if the constraints imposed on it are too great. The desire to move within a country is generally inhibited only by economic and social factors. The desire to move to another country is now usually constrained by political factors in the form of strict immigration laws.

Migration typologies

Various attempts have been made to classify migration (by duration, spatial extent, and cause). L. Smith, after distinguishing between international migration and national migration, subdivided the latter into: rural to urban, urban to rural, state to state (region to region), and local movement. The Swedish demographer Hagerstrand arrived at a similar classification of: urban to rural, rural to urban, rural to rural, and urban to urban.

In 1958, W. Peterson noted the following five migratory types: primitive, forced, impelled, free and mass.

■ The nomadic pastoralism and shifting cultivation practised by the world's most traditional societies are examples of primitive migration.

■ The abduction and transport of Africans to the Americas as slaves was the largest forced migration in history. In the 17th and 18th centuries 15 million people were shipped across the Atlantic Ocean as slaves. The forcible movement of people from parts of the former Yugoslavia under the policy of 'ethnic cleansing' is a more recent example. Migrations may also be forced by natural disasters (volcanic eruptions, floods, drought, etc.) or by environmental catastrophe such as nuclear contamination in Chernobyl.

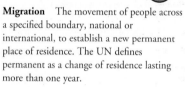

Key Definitions 5

Migration The movement of people across a specified boundary, national or international, to establish a new permanent place of residence. The UN defines permanent as a change of residence lasting more than one year.

Circulatory movements Movements with a time scale of less than a year. This includes seasonal movements which involve a semi-permanent change of residence. Daily commuting also comes into this category.

Mobility An all-embracing term which includes both migration and circulation.

Refugee A person who cannot return to his or her own country because of a well-founded fear of persecution for reasons of race, religion, nationality, political association or social grouping (UN definition).

Push factors Negative conditions at the point or origin which encourage or force people to move.

Pull factors Positive conditions at the point of destination which encourage people to move.

- Impelled migrations take place under perceived threat, either human or physical, but an element of choice lacking in forced migrations remains. Arguably the largest migration under duress in modern times occurred after the partition of India in 1947, when 7 million Muslims fled India for the new state of Pakistan and 7 million Hindus moved with equal speed in the opposite direction. Both groups were in fear of their lives but they were not forced to move by government and small minority groups remained in each country.

- The distinction between free and mass migration is one of magnitude only. The movement of Europeans to North America was the largest mass migration in history.

Within each category Petersen classed a particular migration as either innovating or conservative. In the former the objective of the move was to achieve improved living standards while in the latter the aim was just to maintain existing standards.

Figure 7.44 Japanese sector in São Paulo

International migration

International migration is a major global issue. In the past it has had a huge impact on both donor and receiving nations. In terms of the receiving countries the consequences have generally been beneficial. But today, few countries favour a large influx of outsiders for a variety of reasons.

A number of reasonably distinct periods can be recognised in terms of government attitudes to immigration:

- Prior to 1914 government controls on international migration were almost non-existent. For example, the USA allowed the entry of anybody who was not a prostitute, a convict, a lunatic, and after 1882, Chinese. Thus the obstacles to migration at the time were cost and any physical dangers that might be associated with the journey.

- Partly reflecting security concerns, migration was curtailed between 1914 and 1945. During this period many countries pursued immigration policies which would now be classed as overtly racist.

- After 1945 many European countries, facing labour shortages, encouraged migrants from abroad. In general, legislation was not repealed but interpreted very liberally. The West Indies was a major source of labour for the UK during this period. The former West Germany attracted 'guestworkers' from many countries but particularly from Turkey.

- In the 1970s slow economic growth and rising unemployment in developed countries led to a tightening of policy which, by and large, has remained in force. However, in some countries immigration did increase again in the 1980s and early 1990s, spurring the introduction of new restrictions.

A recent International Labour Organisation publication estimated that 80 million people live in countries they were not born in. Another 20 million are refugees in other countries, having fled from political oppression or natural disaster. About 1.5 million people currently emigrate each year while another million or so, on average, seek temporary asylum abroad. The USA takes as many immigrants each year as the rest of the world put together. Germany is the main receiving country in Europe.

Each receiving country has its own sources, the results of historical, economic and geographical relationships. Earlier generations of migrants form networks that help new ones to overcome legal and other obstacles. Today's tighter rules tend to confine immigration to family members of earlier 'primary' migrants.

The dynamic growth of the newly industrialised countries of Asia in recent decades has created an intricate network of labour migration flows of mutual benefit to both donor and receiving countries. However, the financial crisis that erupted in the latter part of 1997 faced millions of foreign workers with despair. Nations that welcome foreign workers in good times are keen to see the back of them during periods of recession.

Figure 7.45 Chinatown in Toronto

A recognisable recent trend has been the shift towards the migration of highly skilled workers. There are two main reasons for this. The first is that receiving countries prefer highly skilled immigrants and frequently set their immigration criteria accordingly. Another factor is the economic influence of multinational companies. These organisations, as they expand, develop their own internal markets for skilled migrants. Big companies want the freedom to shift employees from country to country as demand requires. If a truly global market for labour ever reappears (it can be argued that if it ever existed it did in the 19th century), it is likely to be for highly skilled workers only.

The costs and benefits of international migration

Much has been written by way of generalisation about the impact of international migration on donor and receiving nations, and Figure 7.46 provides a useful framework for the debate. It must be remembered, however, that each migration situation is unique. For some donor countries a migration that might provide a vital safety valve by relieving pressure on food supply and other resources may for another country drastically reduce its future prospects by skimming off the skilled element of its labour force.

The impact on donor countries can vary at the national and regional scales. Although the Republic of Ireland recorded net immigration in the 1990s this was a reversal of the previous trend. In the past, emigration from the Republic of Ireland was perceived as generally beneficial to the public

purse but its effect on the 'Gaeltacht', the isolated Irish-speaking regions of the west, has been devastating. One community that has suffered severely from this trend is Lettermore, a group of islands off Galway. Most of the young people head abroad after completing their education. In the stony fields stand abandoned cottages, their windows boarded up. Villages lie almost deserted, the remaining population fear that the decline in public and private services caused by a reduced population will trigger off a further outflow.

For receiving countries, previous immigration has almost universally been seen as beneficial but there has been a much higher level of debate about the overall impact of recent immigration. It appears that the economic benefits may take much longer to filter through than previously thought.

Figure 7.47 Vietnamese immigrants, Toronto

Figure 7.46 Migration: Short-term costs and benefits

	Benefits		Costs	
	Individual	**Social**	**Individual**	**Social**
Emigrant countries	1 Increased earning and employment opportunities 2* Training (human capital) 3* Exposure to new culture, etc.	1* Increased human capital with return migrants 2 Foreign exchange for investment via migrant remittances 3 Increased output per head due to outflow of unemployed and underemployed labour 4 Reduced pressure on public capital stock	1 Transport costs 2 Adjustment costs abroad 3 Separation from relatives and friends	1 Loss of social investment in education 2 Loss of 'cream' of domestic labour force 3* Social tensions due to raised expectations of return migrants 4* Remittances generate inflation by easing pressure on financing public sector deficits
Immigrant countries	1 (*) Cultural exposure, etc.	1 Permits growth with lower inflation 2 Increased labour force mobility and lower unit labour costs 3 Rise in output per head for indigenous workers	1 Greater labour market competition in certain sectors	1* Dependence on foreign labour in particular occupations 2 Increased demands on the public capital stock 3* Social tension with concentration of migrants in urban areas

* indicates uncertain effects

Source: *Economist*, 15 November 1988

Migration: theoretical aspects

Although each individual or family decision to migrate is unique, analyses of migration dating from the latter part of the 19th century have identified general patterns and processes. Since then a large body of theory has accumulated of which only the briefest of reviews is given here.

In 1885, E. G. Ravenstein proposed the following laws of migration based on his study of movements within the UK:

1 Most migrants move only a short distance. As distance increases from a particular place the number of migrants from that place decreases.

2 Migration occurs in a series of waves or steps. For example, the 'space' left by people moving from a market town to a city will be filled by people moving into the market town from its rural hinterland.

3 The process of dispersion (emigration) is the inverse of that of absorption (immigration) and exhibits similar features.

4 Each significant migration stream (flow) produces, to a degree, a counterstream.

5 The longer the distance travelled the greater the likelihood of the destination being a major industrial and commercial centre.

6 Town dwellers are less migratory than those living in rural areas.

7 Females are more migratory over short distances while males are more likely to move further.

Later theoretical developments in some cases reinforced aspects of Ravenstein's work but in others, modified it. G. K. Zipf, using the concept of distance decay, presented the Inverse Distance Law, stating that 'the volume of migration is inversely proportional to the distance travelled by migrants.' This is expressed mathematically as:

$$N_{ij} \propto 1/D_{ij}$$

Here N_{ij} is the number of migrants from town i to town j and D_{ij} is the distance between the two towns.

The Gravity Model went a stage further by linking distance to the relative attractiveness of two places of different population size:

$$N_{ij} = k\frac{P_i P_j}{D_{ij}^2}$$

Here N_{ij} and D_{ij} are as above while P_i and P_j are the populations of towns i and j respectively; K is a constant.

In 1940, S. A. Stouffer presented his Theory of Intervening Opportunities in which he stated 'the number of persons going a given distance is directly proportional to the number of opportunities at that distance and inversely proportional to the number of intervening opportunities.' The formula here is:

$$N_{ij} \propto O_j/O_{ij}$$

N_{ij} is the number of migrants from town i to town j, O_j the number of opportunities at j, and O_{ij} the number of opportunities between i and j.

E. S. Lee in 1966 produced a series of Principles of Migration, bringing together all aspects of migration theory at that time. Of particular note was his origin-obstacles-destination model (Figure 7.48) which emphasised the role of push and pull factors.

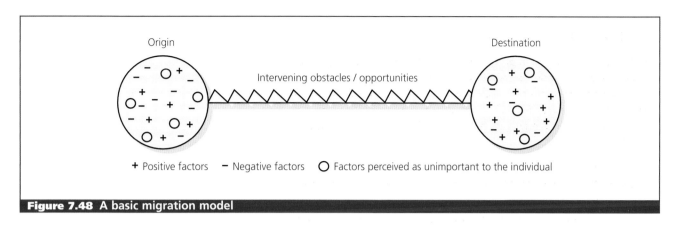

Figure 7.48 A basic migration model

Internal migration

Migration flows occur to varying degrees within all countries. On a regional level movement is invariably from poorer peripheral to richer core regions. This holds for both developed and developing countries. However, this is as far as similarities go. Within developing countries, the strong rural–urban shift that began in the 1950s is still very discernible, causing considerable problems in the rapidly growing urban areas and also at times impacting negatively on the rural donor regions. In contrast, the developed world, which underwent its period of major rural–urban migration in the 19th century, now exhibits the process of counter-urbanisation.

The developing world: rural–urban migration

In general, the intensity of rural–urban movements has been strongly related to the pace of change in both rural and urban regions. In rural areas mechanisation and other aspects of capital intensity have reduced the demand for farm labour. Although not a desirable trend from the individual employment point of view, the resultant increase in productively benefits a country as a whole. On the other hand, movement from rural areas may also be induced by a sudden natural disaster or by a more gradual deterioration of the physical environment such as the spread of desertification. The destinations of migrants from the countryside are those urban areas with the greatest range of employment and other opportunities. This issue is discussed further in the spatial focus on Brazil in Chapter 8.

China: managing migration

Recent economic change in China has unleashed a huge rural to urban migration probably on a scale never before experienced anywhere in the world. The full consequences of such a vast movement will become more apparent in the coming years.

Under China's next national five-year plan, due to begin in 2002, it is planned to move 300 million rural dwellers into 10 000 new towns. This policy is mainly the result of the government's failure to control the huge influx of poor peasants into the bulging cities. The government also hopes the new move will boost the stagnant economy by reviving consumer demand.

In the past the government has tried to restrict migration from the countryside by maintaining the system of residential permits established in the 1950s. It was feared that a deluge of job-seekers would create an unemployment crisis leading to social unrest. However, the shortcomings of the system are now widely recognised. Residential permits are unavailable for many skilled people who could make a legitimate contribution to city life while municipal councils are forced to tolerate a large illegal population of ill-educated peasants. Another problem is that skilled urbanites are often prevented from accepting jobs in other cities because they cannot gain a permit to move there.

The construction of the new towns would result in the loss of millions of hectares of arable land. The scheme would entail spending an estimated £200 billion on housing and infrastructure development. Many of the people likely to be encompassed by the scheme are far from happy. Large-scale protests are expected from those forced to leave rural homes to make way for the new towns. The environmental problems that such large-scale construction will bring are also formidable.

Population movements in Brazil

The Brazil of today is the result of population movements and resultant human activity over a long period of time. Following initial colonisation, the Northeast was the first area to experience significant Portuguese settlement. But, after the demise of the sugar economy in this region, the focus of attention switched to the Southeast. The mineral, agricultural and other resources of the Southeast made it the focus of settlement and investment and by the beginning of the 20th century it was the undisputed economic core of Brazil.

Although earlier population movements provide a fascinating area for study the focus here is on recent movement. The 1991 census found that over 3 million people lived in a different region compared to 1986 (Figure 7.49). In terms of destination the Southeast accounted for 46% of in-migrants, followed by the Centre-West with 21%. The Northeast was the region of origin for over 53% of migrants. Figure 7.49 does not, however, present the total migration picture as a certain number of people will have moved to another region but returned home within the limits of the period under consideration.

In the following five-year period recorded internal migration fell to under 2.7 million (Figure 7.50). During this time 57.4% of migrants originated from the Northeast, relatively higher than in the preceding period. The most substantial change in out-migration occurred in the South. The state of Paraná was mainly

Figure 7.51 Settlers from Northeast Brazil growing crops in the Amazon

responsible for this change in pattern, curtailing the trend established since the beginning of the 1970s by retaining people who formerly would have moved to agricultural areas in the North and Centre-west.

Despite a slight decline, north-eastern migration to the North and Centre-west regions continues to be significant. Here, most migrants head for the state of Pará, the Tocantins river, and the outskirts of Brasília and Goiânia. The data also shows a certain 'returning home' of 'Nordestinos'. Again most in-migrants headed for the Southeast, with the state of São Paulo being the main focus within this region.

Within the Northeast movement from rural areas is greatest in

Figure 7.49 Migrants per region of origin/destination, 1986–91

Region of origin	Region of destination 1991					
1986	North	Northeast	Southeast	South	Centre-west	Total
North	–	72 913	73 280	29 176	95 364	270 733
Northeast	216 995	–	917 464	21 562	198 428	1 354 449
Southeast	78 931	218 206	–	170 416	203 018	670 571
South	41 428	9410	310 580	–	148 294	509 712
Centrewest	71 162	36 304	125 607	64 110	–	297 183
TOTAL	**408 516**	**336 833**	**1 426 931**	**285 264**	**645 104**	**3 102 648**

Source: IBGE, *Censo Demográfico de 1991*

Figure 7.50 Migrants per region of origin/destination, 1991–6

Region of origin	Region of destination 1996					
1991	North	Northeast	Southeast	South	Centre-west	Total
North	–	60 965	78 955	22 978	86 628	249 526
Northeast	182 999	–	835 562	24 914	194 097	1 237 572
Southeast	54 995	262 331	–	156 372	153 307	627 005
South	20 799	17 592	176 532	–	71 852	286 775
Centrewest	60 059	43 403	128 850	50 454	–	282 766
TOTAL	**318 852**	**384 291**	**1 219 899**	**254 718**	**505 884**	**2 683 644**

Source: IBGE, *Censo Demográfico de 1996*

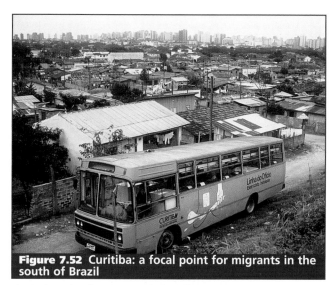

Figure 7.52 Curitiba: a focal point for migrants in the south of Brazil

Figure 7.53 Refugee scene

the Sertão, the dry interior which suffers intensely from unreliable rainfall. However, poor living standards and general lack of opportunity in the cities have also been a powerful incentive to move. Explaining the attraction of urban areas in the Southeast demands more that the 'bright lights' scenario that is still sometimes bandied about. The Todaro model presents a more realistic conceptual framework. According to this model, migrants are all too well aware that they may not find employment by moving to, say, São Paulo. However, they do so because they calculate that the probability of employment, and other factors that are important to the quality of life of the individual and the family are greater in the preferred destination than at their point of origin.

Refugees

Natural disasters, political upheaval and armed conflict has resulted in a huge number of refugees on the move in recent years. According to the United Nations Convention of 1951, refugees are people who have been forced to leave home and country because of 'a well-founded fear of persecution' on account of their race, religion, their social group or political opinions.

A report published by the United Nations High Commissioner for Refugees (UNHCR) in 1997 concluded that although the refugee population worldwide was smaller than two years previously, the overall situation had become

worse. This was primarily due to the way that conflicts were being pursued. Civilians are being targeted more than ever before, partly as the policy of 'ethnic cleansing' spreads from one zone of conflict to another. In certain areas refugees have been pushed from one country to another, fleeing from the very camps they had headed to for sanctuary.

The most desperate conditions are often faced by 'internally displaced people' – those who are forced to abandon their homes, but who don't actually cross an international border. A newspaper report in January 1998 on the six-year Islamic insurgency in Algeria was entitled 'Massacre refugees flee to cities'. The report described how terrified villagers had fled their homes, flooding public squares in the large urban areas, seeking a safe haven from attack.

The human suffering involved is all too apparent when the movement of large numbers is shown on global television. The impact on the human infrastructure of receiving regions can be overwhelming. But what is perhaps less evident at first is the environmental impact of refugees on the move. Refugees often concentrate in marginal and vulnerable environments where the potential for environmental degradation is high. Apart from immediate problems concerning sanitation and the disposal of waste, long-term environmental damage may result from deforestation associated with the need for firewood and building. Increased pressure on the land can result in serious soil degradation.

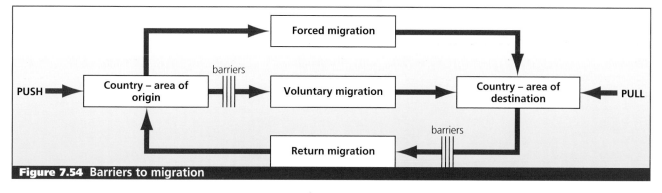

Figure 7.54 Barriers to migration

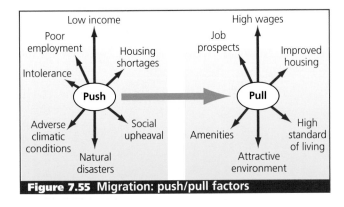

Figure 7.55 Migration: push/pull factors

Population and resources

Defining resources

Resources can be classed as either natural or human. Figure 7.58 provides a classification of natural resources. The traditional distinction is between renewable or flow resources and non-renewable or stock resources. However, the importance of aesthetic resources is being increasingly recognised. Further subdivision of the non-renewable category is particularly relevant to both fuel and non-fuel minerals. Renewable resources can be viewed as either critical or non-critical. The former are sustainable if prudent resource management is employed while the latter can be seen as everlasting.

Figure 7.56 The UK: inter-regional and international migration

	Inflow					Outflow				
	1981	1986	1991	1996	1997	1981	1986	1991	1996	1997
Inter-regional migration[1]										
North East	31	36	40	39	39	39	46	41	45	45
North West	102	112	114	125	107	122	138	123	134	117
Yorkshire and the Humber	68	79	85	91	93	73	91	85	98	100
East Midlands	77	102	90	102	108	72	85	81	94	97
West Midlands	67	87	83	91	93	79	95	88	101	104
East	121	145	122	139	145	104	128	113	121	125
London	155	183	149	168	167	187	232	202	213	222
South East	202	243	198	228	230	166	204	185	199	206
South West	108	149	121	139	144	88	103	99	110	112
England	94	116	96	111	111	93	101	112	105	115
Wales	45	55	52	55	59	42	50	47	53	54
Scotland	47	44	56	47	55	48	58	47	54	53
Northern Ireland	7	9	13	11	10	10	15	9	12	13
International migration[2,3]										
United Kingdom	153	250	267	272	285	233	213	239	216	225
North East	4	9	7	3	12	14	7	4	4	11
North West	15	26	14	17	21	24	17	19	18	9
Yorkshire and the Humber	9	13	20	13	14	14	14	14	9	12
East Midlands	5	9	12	14	15	10	5	7	9	12
West Midlands	11	11	14	24	16	13	8	18	17	10
East	9	21	26	22	19	18	20	22	13	18
London	49	78	79	92	101	55	51	67	55	68
South East	26	38	45	42	39	35	43	37	50	37
South West	10	18	18	17	21	13	16	19	14	16
England	138	223	233	245	257	196	182	207	189	192
Wales	3	8	8	7	11	11	6	6	6	4
Scotland	10	16	22	18	15	21	21	23	20	25
Northern Ireland	2	2	3	3	2	4	4	2	1	3

1 Based on patients re-registering with NHS doctors in other parts of the United Kingdom. See Notes and Definitions.

2 Subject to relatively large sampling errors where estimates are based on small numbers of contacts. See Notes and Definitions.

3 Figures for all years exclude migration to and from the Irish Republic. Data for the South East prior to 1988 include migration via the UK mainland between the Channel Islands and the Isle of Man and the rest of the world. Adjustment of the figures shown are required for 'visitors switchers' and migration to and from the Irish Republic. See Notes and Definitions.

Sources: National Health Service Central Register and International Passenger Survey, Office for National Statistics; General Register Office for Scotland; Northern Ireland Statistics and Research Agency

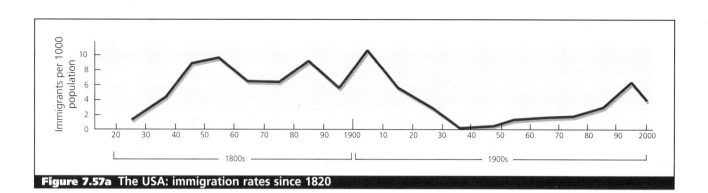

Figure 7.57a The USA: immigration rates since 1820

Figure 7.57b Sources of immigration into the USA

	Immigrants (1000)			
	1820–1979[a]	1971–80[b]	1981–90[b]	1991–95[b]
Europe	36 267	801	706	728
Asia	3038	1634	2817	1634
The Americas	9248	1929	3581	2682
Africa	142	92	192	160
Total (including others)	49 124	4493	7338	5230

[a] by country of last permanent residence
[b] by country of birth

Figure 7.57c America's dividend. Net present value In US$ of gains/losses per immigrant on government finances

Education level	Immigrants	Their descendents	Total
Educated below high school	−89 000	76 000	−13 000
Educated at high school	−31 000	82 000	51 000
Educated above high school	105 000	93 000	198 000
Average	−3 000	83 000	80 000

Source: *Economist*, 29 November 1997

Key Definitions ⑥

Resource   Any aspect of the environment which can be used to meet human needs.

Resource depletion   The consumption of non-renewable, finite resources which will eventually lead to their exhaustion.

Resource management   The control of exploitation and use of resources in relation to economic and environmental costs.

Sustainable development   A carefully calculated system of resource management which ensures that the current level of exploitation does not compromise the ability of future generations to meet their own needs.

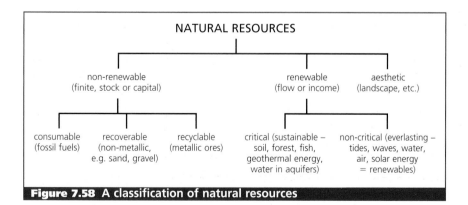

Figure 7.58 A classification of natural resources

Ecological footprints

In the USA it takes 12.2 acres to supply the average person's basic needs (Figure 7.61); in the Netherlands 8 acres; in India, 1 acre. The Dutch ecological footprint covers 15 times the area of the Netherlands, whereas India's footprint exceeds its area by only about 35%. Clearly, the Dutch can only realise such a high rate of resource use by substantial importing. Most alarmingly, if the entire planet lived like Americans, it would take three of planet earth to support the present global population.

Early views on the relationship between population and resources

The relationship between population and resources has concerned those with an understanding of the subject for thousands of years. However, the assumptions made by earlier writers were based on very limited evidence as few statistical records existed more than two centuries ago. These are just some of the views that have been expressed through time:

■ Confucius, the ancient Chinese philosopher, said that excessive population growth reduced output per worker, depressed the level of living and produced strife. He discussed the concept of optimum numbers, arguing that an ideal proportion between land and numbers existed and any major deviation from this created poverty. When imbalance occurred he believed the government should move people from overpopulated to underpopulated areas.

■ Medieval writers generally favoured a high birth rate because of the constant threats of sudden depopulation through wars, famine and epidemics.

■ Thomas Malthus, who was concerned that population was rising too rapidly, wrote his first essay in 1798 entitled 'An essay on the principle of population as it affects the future improvement of society'.

■ In the nineteenth century Karl Marx made the most powerful attack of any on the work of Malthus, stating 'an abstract law of population exists for plants and animals only.' Socialist and Marxist writers believed that any population problems would be solved through the reorganisation of society.

■ In the twentieth century demographic debate was based on the availability of increasingly sophisticated data in terms of both depth and breadth of coverage. Concern about the 'population explosion' developed in the 1960s.

The ideas of Thomas Malthus

The Rev. Malthus (1766–1834) wrote that the crux of the population problem was 'the existence of a tendency in mankind to increase, if unchecked, beyond the possibility of an adequate supply of food in a limited territory'. Malthus thought that an increased food supply was achieved mainly by bringing more land into arable production, and maintained that while the supply of food could, at best, only be increased by a constant amount in arithmetical progression (1 – 2 – 3 – 4 – 5 – 6), the human population tends to increase in geometrical progression (1 – 2 – 4 – 8 – 16 –32), multiplying itself by a constant amount each time. In time, population would outstrip food supply until a catastrophe occurred in the form of famine, disease or war. The latter would occur as human groups fought over increasingly scarce resources. These limiting factors maintained a balance between population and resources in the long term. In a later paper Malthus placed significant emphasis on 'moral restraint' as an important factor in controlling population.

Questions

1 Study Figure 7.54.
   (a) Define the terms (i) forced migration, (ii) voluntary migration, (iii) return migration.
   (b) Briefly describe one example of each of these types of migration.
   (c) How have the barriers to voluntary migration changed over time?
   (d) Discuss two barriers to return migration.
   (e) Why do the demographic characteristics of forced migrants vary from those of voluntary migrants?
   (f) How can voluntary migration benefit (i) a country of origin, (ii) a country of destination?

2 Study Figure 7.55.
   (a) Define the terms (i) push factor, (ii) pull factor.
   (b) Explain, with reference to examples you have studied two of the push factors identified in Figure 7.55.
   (c) Exemplify the influence of pull factors in one country with a high rate of immigration.
   (d) How can the balance of push and pull factors change over time?

3 Study Figure 7.56.
   (a) Describe the pattern of interregional migration in the UK in 1997.
   (b) To what extent had the pattern changed since 1981?
   (c) Suggest reasons for the pattern of interregional movement within the UK.
   (d) Describe the regional differences in international migration shown in Figure 7.56.
   (e) Discuss the factors responsible for such significant variations.

4 Study Figure 7.57.
   (a) Describe the changes in the USA's immigration rate since 1820.
   (b) Discuss possible reasons for these changes.
   (c) Comment on the changing origin of immigration into the USA.
   (d) What does the data in Figure 7.57(c) show about the economic impact of immigration on the USA?
   (e) In 1995, just four of the 50 states, California, New York, Florida and Texas, accounted for 55% of all immigrants. Suggest why immigration is so spatially selective in the USA.

Figure 7.59 Harvesting forest resources, Algonquin Provincial Park, Ontario, Canada

Figure 7.60 Resources from the sea: small fishing boats, Tunisia

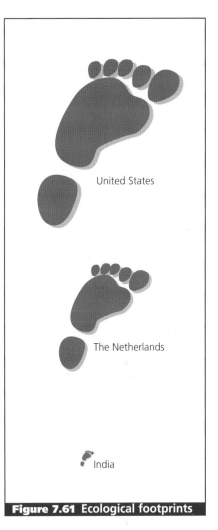

United States

The Netherlands

India

Figure 7.61 Ecological footprints

Clearly Malthus was influenced by events in and before the 18th century and could not have foreseen the great advances that were to unfold in the following two centuries which have allowed population to grow at unprecedented rates alongside a huge rise in the exploitation and use of resources.

Even so, despite technological progress and the increasing production of resources more recent writers have also voiced concern about the population/resource balance. In his book, *The Population Bomb*, published in 1969, Paul Erlich mirrored the concerns of Malthus. He predicted that in the coming decades natural resources such as oil and coal would run out, and that food production would fail to keep pace with population growth.

Optimum population: theory and practice

The idea of optimum population has been mainly understood in an economic sense (Figure 7.62(a)). At first, an increasing population allows for a fuller exploitation of a country's resource base causing living standards to rise. However, beyond a certain level rising numbers place increasing pressure on resources and living standards begin to decline.

In history, the power of the ruling elite or government has often had the edge over individual welfare as an aim. Here power represents a collective aim which may or may not take the form of armament (Figure 7.62(b)). The power optimum is obviously smaller than the maximum population but is always higher than the economic optimum.

Key Definitions ⑦

Underpopulation   When there are too few people in an area to use the resources available efficiently.

Overpopulation   When there are too many people in an area relative to the resources and the level of technology available.

Optimum population   The one that achieves a given aim in the most satisfactory way.

The economic optimum   The level of population which, through the production of goods and services, provides the highest average standard of living.

The power optimum   The population which achieves the greatest level of production above that which is required for its own subsistence.

Optimum rhythm of growth   The level of population growth that best utilises the resources and technology available. Improvements in the resource situation or/and technology are paralleled by more rapid population growth.

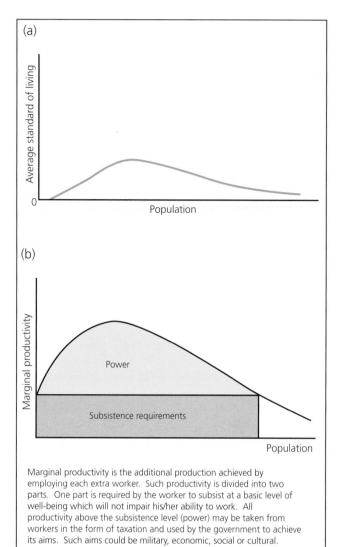

Figure 7.62 The power and economic optimums

(a) Average standard of living / Population

(b) Marginal productivity / Population — Power / Subsistence requirements

Marginal productivity is the additional production achieved by employing each extra worker. Such productivity is divided into two parts. One part is required by the worker to subsist at a basic level of well-being which will not impair his/her ability to work. All productivity above the subsistence level (power) may be taken from workers in the form of taxation and used by the government to achieve its aims. Such aims could be military, economic, social or cultural.

Questions

1 **(a)** Exemplify the distinction between renewable and non-renewable resources.
(b) Why are some renewable resources considered critical while others are said to be non-critical?
(c) What do you understand by the term 'ecological footprint'?
(d) Why do the ecological footprints of countries vary to such an extent?
(e) Suggest ways in which the ecological footprint of a country could be reduced.

2 **(a)** Define the terms (i) economic optimum, (ii) power optimum.
(b) On a copy of Figure 7.62(a) label the position of the economic optimum population, and on a copy of Figure 7.62(b) label the position of the power optimum population.
(c) Briefly discuss another way of considering the optimum population.
(d) On Figure 7.63, suggest why the population initially started to increase.
(e) Suggest reasons for the changing relationship between population and carrying capacity.
(f) Why do question marks appear after Time B?

Thus when a population increases and other factors are constant the following successive positions can be recognised:

■ the minimum population

■ the population resulting in the maximum marginal productivity (fastest rate of growth in total output)

■ the economic optimum population

■ the power optimum population

■ the maximum population.

There is no historical example of a stationary population having achieved appreciable economic progress, although this may not be so in the future. In the past it is not coincidental that periods of rapid population growth paralleled eras of technological advance that have increased the carrying capacity of countries and regions. Thus, we are led from the idea of optimum population as a static concept to the dynamic concept of optimum rhythm of growth (Figure 7.63) whereby population growth responds to substantial technological advances. For example, Abbé Raynal (*Révolution de l'Amerique*, 1781) said of the United States 'If ten million men ever manage to support themselves in these provinces it will be a great deal'. Yet today the population of the USA is over 260 million and hardly anyone would consider the country to be overpopulated.

The most obvious examples of population pressure are in the developing world, but the question here is – are these cases of absolute overpopulation or the results of underdevelopment that can be rectified by adopting remedial strategies over time?

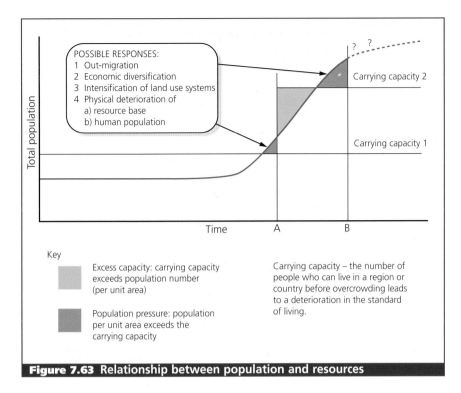

POSSIBLE RESPONSES:
1 Out-migration
2 Economic diversification
3 Intensification of land use systems
4 Physical deterioration of
   a) resource base
   b) human population

Carrying capacity 2

Carrying capacity 1

Total population

Time   A   B

Key

Excess capacity: carrying capacity exceeds population number (per unit area)

Population pressure: population per unit area exceeds the carrying capacity

Carrying capacity – the number of people who can live in a region or country before overcrowding leads to a deterioration in the standard of living.

Figure 7.63 Relationship between population and resources

Malaysia: setting a target population

The concept of an optimum population is one that has generally fallen out of favour, and few countries have adopted a particular population size as a target. Malaysia appears to be unique in recent decades in aiming to achieve an almost five-fold increase in population, and to set a target date for reaching this population as far ahead as 115 years (Figure 7.64). The Prime Minister announced in 1982 that Malaysia should aim for an ultimate population of 70 million. Child tax relief allowances were revised to offer greater reductions for every subsequent child instead of the normal downward scale, and maternity benefits were extended up to the fifth child. The Prime Minister exhorted Malaysians to 'go for five' children, and other politicians have lauded the benefits of large families and early marriage. Malaysia has for long been conscious of the fact that most neighbouring countries have much larger populations. The population of Malaysia was estimated at 21 million in mid-1997, compared to 204 million for Indonesia, 60 million for Thailand, 73 million for the Philippines, and 75 million for Vietnam.

   While there was much support for the plan there was also vociferous opposition. The arguments for and against are summarised in Figure 7.65.

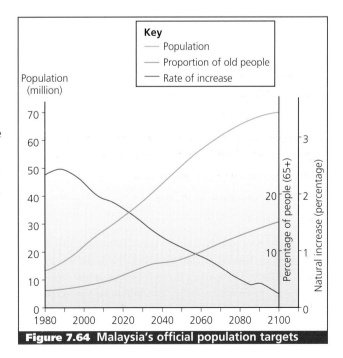

Key
— Population
— Proportion of old people
— Rate of increase

Figure 7.64 Malaysia's official population targets

Figure 7.65 Arguments for and against significant population growth

For – The Malaysian Government	Against – Friends of the Earth, Malaysia
■ Malaysia has achieve impressive economic growth in recent decades. ■ A larger population will provide a bigger domestic market, enabling greater economics of scale and diversification of the manufacturing base. ■ Malaysia would be a more attractive location for foreign multinationals. ■ Malaysia's population would not appear as small in a regional context as it does today.	■ Many of Malaysia's natural resources are being rapidly depleted. ■ Unemployment (6.2 per cent in 1985) is a problem even with the present level of population. ■ Medical and health systems are deteriorating due to financial cutbacks. ■ It will cost $4 billion to provide water to every household in the country between 1995 and 2005. ■ It will cost $1.5 billion to provide all villages with electricity. ■ Food production for domestic consumption has declined in recent years, as emphasis is placed on cash crops which can earn foreign exchange.

Source: *People* Vol. 12 No. 4 1985

Population and food supply

World food summits

Forecasts of famine tend to appear every few decades or so. In 1974, a world food summit held in Rome met against a background of rapidly rising food prices and a high rate of global population growth. The major concern was that the surge in population would overwhelm humankind's ability to produce food in the early 21st century. The possibility that the predictions of Thomas Malthus were going to come true was very real in the opinion of some experts. These neo-Malthusians began to issue dire warnings.

The next world food summit, again hosted by Rome, was held in 1996. It too met against a background of rising prices and falling stocks (Figures 7.66 and 7.67). But new concerns, unknown in 1974 had appeared. Global warming threatened to reduce the productivity of substantial areas of land and many scientists were worried about the long-term consequences of genetic engineering. The errors of past strategies had also become all too apparent in many parts of the world. Across Asia vast areas of irrigated land had become waterlogged and rendered almost totally unproductive. In many regions the intensive use of chemical fertilisers was taking a heavy toll in terms of both runoff into rivers and lakes and the re-emergence of crop diseases, such as the virulent fungus responsible for the Irish potato famine, having developed resistance to traditional farm chemicals.

Lester Brown, president of the environmental organisation the Worldwatch Institute and seen by some as the world's leading modern Malthusian, argued that the world was entering an era of food scarcity. He contended that growing demand for grain, from China in particular, could soon overwhelm the capacity of all the world's grain-producing countries. However, the modern anti-Malthusians counselled

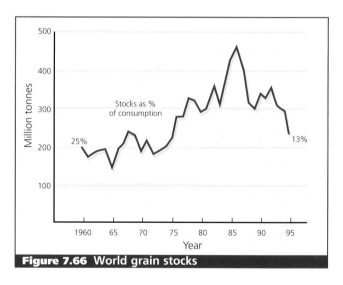

Figure 7.66 World grain stocks

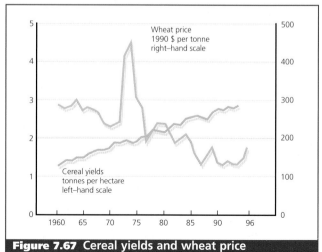

Figure 7.67 Cereal yields and wheat price

against panicking over very short-term fluctuations, pointing in particular to the way in which food production has grown significantly faster than population in the second half of the 20th century (Figure 7.69). Figure 7.70 summarises the opposing views of the neo-Malthusians and the resource optimists such as Boserup who suggested that in a pre-industrial society an increase in population stimulated a change in agricultural techniques resulting in higher food production.

Although it is true that the world grain harvest tripled between 1950 and 1990, production has tapered off significantly in the 1990s. The neo-Malthusians stress the decline in the amount of grain area per person from a global average of 0.23 hectares in 1950 to 0.13 hectares in the mid-1990s. This is a consequence of (a) population growth and (b) loss of cropland due to urban expansion, soil degradation and a number of other factors. Population growth is concentrated in those developing countries least able to cope with the resource and food consequences of such growth.

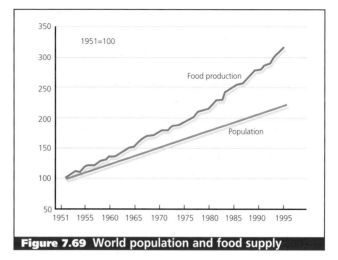

Figure 7.69 World population and food supply

Three agricultural worlds

In terms of agricultural production the nations of the world can be placed into three groups:

- the haves – Europe, North America, Australia and New Zealand – have sufficient cropland to meet most of their food needs and efficient farm production systems enabling the production of more food from the same amount of land.

- the rich have-nots – a mixed group of countries that includes land-short Japan and Singapore, along with rapidly developing countries such as Indonesia, China, and Saudi Arabia. These countries are unable to grow enough food for their populations but can afford to purchase imports to make up the deficit.

- the poor have-nots – consisting of the majority of the developing world. These countries with over 3 billion people are unable to produce enough food for their populations and cannot afford the imports to make up the deficit.

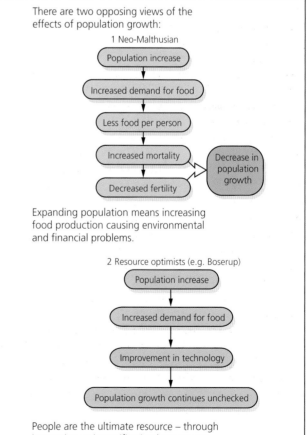

There are two opposing views of the effects of population growth:

1 Neo-Malthusian

Population increase → Increased demand for food → Less food per person → Increased mortality / Decreased fertility → Decrease in population growth

Expanding population means increasing food production causing environmental and financial problems.

2 Resource optimists (e.g. Boserup)

Population increase → Increased demand for food → Improvement in technology → Population growth continues unchecked

People are the ultimate resource – through innovation or intensification humans can respond to increased numbers.

Figure 7.70 Views of population growth

Figure 7.68 Market gardening near Brasília

Figure 7.71 Green Revolution scene - a woman in a padi field in India

Key Definitions

Malthusians (or neo-Malthusians) The pessimistic lobby who fear that population growth will outstrip resources leading to the consequences predicted by Thomas Malthus.

Anti-Malthusians The optimists who argue that either population growth will slow well before the limits of resources are reached or that the ingenuity of humankind will solve resource problems when they arise.

Green Revolution The introduction of high yielding seeds and modern agricultural techniques in developing countries.

Sustainable agriculture Agricultural systems, emphasising biological relationships and natural processes that maintain soil fertility thus allowing current levels of farm production to continue indefinitely.

The Green Revolution: a reassessment

The package of agricultural improvements generally known as the Green Revolution was seen as the answer to the food problem in many parts of the developing world. India was one of the first countries to benefit when a high-yielding variety seed programme (HVP) commenced in 1966–67. In terms of production it was a turning point for Indian agriculture which had virtually reached stagnation. The HVP introduced new hybrid varieties of five cereals, wheat, rice, maize, sorghum and millet. All were drought resistant (with the exception of rice), were very responsive to the application of fertilisers, and had a shorter growing season than the traditional varieties they replaced. Although the benefits of the Green Revolution are clear, serious criticisms have also been made. The two sides of the story can be summarised as follows:

Advantages:	Disadvantages:
■ yields are twice to four times greater than traditional varieties	■ high inputs of fertiliser and pesticide are required to optimise production. This is costly in both economic and environmental terms. In some areas rural indebtedness has risen sharply
■ the shorter growing season has allowed the introduction of an extra crop in some areas	
■ farming incomes have increased allowing the purchase of machinery, better seeds, fertilisers and pesticides	■ HYVs require more weed control and are often more susceptible to pests and disease
■ the diet of rural communities is now more varied	■ middle- and higher-income farmers have often benefited much more than the majority on low incomes, thus widening the income gap in rural communities. Increased rural to urban migration has often been the result
■ local infrastructure has been upgraded to accommodate a stronger market approach	
■ employment has been created in industries supplying farms with inputs	■ mechanisation has increased rural unemployment
■ higher returns have justified a significant increase in irrigation.	■ some HYVs have an inferior taste
	■ the problem of salinisation has increased along with the expansion of the irrigated area.

Food and sustainable farming in Africa

As Figure 7.72 shows, the only world region to record falling food production is Africa. In Africa the downward trend has been continuous, decade by decade, since the 1960s. In 1990 the food consumption of the average sub-Saharan African was 2053 Kcal, just under three-quarters of the consumption of an average person in the developed world. According to recent projections by the International Food Policy Research Institute (IFPRI), Africans can expect their food availability to increase only slightly in the next quarter century (Figure 7.73).

The farming and grazing lands which comprise Africa's arid, semi-arid and dry sub-humid zones cover about one-fifth of the continent and are inhabited by a quarter of its population. Although they may be considered marginal in one sense they also provide major crops such as cotton, groundnuts, millet and sorghum. These lands are also grazed by millions of livestock. Dryland soils are low in humus, poor in nutrients, and vulnerable to wind and water erosion, particularly at the start of the rains. To make the situation even worse, these parts of Africa have experienced reduced precipitation in recent decades. It is not clear how global warming will affect the drylands in the future. However, even though rainfall may well increase, it is likely to fall in fewer more intense storms, with much lost through higher levels of evaporation and runoff.

During the 1960s and 1970s mechanisation, high yield varieties and capital improvements (such as irrigation schemes) were used in the push to increase production. The poor performance of such capital investments, compounded by frequent droughts, necessitated a radical reassessment. Now the search is for more sustainable systems to assure a harvest rather than spectacular yields, utilising the knowledge of local farmers rather than relying on imported technologies. A recent article entitled 'Sustaining Africa's Soil' (*People and Planet*, Vol. 7, No. 1) quoted the following examples:

■ In the north-west of Burkina Faso, Aly Ouedraogo from the village of Gourcy has reclaimed many hectares of barren land over a 15-year period, turning a hard pan, gravel surface into productive fields, able to produce dense stands of sorghum. The techniques employed were

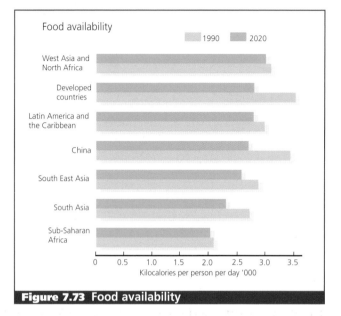

Figure 7.73 Food availability

Figure 7.74 Ploughing in Gambia

Figure 7.72 Index of regional food production per head (1979–81 = 100)				
Region	**Average for**			
	1960s	**1970s**	**1980s**	**early 1990s**
World	92.8	97.4	102.6	105.0
Developing countries	91.2	95.8	106.4	115.4
Asia	86.2	93.8	109.2	122.6
Latin America	95.3	96.2	101.9	106.2
Africa	115.8	109.0	96.6	95.2

Figure 7.75 Major famines and droughts in Africa since the late 1960s

Years	Sub-regions and countries affected
1968–74	Djibouti, Nigeria (Biafra), Somalia, Sudano-Sahelian zone of West Africa and in particular Burkina Faso, Chad, Mali, Mauritania, Senegal
1972–4	Ethiopia, Nigeria (Hausaland)
1973–5	Niger
1974–6	Angola
1977–8	Zaire (Bas-Fleuve)
1980–2	Kenya (Turkana), Uganda (Karamoja)
1982–5	Angola, Burkina Faso, Ethiopia, Malawi, Mali, Mauritania, Mozambique, Niger, Tanzania, Uganda
1984–5	Chad, Mozambique, Sudan
1987	Ethiopia, Mozambique
1988	Somalia, Sudan
1991	Ethiopia, Liberia, Mozambique, Sudan
1992	Eritrea, Ethiopia, Liberia, Mauritania, Mozambique, Sierra Leone, Somalia, Zaire and Southern Africa sub-region
1993	Angola, Burundi, Chad, Liberia, Rwanda, Ethiopia, Eritrea, Kenya, Somalia, Sudan, Zaire and Southern Africa sub-region and in particular Angola, Mozambique, Tanzania
1994	Southern Africa sub-region

Source: Human Development Report 1996

stone lines reinforced by the seeding of perennial grasses and fruit trees, while the areas between the stone lines have been dug with planting pits. The pits trap runoff, drifting soil, dried leaves and other matter which termites break down. At harvest time, Aly leaves at least 20 cm of stubble to slow wind speeds and provide further material on which termites can feed. A compost and manure pit on the edge of his field provides a handful of fertiliser for each pit. A careful watch is kept for straying livestock.

■ In the highlands of northern Shewa, Ethiopia, farmers have traditionally used drainage ditches across their fields as a means of protecting land from being washed away when rains are heavy, and to reduce surface runoff. This contrasts markedly with the huge terracing programmes carried out in the Ethiopian highlands, supported by government and donor agencies in the 1980s. However, having played no role in planning and designing these systems, local farmers felt no sense of ownership or interest in these structures which now lie abandoned.

■ The Dogon people of eastern Mali use a range of soil and water conservation techniques to produce high yields in a very difficult environment. Every square centimetre of soil is used. The crops are carefully selected to make the most of each site. Drier fields grow the fine-grained grassy cereal fonio; damp spots grow rice; patches near water are for irrigated vegetables. On bare, baking rock ledges, crevices only 20 cm wide, as long as they have a little soil, are planted with millet and beans. Narrow hollows among huge boulders are planted with millet, thatching grass and gourds. When weeding occurs, the waste is raked into mounds 15–20 cm high in between the stalks, creating a trellis of mounds and hollows which slows runoff, allowing it to filter slowly

without eroding the soil. The mounds, rich in humus from decaying weeds, are used to plant next year's seed. Along the downhill edge of fields, rocks are piled in long lines level with the contour.

The food situation in many parts of Africa is critical. The continent has been affected by famine and drought on a regular basis (Figure 7.75). However, the difficulties of the physical environment alone are far from fully responsible for the food problem. Human conflict and the host of pathologies that trail in its wake have dogged the continent in the latter part of the 20th century. Sustainability therefore is not just about an approach to natural resource exploitation. This cannot happen unless peaceful human interaction is sustained first.

Questions

1 (a) Describe and explain the Neo-Malthusian view of population growth.
   (b) Outline the view of the resource optimists.
   (c) Describe the relationship between food and population shown in Figure 7.69.
   (d) Suggest reasons why food production could possibly decline at some time in the future.
   (e) Discuss how a resource other than food supply is being pressurised by population growth.
2 (a) Examine the trends in Figure 7.72.
   (b) Describe the differences in food availability illustrated by Figure 7.73.
   (c) What is sustainable agriculture?
   (d) How has sustainable agriculture been put into practice in Africa?
   (e) Conduct your own research to update Figure 7.75.

8 urban and rural *environments*

Urban and rural: the problem of definition

A number of criteria have been used to distinguish between urban and rural (Figure 8.1), of which the most important are:

- **Population size** There are no standard international definitions of urban and rural, and the official minimum population size of an urban area can vary considerably from country to country. In Norway it is 'localities of 200 or more', while in Malaysia it is 'areas with a population of 10 000 or more'. The rural–urban divide for most countries falls between 1500 and 5000. In England the population ranges shown in Figure 8.2 provide a reasonable classification.

- **Land use** The density of buildings in rural settlements is low and they are rarely above two floors in height. Urban areas have much higher residential densities and exhibit a much greater variety of land use.

- **Employment** Traditionally, rural settlements have been dominated by farmers, farm workers and those in other primary occupations. While this is still generally true of the developing world, many villages in the developed world now contain very few people associated with primary activities. Instead they commute to work in urban areas.

Figure 8.2 A possible classification of settlements in Britain by population size

Hamlet	2–50 in more than one household
Village	50–2000
Small town	2000–10 000
Large town	10 000–100 000
City	100 000–1 000 000
Conurbation	1 000 000–10 000 000

Figure 8.3 Rural settlement of Paranapiacaba, Brazil

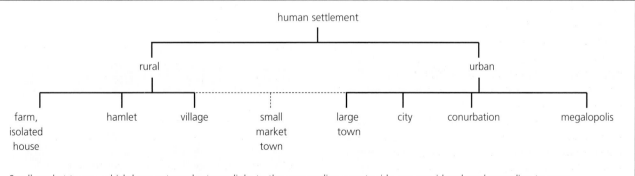

Small market towns, which have extremely strong links to the surrounding countryside, are considered rural according to some classifications but urban according to others.

Figure 8.1 The hierarchy of settlement

Figure 8.4 Urban expansion in Istanbul

- Functions   Rural settlement is characterised by its small number and low level of services. In developed nations the decline of service provision in rural areas is a major issue.

- Social characteristics   In general, rural communities have a longer history than their urban counterparts and are more close-knit. However, in remote rural areas depopulation has resulted in a high level of aged dependency.

In some countries definitions incorporate two measures of rurality. For example, in the Netherlands rural settlements are municipalities with a population of less than 2000 but with more than 20% of their economically active population engaged in agriculture.

Rural settlements form an essential part of the human landscape. In the past rural society was perceived to be

Figure 8.5 Principal characteristics of traditional rural society

1. Close-knit community with everybody knowing and interacting with everyone else.
2. Considerable homogeneity in social traits: language, beliefs, opinions, mores, and patterns of behaviour.
3. Family ties, particularly those of the extended family, are much stronger than in urban society.
4. Religion is given more importance than in urban society.
5. Class differences are less pronounced than in urban society. Although occupational differentiation does exist, it is not as pronounced as in towns and cities. Also the small settlement size results in much greater mixing which in turn weakens the effects of social differentiation.
6. There is less mobility than in urban society, both in a spatial sense (people do not move house so frequently) and in a social sense (it is more difficult for a farm labourer to become a farmer or farm manager than for a factory worker to become a manager).

Source: *The Geography of Rural Resources* by C. Bull, P. Daniel and M. Hopkinson, Oliver & Boyd, 1984

distinctly different from urban society. The characteristics upon which this idea was based are shown in Figure 8.5. However, rapid rural change over the past 40 years or so in Britain and other developed countries has seen the idea of a rural–urban divide superseded by the notion of a rural–urban continuum. The latter is a wide spectrum which runs from the most remote type of rural settlement to the most highly urbanised. A number of the intermediate positions exhibit both rural and urban characteristics. Paul Cloke (1979) used 16 variables (including population density, land use and remoteness) to produce an 'index of rurality' for England and Wales (Figure 8.6). Urban areas now make substantial demands on the countryside, the evidence of which can be found in even the remotest of areas (Figure 8.7).

The development of urbanisation

During the 1990s the world's population became more urban than rural. It seems likely that 1996 was the watershed year but national variations in the quality of data and the way in which urban areas are defined make it difficult to be precise. It has taken about 8000 years for half of the world's population to become urban but it is predicted that it will take less than 80 years for urbanisation to encompass most of the remainder.

The first cities

Gordon Childe used the term 'urban revolution' to describe the change in society marked by the emergence of the first cities some 5500 years ago. The areas which first witnessed this profound social-economic change were (a) Mesopotamia – the valleys of the Tigris and Euphrates rivers, (b) the lower Nile valley and (c) the plains of the River Indus. Later, urban civilisations developed around the Mediterranean, in the Yellow River valley of China, in South East Asia and in the Americas.

These early urban communities appeared when the material foundations of life were such as to create a surplus of food greater than that required by those who produced it and when the means were also available to concentrate this surplus at particular locations. In each case, the volume of surplus production imposed a ceiling on urban development. The most talented in society were freed to perform specialised functions which the newly acquired agricultural techniques, based on irrigation, not only made possible but even demanded for their full application. Such people gathered in clusters to organise and discharge these special services. Compared with anything that had gone before these new settlements were distinctive in size, function and appearance. However, the basis of these earliest urban centres was

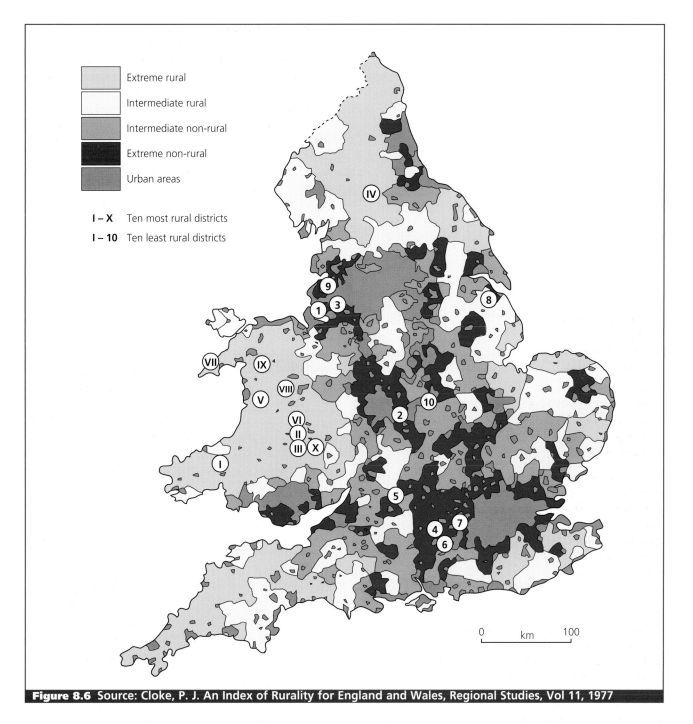

Figure 8.6 Source: Cloke, P. J. An Index of Rurality for England and Wales, Regional Studies, Vol 11, 1977

relatively local. For example, the population of Sumerian cities ranged from 7000 to 20 000. The catalyst for this period of rapid change, which occurred about 8000 BC, was when sedentary agriculture, based on the domestication of animals and cereal farming, steadily replaced a nomadic way of life. As farming advanced irrigation techniques were developed. Other major advances which followed included the ox-drawn plough, the wheeled cart, the sailing boat and metallurgy. However, arguably the most important development was the invention of writing about 4000 BC; for it was in the millennium after this that some of the villages

on the alluvial plains between the Tigris and Euphrates rivers increased in size and changed in function so as to merit the classification of urban. Childe and others also stress the importance of social processes. A level of social development had been achieved that allowed large communities to be socially viable and stable. In this context religious activity, centred around the construction of temples, was undoubtedly an important force in the process of urbanisation.

Trading centres began to develop considerably later than the first cities. The Minoan civilisation cities of Knossos (Figure 8.8) and Phaistos, which flourished in Crete during

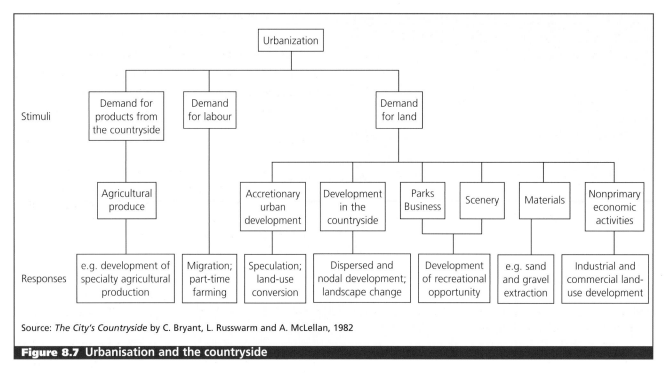

Source: *The City's Countryside* by C. Bryant, L. Russwarm and A. McLellan, 1982

Figure 8.7 Urbanisation and the countryside

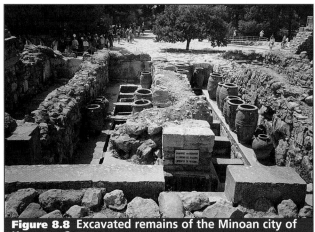
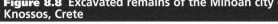

Figure 8.8 Excavated remains of the Minoan city of Knossos, Crete

Figure 8.9 Roman amphitheatre at El Gem, Tunisia

the first half of the second millennium BC, derived their wealth from maritime trade. Later it was the turn of the Greeks and then the Romans to develop urban and trading systems on a scale larger than ever before. For example, the population of Athens in the fifth century BC has been estimated at a minimum of 100 000. The fall of the Roman Empire (Figure 8.9) in the 5th century AD led to a major decline in urban life in western Europe that did not really revive until medieval times.

The medieval revival was the product of population growth and the resurgence of trade with the main urban settlements of this period located at points of greatest accessibility. While there were many interesting developments in urban life during the medieval period it required another major technological advance to set in train the next urban revolution.

The urban industrial revolution

The second 'urban revolution', based on the introduction of mass production in factories, commenced in Britain in the late 18th century (Figure 8.11). This was the era of the Industrial Revolution when industrialisation and urbanisation proceeded hand in hand. The key invention, among many, was the steam engine, which in Britain was applied first to industry and only later to transport. The huge demand for labour in the rapidly growing coalfield towns and cities was satisfied by the freeing of labour in agriculture through a series of major advances. The so-called 'agricultural revolution' had in fact begun in the early 17th century.

By 1801, nearly one-tenth of the population of England and Wales was living in cities of over 100 000 people. This proportion doubled in 40 years and doubled again in another

Figure 8.10 Primitive rural settlement in Tunisia

- the mass production of a very much wider range of goods and services than previously
- the ruthless exploitation of peripheral areas.

The initial urbanisation of much of the developing world was restricted to concentrations of population around points of supply of raw materials for the affluent developed countries. For example, the growth of São Paulo was firmly based on coffee, Buenos Aires on meat, wool and cereals, and Calcutta on jute.

By the beginning of the most recent stage of urban development, in 1950, 27% of the world's population lived in towns and cities (Figure 8.12), with the vast majority of those still living in the developed world where the cycle of urbanisation was nearing completion.

Current patterns

Current levels of urbanisation, as in the past, vary considerably across the globe (Figure 8.13). South America, perhaps surprisingly, is the most urbanised continent with only one of its 13 countries, Guyana, having more rural than urban dwellers. North America, most of Europe, Australia and New Zealand, parts of the Middle East and Japan and Korea in North East Asia are all highly urbanised. In Britain, 89% of people live in towns and cities. In contrast, levels of urbanisation are low through most of Africa, and South and East Asia. The country credited with the lowest urban population of all is Bhutan (6%).

In the developed world around 75% of people live in urban areas compared to only 38% in the developing nations as a whole. Tokyo remains the world's most heavily populated city with 27.2 million people (Figure 8.14), followed by Mexico City with 16.9 million. Of the world's ten largest cities, seven are in less developed countries. The UN expects that between 1990 and 2025 the number of people living in urban areas will double to more than 5

60 years. The 1801 census recorded London's population at one million, the first city in the world to reach this figure. By 1851 London's population had doubled to two million. However, at the global scale fewer than 3% of the population lived in urban areas at the beginning of the 19th century.

As the processes of the Industrial Revolution spread to other countries the pace of urbanisation quickened. The change from a population of 10% to 30% living in urban areas of 100 000 people and more took about 80 years in England and Wales; 66 years in the USA; 48 years in Germany; 36 years in Japan and 26 years in Australia. The transition from industrial capitalism to monopoly capitalism (Figure 8.11) marked the next of the principal stages in urbanisation in recent history. This transition was characterised by:

- a much greater scale of economic activity
- the consolidation of firms into multinational corporations
- the domination of newly created international markets by a small number of producers in each sector

Figure 8.11 Principal stages in global urban development

	1780–1880	1880–1950	1950–
Mode of accumulation			
Economic formation	Industrial capitalism	Monopoly capitalism	Corporate capitalism
Source of wealth	Manufacturing	Manufacturing	Manufacturing and services
Representative unit of production	Factory	Multinational corporation	Trans-national corporation, global factory
World-system characteristics			
Space relations	Atlantic basin	International	Global
System of supply	Colonialism/imperialism	State imperialism	Corporate imperialism
Hegemonic powers	Britain	Britain, USA	USA
Urban consequences			
Level of urbanisation at start of period (%)	3	5	27
Areas of urbanisation during period	Britain	North-western Europe, the Americas, coasts of Empires	Africa and Asia
Dominant cities	London	London, New York	New York, London, Tokyo

Source: *The Geographical Journal* Vol. 164, No. 1, March 1998

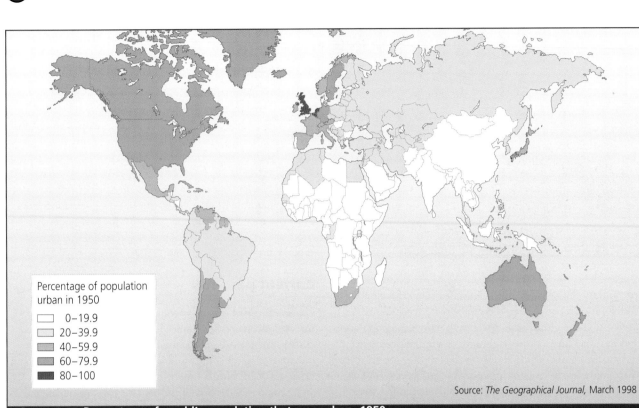

Source: *The Geographical Journal*, March 1998

Percentage of population urban in 1950

- 0–19.9
- 20–39.9
- 40–59.9
- 60–79.9
- 80–100

Figure 8.12 **Percentage of world's population that was urban, 1950**

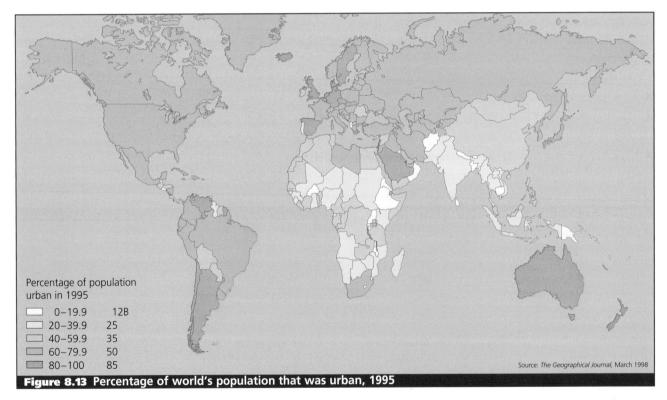

Percentage of population urban in 1995

- 0–19.9   12B
- 20–39.9   25
- 40–59.9   35
- 60–79.9   50
- 80–100   85

Source: *The Geographical Journal*, March 1998

Figure 8.13 **Percentage of world's population that was urban, 1995**

billion, and that 90% of this growth will be in developing countries. Particularly significant will be the growth of very large cities, so-called 'mega-cities'. It has been estimated that 36 cities will have over 8 million inhabitants by 2015, with most of them located in the developing world (Figure 8.15).

In Asia and Africa more than half the population remains in the countryside, compared with only a fifth in North America and Europe. In many developed nations the proportion of people living in large urban areas is actually declining, a process known as 'counterurbanisation'.

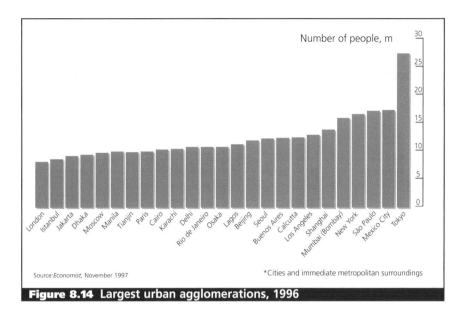

Source:*Economist*, November 1997

*Cities and immediate metropolitan surroundings

Figure 8.14 Largest urban agglomerations, 1996

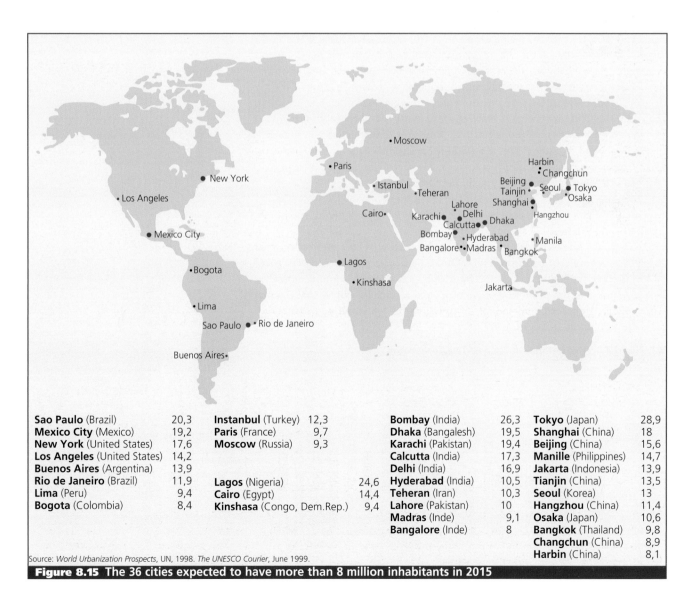

Sao Paulo (Brazil)	20,3	**Instanbul** (Turkey)	12,3	**Bombay** (India)	26,3	**Tokyo** (Japan)	28,9
Mexico City (Mexico)	19,2	**Paris** (France)	9,7	**Dhaka** (Bangalesh)	19,5	**Shanghai** (China)	18
New York (United States)	17,6	**Moscow** (Russia)	9,3	**Karachi** (Pakistan)	19,4	**Beijing** (China)	15,6
Los Angeles (United States)	14,2			**Calcutta** (India)	17,3	**Manille** (Philippines)	14,7
Buenos Aires (Argentina)	13,9			**Delhi** (India)	16,9	**Jakarta** (Indonesia)	13,9
Rio de Janeiro (Brazil)	11,9	**Lagos** (Nigeria)	24,6	**Hyderabad** (India)	10,5	**Tianjin** (China)	13,5
Lima (Peru)	9,4	**Cairo** (Egypt)	14,4	**Teheran** (Iran)	10,3	**Seoul** (Korea)	13
Bogota (Colombia)	8,4	**Kinshasa** (Congo, Dem.Rep.)	9,4	**Lahore** (Pakistan)	10	**Hangzhou** (China)	11,4
				Madras (Inde)	9,1	**Osaka** (Japan)	10,6
				Bangalore (Inde)	8	**Bangkok** (Thailand)	9,8
						Changchun (China)	8,9
						Harbin (China)	8,1

Source: *World Urbanization Prospects*, UN, 1998. *The UNESCO Courier*, June 1999.

Figure 8.15 The 36 cities expected to have more than 8 million inhabitants in 2015

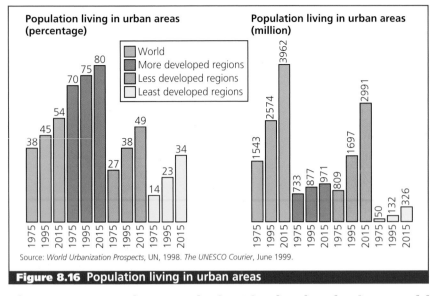

Source: *World Urbanization Prospects*, UN, 1998. *The UNESCO Courier*, June 1999.

Figure 8.16 Population living in urban areas

Key Definitions

Rural Belonging to or relating to life in the countryside in contrast to an urban lifestyle.

Rural landscape A mental or visual picture of countryside scenery which is difficult to define as rural areas are constantly changing and vary from place to place.

Rural population People living in the countryside in farms, isolated houses, hamlets and villages. Under some definitions small market towns are classed as rural.

Urbanisation The process whereby an increasing proportion of the population in a geographical area lives in urban settlements.

Urban growth The absolute increase in physical size and total population of urban areas.

Urbanism The tendency for people to lead increasingly urban ways of life.

The post-1945 urban 'explosion' in the developing world

Throughout history urbanisation and significant economic progress have tended to occur together. In contrast, the rapid urban growth of the developing world in the latter part of the 20th century has in general far outpaced economic development, creating huge problems for planners and politicians. Because urban areas in the developing world have been growing much more quickly than the cities of the developed world did in the 19th century the term 'urban explosion' has been used to describe contemporary trends. Between 1990 and 1995, the world's urban population grew by nearly 60 million, with four-fifths of this increase in the developing world. Singapore tops the world urbanisation league as everyone in this small city-state is classed as an urban resident. However, the clear distinction between urbanisation and urban growth should be kept in mind as some of the least urbanised countries, such as China and India contain many of the world's largest cities. China's urban population at 37% in 1995 is predicted to rise to 70% by 2040 and 89% by 2087 (Figure 8.17).

An approach known as dependency theory has been used by a number of writers to explain the urbanisation of the Third World, particularly the most recent post-1950 phase. According to this approach, urbanisation in the developing world has been a response to the absorption of countries and regions into the global economy. The capitalist global economy induces urbanisation by concentrating production and consumption in locations that:

■ offer the best economies of scale and agglomeration

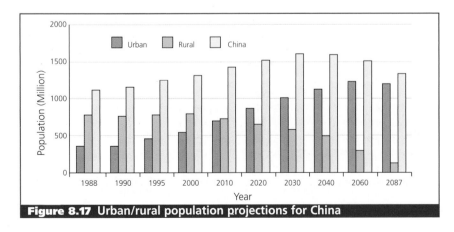

Figure 8.17 Urban/rural population projections for China

- provide the greatest opportunities for industrial linkage
- give maximum effectiveness and least cost in terms of control over sources of supply.

Thus urban development is one of the spatial outcomes of the capitalist system. Trans-national companies (TNCs) are the major players in this economic process which enables and encourages people to cluster in geographical space. The actions of TNCs encourage urbanisation directly in response to localised investment. However, TNCs also influence urbanisation indirectly through their impact on traditional patterns of production and employment. For example, the advance of export-oriented agriculture at the expense of traditional food production has reduced employment opportunities in the countryside and encouraged rural to urban migration.

Other factors which have encouraged urbanisation in the developing world include:

- the investment policies of central governments which have generally favoured urban over rural areas, often in an attempt to enhance their prestige on the international stage

- the higher wage rates and better employment protection in cities
- greater access to health care and education
- the decline in the demand for locally produced food as consumers increasingly favour imported food.

The combined result of these factors has been 'backwash urbanisation', destroying the vitality of rural areas and placing enormous pressure on cities. In the longer term, the rate of urban growth should eventually slow as a result of falling fertility rates and a deceleration in the urbanisation process itself, as a growing share of the population becomes urbanised. In West Africa the growth of the urban population should slow from 6.3% per year from 1960–1990, to 4.2% between 1990 and 2020. However, even with a reduction in the rate of urbanisation the problems faced by large cities in the developing world are going to be severe (Figure 8.18).

Urban future

The population explosion in the world's cities will lead to serious social problems, with widespread poverty, exclusion and destitution. Remedial action will be required, with comprehensive social measures to provide basic services in education, health, nutrition, family planning and vocational training. As many as a billion new jobs may be needed between 1990 and 2025, most of them in cities.

The environmental risks in developing world cities will be legion. Insufficient shelter, inadequate sanitation, poor water supply, air pollution and congestion on the streets are all to be expected. Huge investments could help prevent the worst excesses of this, perhaps to the tune of $100–$150 billion a year world-wide. The lion's share will come from the revenues of national and local governments. Some will come from the private sector, including foreign direct investment. Official development assistance will at best act as a top-up, unless recent declining trends are radically reversed.

City opportunity

Problems like these have led many to dismiss the city's role in the future world economy, almost to the point of predicting its demise at the hands of new technologies. However, the potential of cities will be enormous, with urbanisation representing a major opportunity, for developing countries in particular in terms of overall spatial planning. Channelling rural migration into cities can help relieve damaging population pressure on marginal rural land. Focusing development in cities themselves will, on balance, continue to bring economies of scale, in terms of transport, waste

treatment and, of course, business. And even if there are countering forces of decentralisation enabled by new technologies, other technological innovations, such as in emissions control and communications, may make cities more attractive. So, though information and communications technology may generate a greater geographic dispersion of economic activity in the future, urbanisation will continue to be a major force of economic development. In fact, the importance of cities could well grow in the 21st century. One reason is that they will sharpen their profile in the open and competitive international space of the new global economy. The importance of cities is likely to be strengthened by the creation of global city networks. These networks will give cities greater autonomy of action to tap into international markets and forge new economic links across national boundaries. The city will be conditioned by those networks, both politically and architecturally, and often to a greater extent than by their regional or national hinterlands. These networks will act as highways for the transfer of knowledge and best practice, for stimulating innovation in policy and project development; and as catalysts in economic co-operation. Cities in developing countries will have to be particularly determined that they are 'plugged' into these networks so as to profit from them and to avoid being sidelined. Sao Paulo or Beijing should therefore see to it that they are as much part of the network as Paris or London, but so too should medium cities, from Bahia Blanca to Bordeaux. Networked cities are rich in promise and diversity, giving us all an urban future to look forward to.

Source: OECD Observer No. 217/218 Summer 1999

Figure 8.18 The future for cities

Figure 8.19 Urban poverty: Manaus, Brazil

SPATIAL FOCUS SPATIAL

A summary of the evolution of settlement in South East England

The human landscape may be viewed as a palimpsest, consisting of the legacy of layer upon layer of human occupation, with each successive human imprint obscuring at least partially the evidence of earlier periods of settlement and economic activity.

The following are significant periods in the settlement of South East England. The grid references refer to the I: 50 000 Dorking, Reigate and Crawley Ordnance Survey map [No 187]. Analyses of OS maps can confirm a considerable amount about the evolution of settlement in an area but an examination of other sources is also required to obtain a fuller picture.

■ Between 4000 and 2000 BC woodland was being cleared for grazing and cultivation as more permanent settlements than had existed beforehand were established. By the late Bronze Age, some enclosed hilltop sites, such as that now covered by Queen Mary's Hospital, Carshalton (280625) were occupied. Iron implements began to appear about 800 BC which have been found at the sites of earthen-rampart forts crowning St George's Hill in the Thames Valley (085615) and Cardinal's Gap on the downs south of Caterham (328533). However it is uncertain whether any of these hillforts were occupied on a long term basis.

■ The Roman invasion found a fairly densely populated countryside of agricultural villages interspersed with a few larger centres, largely established by the Belgae who arrived in Britain early in the first century BC. The influence of the Belgae extended over most of South East England.

■ The most important Roman town in England was London (Londinium) with a population of 25 000. In regional terms the South East had the greatest concentration of Roman towns. Associated with the towns were the roads (i.e., 357475) which linked the towns to their hinterlands.

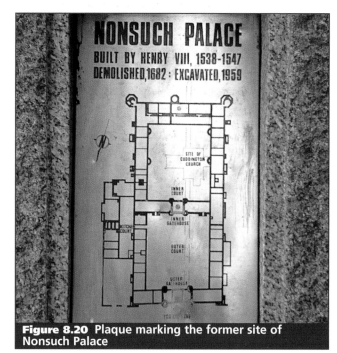

Figure 8.20 Plaque marking the former site of Nonsuch Palace

Outside of the towns and developed later was a scattering of villas – large farmsteads belonging to Romanised folk. One of the best known examples is the Roman palace at Fishbourne, Sussex. Others were located along the gault clay vale in Sussex (i.e., 410545) and further south at Ewhurst (083417). However, much of the countryside was little affected by Romanisation.

■ Recognisable towns all but vanished during the post-Roman period and a resurgence did not occur until the 8th and 9th centuries. Again the South East was at the forefront. There is only limited evidence of the Germanic

Figure 8.21 Remains of the Banqueting House at Nonsuch Palace

settlers (Angles, Saxons, Jutes) who penetrated lowland England during the 5th and 6th centuries. Much of our understanding of this colonisation is through place name study. Map evidence can be seen in the tumuli (burial grounds) at Banstead (248608) and Coulsdon (300582), both of which probably date from the mid-seventh century. Other pagan cemeteries have been found at Tattenham Corner (227582) and Leatherhead and Guildford.

- By the time of Alfred, the settlement pattern that was to prevail until comparatively modern times had probably become firmly established. Villages thrived in the Thames valley and along the dip-slope spring line of the North Downs. Smaller settlements were sited at favourable places on the chalk downs themselves. On the gault clay to the south this limited area of fertile soils was farmed by a long line of closely spaced settlements.

- The Normans built strong castles at strategic places like Guildford and manor houses at fertile and accessible locations. The parks which usually accompanied Norman castles and manor houses were generally broken up for farmland in the Tudor period but, in some cases, such as that of the former park at Lagham in south Godstone (370474), the distinctive curved outline of their boundaries can still be traced among the modern fields.

- The settlement pattern was consolidated during the Middle Ages with only limited change. Only Southwark, Guildford, Kingston and to a lesser extent Farnham could really be called towns in medieval Surrey. Here, the proximity of London appears to have inhibited the growth of urban centres. Henry VIII was an enthusiastic builder and built Nonsuch (230630) and Oatlands Palaces in Surrey. Industries appeared in some towns, including the spinning and weaving of wool (Guildford), leather-working (Leatherhead) and metal-working.

- The construction of turnpikes and then railways brought suburban London to the north-eastern part of Surrey,

followed by the expansion of Croydon, Guildford and Horley and the creation of 'railway towns at New Woking, Redhill and Surbiton. Between the two world wars motor cars and electric trains brought a further exodus out of London into Surrey. Meanwhile the destructive ugliness of the speculative builder was threatening some of the finest scenery in the county. This process was to continue, checked only by philanthropy and the National Trust, until the Green Belt Act was passed.

- In spite of increasing legislation to protect the landscape, in particular the establishment of London's Green Belt, post-war development has had a significant impact. Crawley (270364) is one of the eight New Towns built around London following the New Towns Act of 1946. To the north of Crawley, the construction and subsequent expansion of Gatwick airport has had a considerable effect on the demand for housing in the surrounding area.

- The villages and towns of the region have increasingly become the preserve of commuters, considerably changing the socio-economic character of these settlements. The construction of motorways, particularly the M25 and the M23 has had a major effect. The commuter belt now extends right to the south coast with Brighton being the most significant example. The retirement function of many coastal towns has grown significantly in recent decades. The South-East exhibits the most advanced landscape of counterurbanisation in Britain.

Settlement patterns

Site and situation

The **site** of a settlement is the characteristics of the actual point at which the settlement is located and its immediate surroundings. These characteristics would have been of major importance in the initial establishment of the settlement and its subsequent growth. When trying to identify site characteristics from Ordnance Survey maps we need to look at the land covered by the built environment and a zone about three to four kilometres around it. The latter would be the reasonable daily walking distance to grow crops, tend cattle, collect wood and so on when the settlement was first established. Figure 8.22 shows the main site characteristics of the village of Alfriston in the South Downs.

The term **situation** refers to the location of a settlement in relation to its wider surroundings. Relevant factors are other settlements, rivers, relief, major roads and railways.

The original decision to locate a settlement at a particular point would have involved the consideration of a range of factors. It is unlikely that all of the characteristics of the site selected were ideal but that it was the best of all the realistic alternatives. The following are the main factors which have influenced the location of settlements:

- Relief: flat, low-lying land with fertile soils such as the clay vales of southern England was preferable to the steeper chalk downlands for a number of obvious reasons. However, hazards such as flooding and potential hostility from other human groups sometimes persuaded early settlers to opt for sites on higher land.

- Defence: many early settlements in Britain were on hilltops, along recognised trade routes and in commanding positions, providing defence from attack. Of the larger settlements in Britain, Edinburgh occupies the most commanding hilltop site. Another classic defensive site was to be surrounded by water on three sides. The site of Durham on high ground in the neck of a meander of the River Wear is a prime example.

- Water supply: in pre-industrial times Britain's rivers provided a clean and safe water supply. This was the main reason why so many settlements developed on river systems. Water transport was often an added bonus. Spring-line settlement along the foot of chalk escarpments is a clear example of the importance of water supply in settlement location. Sites located in immediate proximity to a supply of water are known as wet-point sites.

- Flood avoidance: where the likelihood of flooding was significant, settlements were usually built above the level of the perceived threat. Good examples of such dry-point (or water-avoiding) sites are the Fenland settlements such as Ely which were built on mounds of land that formed natural islands above the surrounding marsh.

- Bridging points: wherever routes could cross a river by a ford in earlier times (e.g., Oxford, Guildford), and later by a bridge a settlement often developed. Of particular importance was the lowest bridging point. Downstream of this crossing point the tidal waters of an estuary would have been too wide and turbulent for bridge construction.

- Food supply: because all food was produced locally sites were sought which were suitable for both arable and pastoral farming. Spring-line settlements at the foot of chalk escarpments in South East England grazed sheep on the higher and steeper chalk and utilised the fertile clay vales for arable farming and for raising cattle. In upland landscapes, such as the Pennines and the Lake District, agricultural land use also changed from flat valley floors to steeper windswept slopes.

- Building materials: locally available wood, stone and clay were major influences on site selection. Wood was of course more commonly available than it is today.

- Fuel supply: for early settlement in Britain firewood was the dominant source of fuel as it is in much of the developing world today.

- Accessibility: settlements frequently developed where natural routeways converged. Confluence settlements developed where two or three rivers met and the valley routes converged. Good examples are Monmouth, Salisbury and Reading. Gap towns command key routeways through hills or mountains. For example, Dorking controls the scarp slope end of the gap through the North Downs cut by the River Mole, a tributary of the River Thames, while Leatherhead is at the dip slope entrance to the gap.

- Aspect: where possible early settlers selected south-facing slopes to benefit from maximum insolation and protection from cold northerly airstreams.

- Mineral resources: the distribution of coalfields has had a considerable influence on the location of settlement in Britain. Other minerals such as iron ore, salt and tin have also contributed to site selection in certain areas.

- Natural harbours: the combination of flat land and a sheltered harbour provided the ideal location for fishing communities. Many larger coastal settlements eventually developed important trade functions.

Questions

1 (a) Suggest three ways in which villages can be distinguished from hamlets.

(b) Why are small market towns the most difficult to classify in the rural–urban divide?

(c) Explain why the population ranges given in Figure 8.2 might not be appropriate for highly populated developing countries such as India, China and Indonesia.

(d) How would you distinguish between the four classes of urban settlement shown in Figure 8.1?

(e) Why can the hierarchy of settlement in a country or region be viewed as a dynamic rather than a static phenomenon?

2 (a) Explain two of characteristics of traditional rural society listed in Figure 8.5.

(b) Why are these characteristics weaker today than they were 50 years ago?

(c) Describe the distribution of the four types of rural area illustrated in Figure 8.6.

(d) How would you expect extreme rural and extreme non-rural areas to differ in terms of the following characteristics: (i) demographic, (ii) social, (iii) economic, (iv) environmental?

(e) Use Figure 8.7 to describe how urbanisation has impacted on the countryside for one region you have studied.

3 (a) Describe the expected distribution of the world's largest cities in 2015.

(b) Explain why there will be such a large variation by continent.

(c) Describe the trends illustrated by Figure 8.16.

(d) Discuss the major problems that rapidly growing cities will face in the foreseeable future.

(e) Explain why it is possible to construct optimistic and pessimistic urban future scenarios.

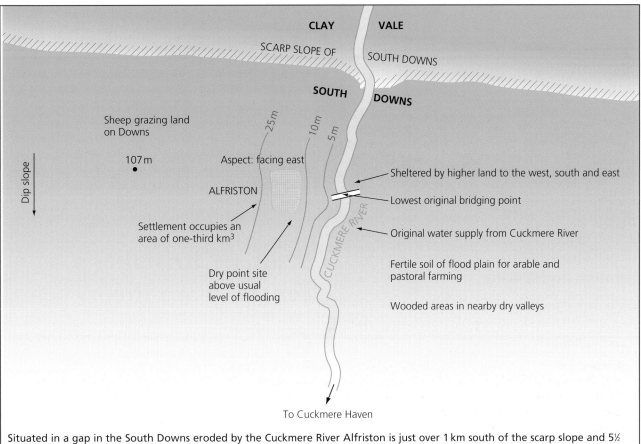

CLAY   VALE

SCARP SLOPE OF   SOUTH DOWNS

SOUTH   DOWNS

Dip slope

Sheep grazing land
on Downs

107 m

25 m   10 m   5 m

Aspect: facing east

ALFRISTON

CUCKMERE RIVER

Settlement occupies an
area of one-third km³

Dry point site
above usual
level of flooding

Sheltered by higher land to the west, south and east

Lowest original bridging point

Original water supply from Cuckmere River

Fertile soil of flood plain for arable and
pastoral farming

Wooded areas in nearby dry valleys

To Cuckmere Haven

Situated in a gap in the South Downs eroded by the Cuckmere River Alfriston is just over 1 km south of the scarp slope and 5½ km north of the coast. The town of Seaford is 5 km to the south west while Eastbourne is 8 km to the south east. The A27T, 2 km north of Alfriston links the village to settlements further east and west such as Newhaven and Brighton.

Figure 8.22 **The site of the village of Alfriston, East Sussex**

Figure 8.23 **The village of Westhumble in the Mole Gap**

Figure 8.24 **Oasis settlement in Tunisia**

Figure 8.25 Natural harbour: Hope Cove, south Devon

Location and spacing

The analysis of any landscape of at least a regional scale will reveal settlements of varying sizes (the vertical component) along with some degree of order and logic in their location and spacing (the horizontal component). The two components are, of course, mutually interrelated, with the largest settlements being fewer in number and spaced farther apart than settlements of a lesser size. Not surprisingly, geographers and those of other related disciplines have tried to explain the reasons behind such regularity.

Settlements above a certain minimum size act as central places. Small central places tend only to provide a narrow range of convenience (low order) services which have a limited range and a low threshold. With ascent of the hierarchy of settlement the range of services on offer increases, in particular the supply of comparison (high order) services. The latter have more extensive ranges and higher thresholds. Settlements provide services on the basis of the threshold population available to utilise them. The range of a service is controlled at the lower level by the minimum threshold population necessary to sustain it. In the regional landscape the largest settlements have wide market or catchment areas within which can be found the more limited market areas of the smaller settlements. The well established relationship between settlement size and function is shown in Figure 8.26.

Walter Christaller's central place theory

One of the earliest attempts to seek an understanding of the order underlying settlement spacing was that of Walter Christaller. Christaller's work, entitled *Die zentralen Orte in Suddeutschland* (Central Places in Southern Germany), was published in 1933. He asserted that the numbers, sizes and spatial patterns of central places can be explained by the operation of the forces of supply and

Key Definitions

Site The characteristics of the actual point at which the settlement is located and its immediate surroundings.

Situation The location of a settlement in relation to its wider surroundings.

Central place A settlement which provides goods and services not only for its own population but also for those living in a surrounding area.

Hierarchy of settlement The grouping together of central places into distinctive levels of functional importance. Settlements at the top of the hierarchy have larger populations, a wider range of functions and more extensive market areas than settlements lower down the hierarchy.

Range The maximum distance people are willing to travel to obtain goods or services.

Threshold The minimum number of customers necessary to support the profitable sale of goods or services.

Market area The spatial area in which the consumers of an enterprise's goods or services are located.

Sphere of influence The area around a settlement that comes under its economic, social and political influence.

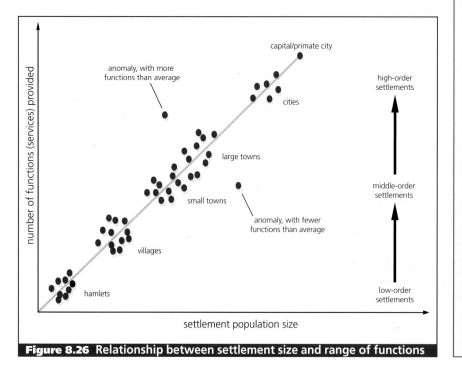

Figure 8.26 Relationship between settlement size and range of functions

demand. But first, as with all theories, certain assumptions had to be made:

- unbounded flat land with an even distribution of resources
- equal ease and opportunity of movement in all directions with only one form of transport and with transport costs proportional to distance
- an evenly distributed population
- consumers with identical needs, tastes and purchasing power whose objective was to minimise the distances they travelled to obtain goods and services. This was done by using the nearest central place
- different orders of central place existed in the landscape with higher order central places providing both high and low order goods
- no excess profit would be made by any central place with all central places located as far away as possible from each other in order to maximise profits
- hexagonal market areas, the hexagon being the shape nearest to the circle which will pack tightly together without leaving gaps or overlapping.

Figure 8.27 A central place: Nabul, Tunisia

The term isotropic surface has been applied to the uniform landscape assumed. Christaller proposed that settlements with the lowest order of specialisation would be equally spaced and surrounded by hexagonal-shaped market areas. For every six lowest order settlements there would be a larger and more specialised settlement which in turn would be situated at an equal distance from other settlements of the same order. These higher order settlements would also be surrounded by hexagonal market areas. Progressively more specialised settlements would be similarly located at an equal distance from each other, with, at each stage, the hexagonal market area becoming larger (Figure 8.28). The six to one relationship is maintained between every pair of levels in the hierarchy. Each level in the settlement hierarchy would provide all the services of lower order settlements as well as a range of high order services which could not be offered on a profitable basis lower down the hierarchy. Christaller referred to this regular progression as the K = 3 hierarchy. His logic was that each lowest order settlement was situated at the corners of the hexagonal market areas and was thus equidistant from three settlements of the next level up the hierarchy. The purchasing power of the smaller settlements for the goods and services of the larger settlements would be split on this basis (Figure 8.29). Thus a larger settlement would benefit from one-third of the custom of each of the six smaller settlements on its hexagonal boundary, equivalent to the full purchasing power of two lower order settlements. Added to this would be the purchasing power of its own population to give a total of 3.

In spatial terms each settlement would serve three times the market area of the next order settlement down the hierarchy. For every settlement of the largest size, there would be three of the second level, nine of the third level, 27 of the fourth level and so on. According to Christaller's study of southern Germany the smallest settlements would be spaced 7 km apart. Centres at the next level up, serving three times the area and population, would be located 12 km apart (sq root 3 times 7). This relationship would continue throughout the hierarchy. The K = 3 network was based on the 'marketing principle' which placed lower order settlements as close as possible to higher order settlements.

Christaller recognised that different hierarchical arrangements would be formed if other factors were more important than the demand for and supply of goods in a

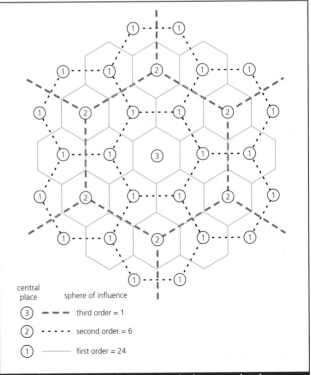

central place · sphere of influence

③ – – – third order = 1

② · · · · second order = 6

① —— first order = 24

Figure 8.28 Christaller's central places and spheres of influence

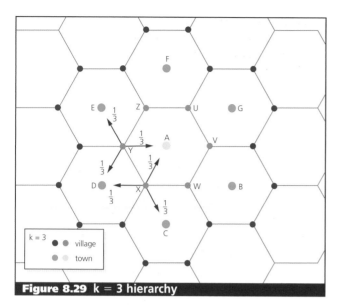

Figure 8.29 k = 3 hierarchy

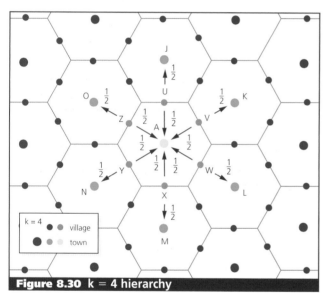

Figure 8.30 k = 4 hierarchy

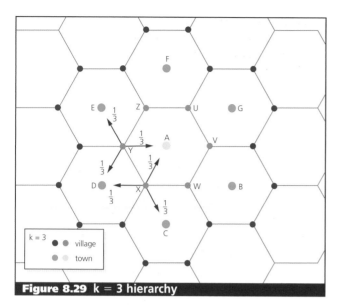

Figure 8.31 k = 7 hierarchy

region. Thus he also devised networks based on the 'traffic principle' (K = 4) and the 'administrative principle' (K = 7). He postulated that a K = 4 hierarchy (Figure 8.30) would develop in regions where transport costs were particularly important since such an arrangement maximises the number of central places on straight-line routes. In regions with a highly developed system of central administration a K = 7 hierarchy (Figure 8.31) would tend to evolve, maximising the number of central places dependent on any one higher order central place and eliminating the shared allegiances of the other K value systems.

Criticisms of central place theory

Nowhere in the world does the settlement pattern exactly match that set out in central place theory. Thus it is not surprising that the model has been criticised. Equally unsurprising is the fact that most criticism is linked to the assumptions on which the theory is based. So:

- perfectly uniform physical landscapes rarely exist in reality

- there is usually competition within and between different modes of transport, and transport costs are not generally proportional to distance

- the distribution of population and purchasing power is not uniform

- for a variety of reasons people do not always use the same central place

- the main function of some settlements is not as a central place, for example industrial towns

- the level of profit tends to vary between businesses and places, thus perfect competition exists only in theory

- as the century has progressed government has played an increasingly important role in the location of settlement.

Despite the criticism, Christaller's central place theory has provided a starting point for planners and others interested in the arrangement of the human landscape. Perhaps the most famous of later works is that of the economist August Losch. Following Christaller, Losch used hexagonal market areas but allowed various hexagonal systems to co-exist. Each system operated at a different level and is superimposed on the other. This more variable and complex system produces a continuum of settlement sizes which more closely relates to reality, rather than the stepped distribution in the Christaller model.

The hierarchical arrangement of settlement

The vertical component in the organisation of settlements has also led to the formulation of a body of theory. Although earlier writers had pointed to a certain regularity in the size of cities when ranked from the largest downward, G. K. Zipf expressed this relationship precisely when he proposed the rank-size rule in 1949. He stated that 'if all the urban settlements in an area are ranked in descending order of population, the population of the nth town will be 1/nth that of the largest town'. Thus, according to the rank-size rule the fifth largest urban settlement in a country would be expected to have a population one-fifth the size of the largest settlement. Plotted on a logarithmic graph a perfect rank-size relationship produces a straight line. The hierarchy of cities in some countries fits this pattern to a reasonable degree but even then significant changes can occur over time. For example, when the rank-size rule was proposed the USA provided quite a good fit but the relationship is not so good today. Near the top of the hierarchy the rapid growth of Los Angeles in the latter part of the 20th century to just overtake Chicago as the second-ranking city means that after New York at the top the next two cities are very similar in population size. Alternative patterns to the rank-size rule are (Figure 8.32):

- a stepped order pattern where there are distinct levels but where a number of settlements occur at each level

- a binary pattern where a number of cities of similar size dominate the upper end of the hierarchy, as is the case in Canada with Toronto and Montreal, and in Spain with Madrid and Barcelona

- a primate pattern where the largest city, usually the capital, is many times the size of the next ranking city. The Law of the Primate City proposed by M. Jefferson in 1939 suggested that once a major city had become larger than its competitors a combination of economic,

social and political factors will tend to create a much faster rate of growth in that city compared to its rivals. Clear examples of urban primacy are London, Paris, Montevideo and Lima.

Interaction between settlements

Reilly's gravity model

W. J. Reilly's retail gravitation model (1931) attempts to predict the degree of interaction between two places. The model states that two centres attract trade from intermediate places in direct proportion to the size of the centres and in inverse proportion to the distance between them. Thus the 'break-point' between two settlements can be calculated by using the following formula:

$$\text{Breaking point AB} = \frac{\text{Distance from A to B}}{1 + \sqrt{\dfrac{\text{Popuation A}}{\text{Population B}}}}$$

However, the basic assumption that the larger the settlement the greater will be its trade area may not always be true because:

- the level of services in a place is not always related to population size. For example tourist resorts have more services than expected in relation to their resident populations

- the accessibility of different places can vary significantly

- consumer perceptions, negative or positive, of a place may override logical considerations

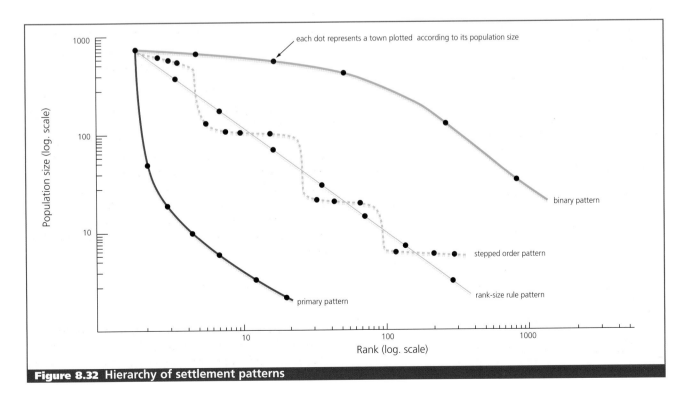

Figure 8.32 Hierarchy of settlement patterns

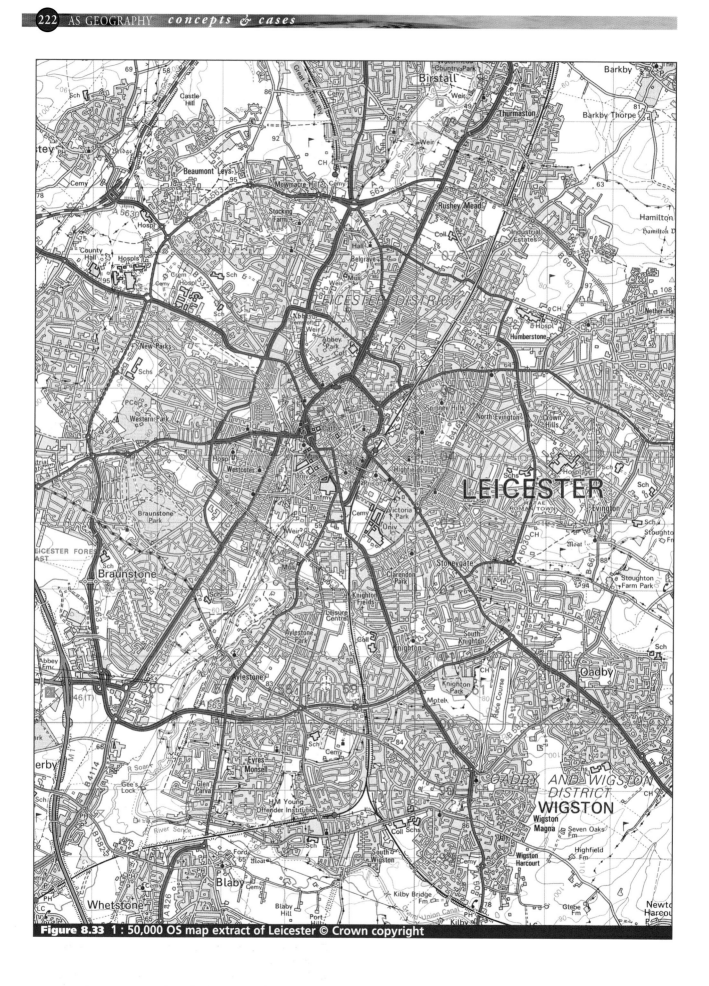

Figure 8.33 1 : 50,000 OS map extract of Leicester © Crown copyright

Testing of the model has shown that it suits agricultural areas where towns are of limited size and fairly evenly spaced better than closely packed urban areas.

Huff's behavioural model

D.L. Huff (1962) was responsible for restarting the gravity model in probabilistic terms on the basis that the likelihood of a consumer going to any centre is based on the relative attractiveness of different centres and the distance that would have to be travelled in each case. The attraction of towns can be measured in various ways but perhaps the best is the total number of shops in each centre.

$$P_1 = \dfrac{\dfrac{\text{Number of shops in centre[1]}}{\text{Distance or time to reach them}}}{\dfrac{\text{Total numbers of shops in study area}}{\text{Total distance or time to reach them}}}$$

If P is calculated for a series of points of origin then isopleths, or equiprobability contours can be drawn for each of the centres in an area.

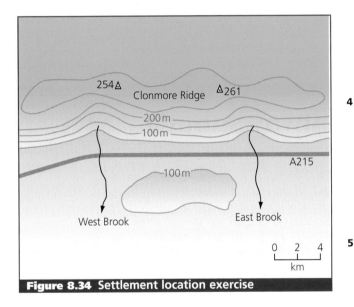

Figure 8.34 Settlement location exercise

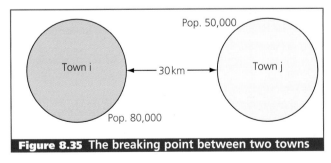

Figure 8.35 The breaking point between two towns

Questions

1   (a) Define the terms (i) site, and (ii) situation.
    (b) On a copy of Figure 8.34, using the letters A and B, mark the likely sites of two villages originally located in the Anglo-Saxon or medieval periods.
    (c) Explain your choice of sites.
    (d) Suggest how these settlements might have expanded with modern population increase.
    (e) If evidence of pre-Roman settlement existed in this area, where would you expect to find it? Give a reason for your answer.

2   (a) Describe the relationship between settlement population size and the number of functions provided.
    (b) Explain the reasons for such a relationship.
    (c) Suggest reasons for the two anomalies shown in Figure 8.26.
    (d) How might a settlement ascend or descend the hierarchy?
    (e) For any settlement that you have studied describe and explain its service provision.

3   (a) According to Christaller, what is the K-value illustrated in Figure 8.28?
    (b) Account for the hexagonal shape of the market areas.
    (c) Comment on the relationship between the number and size of settlements.
    (d) What did Christaller mean by the term 'an isotropic surface'?
    (e) What are the main criticisms of central place theory?

4   (a) Study Figure 8.35. Calculate the breaking point between the two towns.
    (b) Outline the theoretical basis of the theory.
    (c) Discuss the factors which in reality could result in a breaking point different from your answer to (a).
    (d) Describe three ways in which you could attempt to determine the breaking point between two settlements through fieldwork.
    (e) Why would major chain stores be interested in this geographical concept?

5   Use the Ordnance Survey map extract (Figure 8.33) to answer the following questions.
    (a) What is the area (i) covered by the map extract? (ii) of the city contained within the A563 and the A6030?
    (b) Examine the site of Leicester.
    (c) Use an atlas to assess the situation of the city.
    (d) Using OS map evidence only, state the functions of Leicester.
    (e) What is the evidence on the map that settlement has evolved over a long period of time?
    (f) Examine the physical and human factors which might have influenced the growth of Leicester.
    (g) Referring to evidence from the map, identify the location of the CBD. Which land uses not evident on the map would you expect to find in the CBD?
    (h) Using map evidence only, compare the inner city and suburbs of Leicester. How would you expect census data to reveal other contrasts between the inner city and the suburbs?
    (i) Comment on the location and number of (i) open spaces, (ii) churches.
    (j) Discuss the land use in the rural–urban fringe.
    (k) Describe and explain the pattern of communications within the city.
    (l) Select and justify a location for a new industrial estate which would cover an area of approximately 0.5 km[2].

Changing urban environments

The cycle of urbanisation

The development of urban settlement in the modern period can be seen as a sequence of processes, known as the cycle of urbanisation. The key processes and their landscape implications are: suburbanisation, counterurbanisation and reurbanisation. In Britain, suburbanisation was the dominant process until the 1960s. From this decade counterurbanisation impacted increasingly on the landscape. Reurbanisation of some of the largest cities, beginning in the 1990s, is the most recent phenomenon.

Suburbanisation

Although the urban Industrial Revolution in Britain began in the late 18th century it was not really until the 1860s that urban areas began to spread out significantly. The main factor in this development was the construction of suburban railway lines. In south-west London, for example, one arm of the District Line reached Fulham Broadway in 1880, Putney Bridge in 1880, and Wimbledon in 1889. At each location the opening of the railway spurred on a rapid period of house building. Initially the process of suburbanisation was an almost entirely middle-class phenomenon. It was not until after the First World War, with the growth of public housing, that working-class suburbs began to appear.

In the inter-war period about 4.3 million houses were built, mainly in the new suburbs. Just over 30% were built by local authorities. The reasons for such a rapid rate of suburban growth were:

- **government support for house building**
- **the willingness of local authorities to provide piped water and sewerage systems, and gas and electricity**
- **the expansion of building societies**
- **low interest rates**
- **development of public transport routes**
- **improvements to the road network.**

Figure 8.36 describes the development of Stoneleigh, an outer suburb in south-west London. In the latter half of the 20th century suburbanisation was limited by the creation of Green Belts and the introduction of general planning controls.

Stoneleigh: a railway suburb

By the end of the 1930s developments were taking place on the rural–urban fringe. Stoneleigh acquired a railway station in 1932 and witnessed spectacular growth thereafter. The Stoneleigh Estate consisted of three farms. These had been offered for building development in the early 1900s but by the end of the 1920s only a few dozen houses had been built. However, following the arrival of the railway, development intensified. By 1933 a 3500 acre site for 3000 homes existed, and the area had a complete set of drains and sewers. By 1937 all farmland and woodland within a 1 mile radius of the railway station had been destroyed.

The housing density at Stoneleigh was low at eight houses per acre. As well as the railway there was a good bus service to Epsom, Surbiton and Kingston. Further developments followed quickly:
- a block of 18 shops (by 1933);
- a sub-post office (1933) and a bank (1934);
- Stoneleigh's first public house (1934);
- a cinema (1937);
- a variety of churches (1935 onwards);
- schools (from 1934);
- recreational grounds at Nonsuch Park and Cuddington.

Stoneleigh benefited from a strong and dynamic residents' association. The residents were aggrieved that nearby working-class areas in Sutton and Cheam were reducing their own land values. They canvassed successfully for boundaries to be redrawn, raising the values of their properties. There were many social activities too, including dances, whist-drives, cricket, children's parties, choral societies, cycling and tennis. This went a long way to creating a sense of community. The chairman of the residents' association was also the editor of a local newspaper, which helped the residents in their aims.

By 1939 Stoneleigh was a model railway suburb. Over 3000 people used the railway each day for commuting to work and it was also useful for reaching the south coast. However, the railway also split the community in two. There were problems for buses and cars trying to move from one side of the town to the other. Socially, it also split the community.

The development of Stoneleigh shows many similarities with other suburbs:
- a variety of housing styles, reflecting the different building companies;
- a somewhat chaotic road layout;
- complete destruction of the former farming landscape;
- ponderous shopping parades;
- the claim by some that it is dull and soulless.

Yet because of its poor road layout, in particular the lack of railway crossings, and its housing developments right up to the railway line, it does not have the worst trappings of modern suburban development.

Source: *Geography Review*, September 1998

Figure 8.36 Stoneleigh: the development of a London suburb

Figure 8.37 Stoneleigh: the High Street

Figure 8.38 Stoneleigh: semi-detached houses

Counterurbanisation

Urban deconcentration is the most consistent and dominant feature of population movement in Britain today, in which each level of the settlement hierarchy is gaining people from the more urban tiers above it but losing population to those below it. However, it must be remembered that the net figures hide the fact that there are reasonable numbers of people moving in the opposite direction. Figure 8.39 shows the consistent loss of population for metropolitan England in terms of net within-UK migration for the period 1981–94. It does not, however, mean an overall population decline of this magnitude, because population change is also affected by natural increase and international migration. London is the prime example of the counterbalancing effect of the latter two processes.

Figure 8.40, based on the 1991 census, is a summary of change of addresses within Great Britain during the year preceding census night. It focuses on migration between local authority districts grouped according to their degree of 'urbanness' and their distance from the main metropolitan centres. The top six most urban areas all averaged net migration losses while the other seven district types were all net gainers. The trend from most urban to least urban is highly regular.

Around London, where central rents are particularly high, back offices have diffused very widely across South East England. Between 20 and 30 decentralisation centres can be identified in the Outer Metropolitan Area, between 20 and 80 km from central London, especially along the major road and rail corridors. Examples include Dorking, Guildford and Reigate.

Liverpool's population has been falling since 1931 (Figure 8.42) with a current outflow of about 2500 a year. Population loss accelerated during the 1960s when the port of Liverpool and its associated industries went into steep decline. Many people left for new towns such as Runcorn and Skelmersdale, lured by the promise of better housing and new jobs. During the 1990s, however, the decline has slowed. Other cities show similar trends. The population of Glasgow shrank from more than one million in 1960 to 623 000 in the mid-1990s. Manchester has lost up to a third of its population in the last 30 years.

Figure 8.39 Net within-UK migration, 1981–94 for metropolitan England		
All metropolitan England	**Greater London**	**Six metropolitan counties**
	thousands	
1981	−82.6	−32.2	−50.4
1982	−78.2	−33.5	−44.7
1983	−88.3	−33.3	−55.0
1984	−89.2	−33.1	−56.1
1985	−111.1	−55.2	−55.9
1986	−113.7	−49.2	−64.5
1987	−125.1	−72.9	−52.2
1988	−118.8	−74.4	−44.5
1989	−64.8	−36.9	−27.8
1990	−64.9	−37.4	−27.5
1991	−80.6	−52.7	−27.9
1992	−81.3	−51.7	−29.6
1993	−87.4	−52.9	−34.6
1994	−89.9	−45.9	−44.0

Source: NHSCR data. Reproduced from *Population Trends 83*, Spring 1996

Figure 8.40 Net within-Britain migration, 1990–91, by district types

District type	Population 1991	Net migration 1990–91	%
Inner London	2 504 451	−31 009	−1.24
Outer London	4 175 248	−21 159	−0.51
Principal metropolitan cities	3 992 670	−26 311	−0.67
Other metropolitan districts	8 427 861	−6900	−0.08
Large non-metropolitan cities	3 493 284	−14 040	−0.40
Small non-metropolitan cities	1 861 351	−7812	−0.42
Industrial districts	7 475 515	7194	0.10
Districts with new towns	2 838 258	2627	0.09
Resorts, ports and retirement districts	3 591 972	17 637	0.49
Urban–rural mixed	7 918 701	19 537	0.25
Remote urban–rural	2 302 925	13 665	0.59
Remote rural	1 645 330	10 022	0.61
Most remote rural	4 731 278	36 450	0.77

Note: Metropolitan cities and districts include the Central Clydesdale conurbation area

Source: Calculated from the 1991 Census SMS and LBS/SAA (ESRC/JISC purchase). Reproduced from *Population Trends 83*, Spring 1996

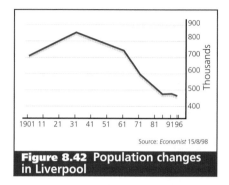

Figure 8.42 Population changes in Liverpool

Source: *Economist* 15/8/98

Figure 8.41 The village of Brockham, Surrey: 1990s private housing development

Reurbanisation

In the last few years British cities have (to a limited extent so far) reversed the population decline that has dominated the post-war period. In fact, Merseyside was the only urban region where population fell between 1991 and 1996. Central government finance, for example the millions of pounds of subsidies poured into London's docklands, Manchester's Hulme wastelands and Sheffield's light railway, has been an important factor in the revival. New urban design is also playing a role. The rebuilding of part of Manchester's city centre after a massive IRA bomb explosion has allowed the planners to add new pedestrian areas, green spaces and residential accommodation.

In London, the City's so-called 'ring of steel' security measures against terrorism have proved so successful in deterring traffic that they are being extended over a much wider area to do just that. Road traffic in the City has been cut by 25%, with pollution down 25% and road accidents

reduced by more than a third. Peak period bus journey times have been reduced by as much as 70%. The popularity of the City's scheme suggests that the advantages of urban traffic restriction measures of this kind may have been underestimated in the past.

The reduction in urban street crime due to the installation of automated closed-circuit surveillance cameras has significantly improved public perception of central areas. In Newcastle police claim that crime in areas covered by cameras was cut by nearly half between 1991 and 1997. Rather than displacing crime to nearby areas as some critics have claimed, a recent Home Office study found that, on the contrary, the installation of cameras had a halo effect, causing reductions in crime in surrounding areas.

Is recent reurbanisation just a short-term blip or the beginning of a significant trend, at least in the medium-term? Perhaps the most important factor favouring the latter is the government's prediction of the formation of 4.4 million extra households over the next two decades. Sixty per cent of these new households will have to be housed in existing urban areas because there is such fierce opposition to the relaxation of planning restrictions in the countryside. Also, as many of the new households will be single-person, the existing urban areas may well be where most would prefer to live.

The rejuvenation of inner London

For the first time in about 30 years London stopped losing population in the mid-1980s and has been gaining people ever since, due to net immigration from overseas and natural increase. Perhaps the most surprising aspect of this trend is the rejuvenation of inner London where the population peaked at 5 million in 1900 (Figure 8.43), but then steadily dropped to a low of 2.5 million by 1983. The Department of the Environment forecasts it will reach 3 million by 2011,

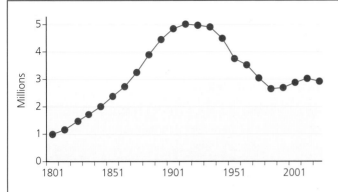

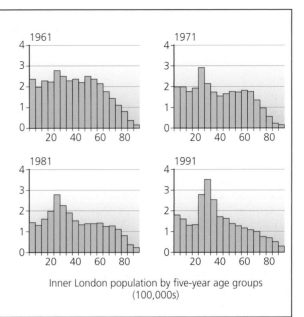

The influx of young migrants to London has brought with it a whole host of new industries and services, not only because of their connections back to their country of origin, but also because migrants tend to be the more imaginative and enterprising of their peers.

London is now a recognised world leader in design, arts, fashion, food, entertainment, music, computer software development, multi-media programming and in a whole range of other new industries. This generates billions of pounds of new revenue for the country.

Inner London population by five-year age groups (100,000s)

Figure 8.43 Population change and age structure in Inner London

Figure 8.44 Reurbanisation in London Docklands

Key Definitions   3

Cycle of urbanisation   The stages of urban change from the growth of a city to counterurbanisation through to reurbanisation.

Suburbanisation   The outward growth of towns and cities to engulf surrounding villages and rural areas.

Counterurbanisation   The process of population decentralisation as people move from large urban areas to smaller urban settlements and rural areas.

Reurbanisation   When, after a clear period of decline, the population of a city, in particular the inner area, begins to increase again.

and subside thereafter. Young adults now form the predominant population group in Inner London, whereas in the 1960s all the inner London boroughs exhibited a mature population structure.

Urban structure

The patterns evident and the processes at work in large urban areas are complex but by the beginning of the 20th century geographers and others interested in urban form were beginning to see more clearly than before the similarities between cities as opposed to laying stress on the uniqueness

of each urban entity. The first generalisation about urban land use to gain widespread recognition emanated from the so-called 'Chicago School'.

The concentric zone model

Published in 1925, and based on American Mid-Western cities, particularly Chicago, E. W. Burgess's model (Figure 8.45) has survived much longer than perhaps it deserves as it has only limited applicability to modern cities. However, it did serve as a theoretical foundation for others to investigate further.

The main assumptions upon which the model was based are:

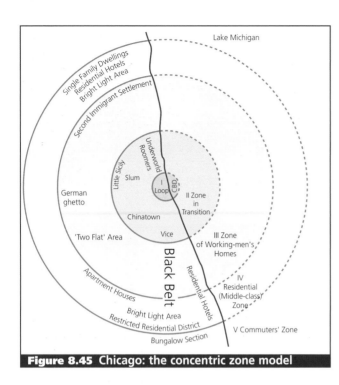

Figure 8.45 Chicago: the concentric zone model

- a uniform land surface

- free competition for space

- universal access to a single-centred city

- continuing in-migration to the city, with development taking place outward from the central core.

Burgess concluded that the city would tend to form a series of concentric zones. The model's basic concepts were drawn from ecology, with the physical expansion of the city occurring by invasion and succession, with each of the concentric zones expanding at the expense of the one beyond.

Business activities agglomerated in the central business district (CBD) which was the point of maximum accessibility for the urban area as a whole. Surrounding the CBD was the 'zone in transition' where older private houses were being subdivided into flats and bed-sitters or converted to offices and light industry. Newcomers to the city were attracted to this zone because of the concentration of relatively cheap, low-quality rented accommodation. In-migrants tended to group in ethnic ghettos and areas of vice could be recognised. However, as an ethnic group assimilated into the wider community, economically, socially and politically its members would steadily move out to zones of better housing, to be replaced by the most recent arrivals. Beyond the zone in transition came the 'zone of working-men's homes' characterised by some of the oldest housing in the city and stable social groups. Next came the 'residential zone' occupied by the middle classes with its newer and larger houses. Finally, the commuters' zone extended beyond the built-up area.

Burgess observed in his paper that 'neither Chicago nor any other city fits perfectly into this ideal scheme.

Complications are introduced by the lake front, the Chicago River, railroad lines, historical factors in the location of industry, the relative degree of the resistance of communities to invasion, etc.'

Bid-rent theory

Alonso's theory of urban land rent (1964) also produces a concentric zone formation, determined by the respective ability of land users to pay the higher costs of a central location (Figure 8.46). The high accessibility of land at the centre, which is in short supply, results in intense competition among potential land users. The prospective land user willing and able to bid the most will gain the most central location. The land user able to bid the least will be relegated to the most peripheral location.

He explained the paradox of poorer people living on expensive land in inner areas and more affluent people living on cheaper land further out as follows:

- With poor personal mobility low income groups prefer to reside in inner locations. They overcome the problem of land costs by living at high densities, each household buying or renting only a small amount of space.

- The more affluent, desiring a large house and garden, seek out cheaper land in the low-density suburbs where they can realise their 'dreams'. Being highly mobile they trade off space against accessibility to the CBD.

The assumptions upon which the theory is based and the criticisms of it are similar to the Burgess model.

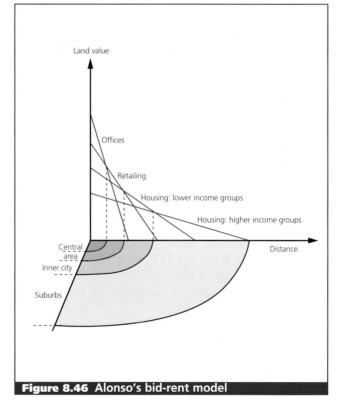

Figure 8.46 Alonso's bid-rent model

The sector model

Homer Hoyt's sector model (1939) was based on the study of 142 cities in the USA (Figure 8.47). Following Burgess, Hoyt placed the business district in a central location for the same reason – maximum accessibility. However, he observed that once variations arose in land uses near to the centre, they tended to persist as the city expanded. High-income housing usually developed where there were distinct physical or social attractions with low-income housing confined to the most unfavourable locations. Middle-income groups occupied intermediate locations. Major transport routes often played a key role in influencing sectoral growth, particularly with regard to industry. As new land was required by each sector it was developed at the periphery of that sector. However, medium- and high-class housing near the centre, the oldest housing in each case, was subject to suburban relocation by its residents, leading to deterioration, subdivision and occupation by low-income groups.

Figure 8.49 New York – zone in transition

The multiple nuclei model

C. D. Harris and E. Ullman (1945) argued that the pattern of urban land use does not develop around a single centre but around a number of discrete nuclei (Figure 8.50). Some nuclei may be long established, for example old villages which have been incorporated into the city by urban expansion. Others, such as industrial estates for light manufacturing, are much newer. Similar activities group together, benefiting from agglomeration while some land uses repel others. Middle- and high-income house buyers can afford to avoid residing close to industrial areas which become the preserve of the poor. A very rapid rate of urban expansion may result in some activities being dispersed to new nuclei, such as a new out-of-town shopping centre.

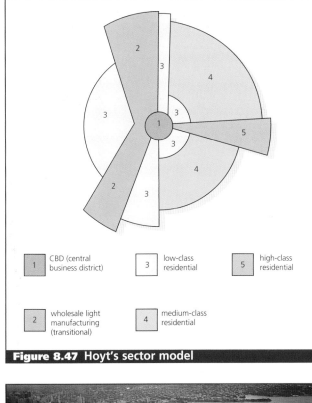

1 CBD (central business district)	3 low-class residential	5 high-class residential
2 wholesale light manufacturing (transitional)	4 medium-class residential	

Figure 8.47 Hoyt's sector model

Figure 8.48 Sydney: CBD and beyond

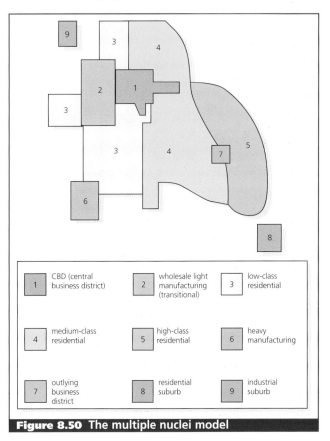

1 CBD (central business district)	2 wholesale light manufacturing (transitional)	3 low-class residential
4 medium-class residential	5 high-class residential	6 heavy manufacturing
7 outlying business district	8 residential suburb	9 industrial suburb

Figure 8.50 The multiple nuclei model

Figure 8.51 Little Italy, inner city, New York

Figure 8.52 The inner suburb of Queen's, New York

Key Definitions

Concentric zone A region of an urban area, circular in shape, surrounding the CBD and possibly other regions of a similar shape, that has common land use/socio-economic characteristics.

Zone in transition (twilight zone) The area just beyond the CBD which is characterised by a mixture of residential, industrial and commercial land use, tending towards deterioration and blight. The poor quality and relatively cheap cost of accommodation makes this part of the urban area a focus for in-migrants, resulting in a rate of population change higher than in other parts of the urban area.

Sector A section of an urban area in the shape of a wedge, beginning at the edge of the CBD and gradually widening to the periphery.

Bid-rent Decreasing accessibility from the centre of an urban area, with corresponding declining land values, allows (in theory) an ordering of land uses related to rent affordability.

Urban density gradient The rate at which population density and/or the intensity of land use falls off with increasing distance from the centre of the city.

Family life cycle Families with children pass through various stages over time (pre-child stage, family building, dispersal, post-child stage) with corresponding changes in housing needs.

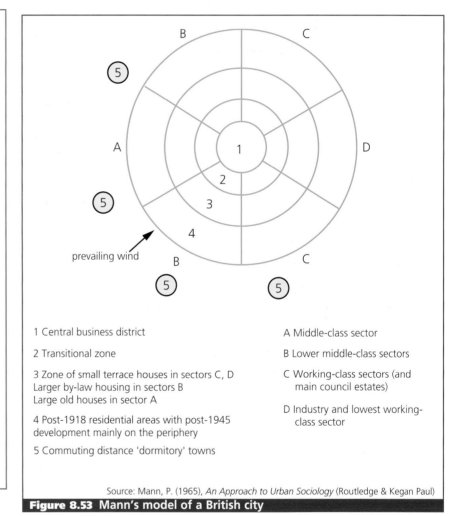

1 Central business district

2 Transitional zone

3 Zone of small terrace houses in sectors C, D
Larger by-law housing in sectors B
Large old houses in sector A

4 Post-1918 residential areas with post-1945 development mainly on the periphery

5 Commuting distance 'dormitory' towns

A Middle-class sector

B Lower middle-class sectors

C Working-class sectors (and main council estates)

D Industry and lowest working-class sector

Source: Mann, P. (1965), *An Approach to Urban Sociology* (Routledge & Kegan Paul)

Figure 8.53 Mann's model of a British city

A British urban land use model

P. Mann based his land use model for a typical British city on the theories of both Burgess and Hoyt (Figure 8.53) which he tried to apply to Sheffield, Nottingham and Huddersfield. The outcome was very much a compromise between the two models which he regarded as being complementary.

Identifying four residential sectors from middle class to lower working class he noted the influence of prevailing winds on the location of industry and the most expensive housing. He also allowed for local authority house building, particularly towards the periphery of the urban area, and for commuter villages.

A model of the modern North American city

Another model which incorporates aspects of both Burgess and Hoyt was produced by David Clark (Figure 8.54) in his book *Post-Industrial America*, although similar diagrams have also been produced by others. Here the CBD is subdivided into core and frame. Outside the low-income inner city are three suburban rings divided into sectors of lower middle, middle and high income. Important elements in the commercial hierarchy are included along with industrial and office parks. Thus, decentralisation is a key element of this model. The central city boundary shows the legal limits of the main city which once contained the whole urban area. In the 20th century the city has sprawled way beyond its legal limits to incorporate other legal entities. The Standard Metropolitan Statistical Area (SMSA) also includes the rural sections of counties which form part of the wider urban area.

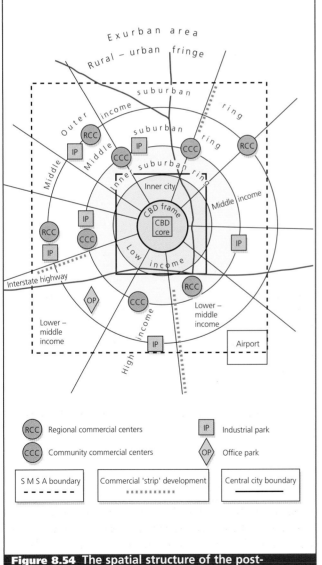

Figure 8.54 The spatial structure of the post-industrial American city

Models of developing world cities

Although the development of urban land use models has favoured western cities some interesting contributions relating to developing cities and socialist cities have appeared at various points in time.

Griffin and Ford's model (Figure 8.55) summarises many of characteristics that they noted in modern Latin American cities:

■ Central areas which had changed radically from the colonial period to now exhibit most of the characteristics of modern Western CBDs.

■ The development of a commercial spine, extending outwards from the CBD, enveloped by an elite residential sector.

■ The tendency for industries with their need for urban services such as power and water to be near the central area.

■ A 'zone of maturity' with a full range of services containing both older, traditional-style housing and more recent residential development. The traditional housing, once occupied by higher income families who now reside in the elite sector, has generally undergone subdivision and deterioration. A significant proportion of recent housing is self-built of permanent materials and of reasonable quality.

Figure 8.55 Low cost government housing.

■ A zone of 'in situ accretion' with a wide variety of housing types and quality but with much still in the process of extension or improvement. Urban services tend to be patchy in this zone with typically only the main streets having a good surface. Government housing projects are often a feature of this zone.

■ A zone of squatter settlements which is the place of residence of most recent in-migrants. Services in this zone are at their most sparse with open trenches serving as sewers and communal taps providing water. Most housing is of the 'shanty' type, constructed of wood, flattened oil cans, polythene and any other materials available at the time of construction. The situation is dynamic and there is evidence of housing at various stages of improvement.

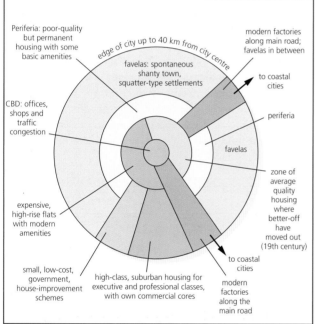

Figure 8.56 Waugh's model of a South American city

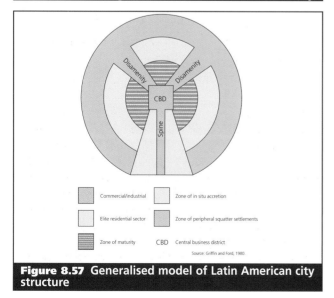
Figure 8.57 Generalised model of Latin American city structure

Of perhaps greater familiarity is Waugh's model based on Brazilian cities. The annotations on Figure 8.56 make the model largely self-explanatory.

Socialist cities

Although communism as a political ideology appears to be in its death throes, the impact of decades of central planning is still very clear in the urban and rural landscapes of the countries of the old Soviet Union, its eastern European satellites and other countries such as China and Cuba, where such a system of government operated. Evidence of the pre-communist era is most widespread in the historic core. Beyond this the communist or socialist city shows a greater degree of uniformity by urban region than other cities because of the absolutely dominant role of central planning and the absence of private enterprise.

F. E. Hamilton's study of East European socialist cities (Figure 8.58) identified the following characteristics:

■ the historic core, generally conserved for the purposes of cultural heritage

■ the CBD where public buildings are much more dominant than in Western cities

■ areas of 1950s housing, mainly in the form of high-rise apartment blocks

■ integrated neighbourhoods and residential districts constructed in the 1970s and 1980s. A reaction to the unpopularity of high-rise, neighbourhoods were planned to contain all necessary services to minimise population movement

■ open or planted 'isolation belts' to contain the expansion of the built-up area

■ industrial zones located at the periphery of the urban area and separated from residential areas by isolation belts to minimise the impact of pollution and other externalities.

Urban density gradients

Contrasting functional zones within urban areas characteristically vary in residential population density. Examination of population density gradients, termed **gradient analysis**, shows that for most cities densities fall with increasing distance from the centre. Gradient analysis of developed cities over time (Figure 8.62) shows the following trends:

■ the initial rise and later decline in density of the central area

■ the outward spread of population and the consequent reduction in overall density gradient over time.

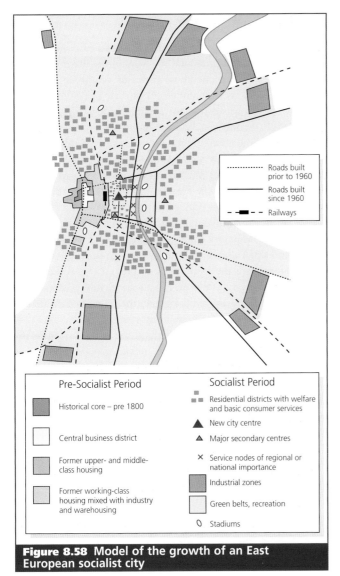

Figure 8.60 CBD, St Petersburg

Pre-Socialist Period

- Historical core – pre 1800
- Central business district
- Former upper- and middle-class housing
- Former working-class housing mixed with industry and warehousing

Socialist Period

- Residential districts with welfare and basic consumer services
- ▲ New city centre
- ▲ Major secondary centres
- ✕ Service nodes of regional or national importance
- Industrial zones
- Green belts, recreation
- ◯ Stadiums

Figure 8.58 Model of the growth of an East European socialist city

Figure 8.61 Public housing, St Petersburg

Figure 8.59 The central public area, St Petersburg

In contrast, analysis of density gradients in developing countries shows:

- **a continuing increase in central area densities**
- **the consequent maintenance of fairly stable density gradients as the urban area expands.**

In developing cities both personal mobility and the sophistication of the transport infrastructure operate at a considerably lower level. Also, central areas tend to retain an important residential function. Both of these factors result in a more compact central area and the transport factor in particular has restricted urban sprawl to levels below that of developed cities. The presence of extensive areas of informal settlement in the outer areas also results in higher suburban densities. However, in the more advanced of the developing nations where car ownership is rising rapidly, significant sprawl is now occurring.

Figure 8.63 shows how the distance–density relationship has changed in London since 1801.

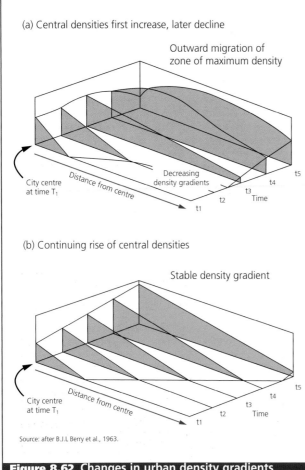

(a) Central densities first increase, later decline

Outward migration of zone of maximum density

City centre at time T₁

Distance from centre

Decreasing density gradients

Time

(b) Continuing rise of central densities

Stable density gradient

City centre at time T₁

Distance from centre

Time

Source: after B.J.L Berry et al., 1963.

Figure 8.62 Changes in urban density gradients through time for (a) 'Western' and (b) 'non-Western' cities

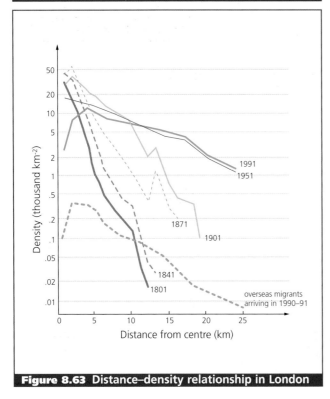

Figure 8.63 Distance–density relationship in London

The family life-cycle

Demographic analysis shows that movements of population within cities are closely related to stages in the life-cycle, with the available housing stock being a major determinant of where people live at different stages in their life. Studies in Toronto show a broad concentric zone pattern (Figure 8.64). Young adults frequently choose housing close to the CBD while older families occupy the next ring out. Middle-aged families are more likely to reside at a greater distance from the central area and farther out still, in the newest suburban areas, young families dominate. This simplified model applies particularly well to a rapidly growing metropolis like Toronto where an invasion and succession process evolves over time.

Toronto's inner city contains a much higher percentage of rented and small unit accommodation than the outer regions which, along with the stimulus of employment and the social attractions of the central area, has attracted young adults to the area. Most housing units built in the inner area in recent decades have been in the form of apartments.

Studies in Britain have highlighted the spatial contrasts in life-cycle between middle and low income groups (Figure 8.65). With life-cycle and income being the major determinants of where people live, residential patterns are also influenced by a range of organisations foremost of which are local authorities, housing associations, building societies and landowners. On top of this is the range of choice available to the household. For those on low income this is frequently very restricted indeed. As income rises the range of choice in terms of housing type and location increases.

Questions

1 Study Figure 8.66.
  (a) State briefly the meaning of the terms (i) suburbanised village, (ii) declining village, (iii) overspill town.
  (b) What has prevented the coalescence of the conurbation and the nearest settlements to it?
  (c) How has distance from the conurbation affected its impact on surrounding smaller settlements?
  (d) Suggest how the number of second homes can affect the character of a village.
2 **(a)** Explain why the CBD is placed at the centre in the concentric zone and sector models.
  (b) Describe the characteristics of the zone in transition.
  (c) How can the growth of an urban area be explained through the use of the concentric zone model?
  (d) To what extent does the bid-rent model support concentric zone theory?
  (e) Discuss three reasons why a concentric zone pattern is unlikely in reality.
3 **(a)** In Figure 8.47, why is land use arranged in sectors?
  (b) Why does the character of the inner areas of middle-class and high-class housing change over time?
  (c) In Figure 8.50 comment on the position of the following:

(i) wholesale light manufacturing, (ii) the outlying business district, (ii) heavy manufacturing.

(d) How is it possible that such different models were proposed to explain the pattern of land use in American cities?

4 (a) On which two previous theories of urban land use is Figure 8.53 based?

(b) Suggest two reasons for the location of the industrial zone.

(c) Which factors might explain the location of the middle-class sector?

5 Study Figure 8.54.

(a) How and why will land use and vertical profile differ between the CBD core and the CBD frame?

(b) State four differences you would expect to find between residential environments in low income inner city areas and high income middle suburban ring neighbourhoods.

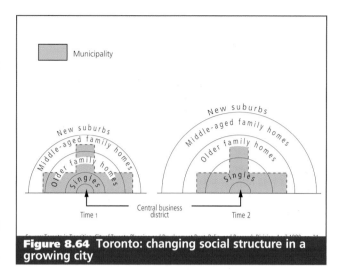

Figure 8.64 Toronto: changing social structure in a growing city

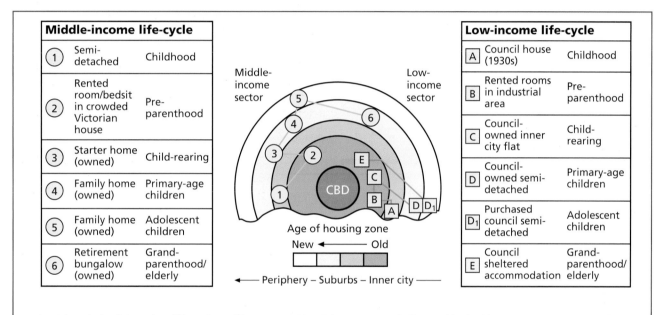

Housing choice is based on life-cycle and income. Residential patterns are influenced by building societies, landowners, local authorities/housing associations, and free choice.

Figure 8.65 Middle and low income models of the family life cycle

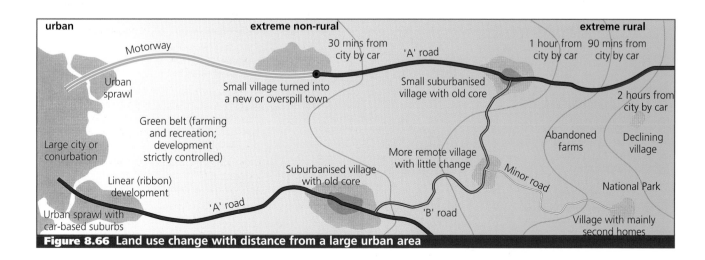

Figure 8.66 Land use change with distance from a large urban area

(c) Account for the location of industrial and office parks.

(d) How would you expect the functions of 'commercial strips' to differ from those of regional commercial centres?

(e) To what extent can earlier models of urban structure be recognised in Figure 8.54?

6 (a) Suggest reasons for the development of a commercial spine (Figure 8.55) extending outwards from the CBD.

(b) How do socio-economic conditions vary between the zone of maturity and the zone of 'in situ' accretion?

(c) Why in Figure 8.56 is there a significant area of higher income adjacent to the CBD?

(d) Discuss the main differences between the periferia and the favelas.

(e) Describe the characteristics of a low-cost government housing scheme for a developing city you have studied.

7 (a) Look at Figure 8.67. Suggest reasons for the location of (i) the hotel zone, (ii) the industrial zone.

(b) Explain the location of the three areas of squatter settlement.

(c) Comment on the location of the medina. How might the street pattern of the medina differ from that of the French and high-class Arab quarter?

(d) Suggest four ways in which the urban environment would differ between the French and high-class Arab quarter and the low-class Arab quarter.

(e) How would you expect population density to vary between the different areas of Sousse?

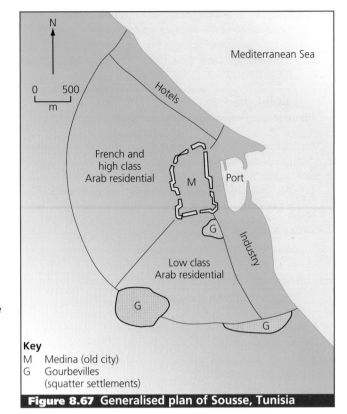

Key
M   Medina (old city)
G   Gourbevilles (squatter settlements)

Figure 8.67 Generalised plan of Sousse, Tunisia

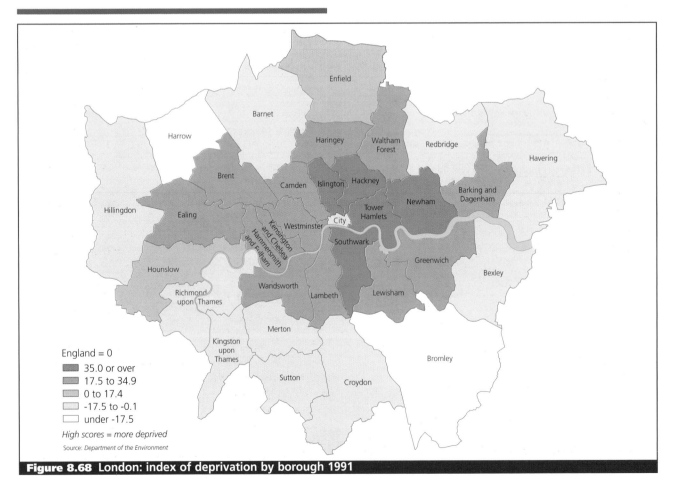

England = 0

- 35.0 or over
- 17.5 to 34.9
- 0 to 17.4
- -17.5 to -0.1
- under -17.5

High scores = more deprived

Source: *Department of the Environment*

Figure 8.68 London: index of deprivation by borough 1991

The quality of life in cities

The quality of life usually varies considerably both between and within cities in the same country. Over time a wide range of different socio-economic indicators have been used to identify such variation. It is not surprising that each indicator, at least to some extent, produces a different pattern as every individual measure has its merits and limitations. Thus most recent attempts to measure spatial variations in the quality of life have combined a range of indicators to form a composite quality of life index. The Department of the Environment used 1991 census data and other information to compile an Index of Deprivation for the 366 local authority districts (including London Boroughs) of England. The index gave equal weight to 13 indicators including level of unemployment, proportion of households without a car, standardised mortality rate, and households lacking basic amenities. The results of this analysis for London are shown in Figure 8.68. The average for the country has been set at zero. Positive scores indicate above average levels of deprivation, negative scores below average levels. Fourteen of the 20 most deprived districts in England were recorded as being in London, with Newham, Southwark, Hackney and Islington being the most deprived boroughs in the country in that order. The most affluent London boroughs were Bromley, Harrow, Sutton and Bexley.

The Department of the Environment, Transport and Regions (DETR) has recently updated and revised the 1991 analysis and issued it as the 1998 Index of Local Deprivation (Figure 8.69). This includes four measures of deprivation:

- the degree of deprivation as expressed by the overall district level score

- the intensity of deprivation as calculated by the average score of the worst three wards in the local authority

- ward extent: the proportion of the local authority population living in wards that are within the most 10% deprived in England

- ED extent: the proportion of the EDs (Enumeration Districts) in the local authority population that fall within the 7% most deprived EDs in England.

Compared with the 1991 analysis, London boroughs are slightly lower down (better-off) in the national rankings, largely due to the removal of children in the unsuitable accommodation indicator. On the overall degree of deprivation measure, 13 of the most deprived districts in England are in London. On the other three measures London boroughs form at least half of the 20 most deprived local authority areas.

Figure 8.70 shows the regional location of deprived local authority housing estates based on the 1991 analysis. The

Figure 8.69 National ranking of London boroughs on the 1998 Index of Local Deprivation (1 = most deprived)

	Degree	Intensity	Ward Extent	ED Extent
City of London	183	287	=158	=253
Barking & Dagenham	15	65	22	89
Barnet	130	106	105	100
Bexley	148	147	127	181
Brent	20	16	21	14
Bromley	179	97	88	148
Camden	17	32	9	23
Croydon	88	63	91	125
Ealing	36	52	48	66
Enfield	70	79	77	116
Greenwich	11	45	16	26
Hackney	4	8	1	1
Hammersmith & Fulham	18	17	14	16
Haringey	13	12	7	9
Harrow	145	237	=158	217
Havering	143	81	117	179
Hillingdon	120	222	=158	228
Hounslow	59	132	138	112
Islington	10	22	3	10
Kensington & Chelsea	63	28	28	24
Kingston	220	196	=158	244
Lambeth	12	9	5	7
Lewisham	14	19	17	20
Merton	122	110	107	106
Newham	2	10	2	3
Redbridge	90	75	78	141
Richmond	156	267	=158	250
Southwark	8	14	6	4
Sutton	284	142	139	178
Tower Hamlets	6	5	4	2
Waltham Forest	22	13	12	25
Wandsworth	30	48	35	57
Westminster, City of	57	20	13	21

Source: Department of the Environment, Transport, and the Regions

boroughs of Hackney, Southwark and Tower Hamlets all contained a hundred or more deprived estates, accounting between them for a fifth of the national total. Local authority and housing association tenants are a more uniformly poor and disadvantaged group than ever before.

Not surprisingly, the distribution of income in London is more unequal than in the UK as a whole and the gap between better and worse off in London has been getting wider (Figure 8.72). One result is that the proportion of Londoners reliant upon means-tested benefits has increased sharply since the early 1980s. Over 1.5 million Londoners were reliant upon Income Support in 1994. In the same year more than one in two pupils in Lambeth, Hackney, Southwark and Tower Hamlets were eligible for free school meals. At 64%, the rate in Tower Hamlets was the highest of any local authority in the country.

Unemployment also shows a strong spatial concentration. Twenty per cent of unemployed people live in the ten most

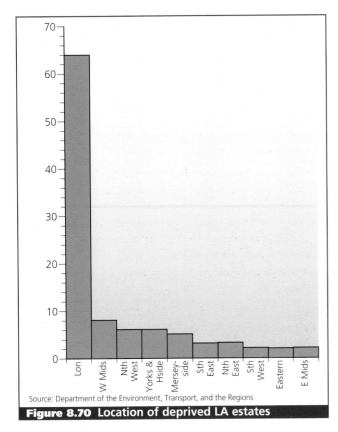

Source: Department of the Environment, Transport, and the Regions

Figure 8.70 Location of deprived LA estates

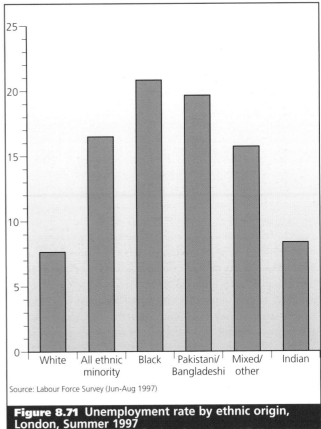

Source: Labour Force Survey (Jun-Aug 1997)

Figure 8.71 Unemployment rate by ethnic origin, London, Summer 1997

deprived wards. Unemployment is two to three times higher for non-white ethnic groups compared with those from white ethnic groups (Figure 8.71).

In terms of individual measures perhaps the standard mortality ratio conveys the contrast within London at its bleakest (Figure 8.73). Mortality is to a significant extent the product of all the other indicators of relative deprivation/affluence. Observed deaths in 1995 were 25% higher than expected in the most deprived wards given their population characteristics. In the least deprived wards they were 20% less than expected. This mortality gap is wider now than it was in 1981. Infant mortality rates in the most deprived inner London boroughs are up to twice as high as within some outer London boroughs.

On all counts the contrast between inner and outer London is striking, with the most intense deprivation in inner London being concentrated towards the east. Even so, significant contrasts exist within virtually all boroughs so that the better-off wards in some inner London boroughs often record a higher quality of life than the least affluent wards in outer London boroughs. The pattern found within boroughs is often quite intricate, forming the 'residential mosaic' that social geographers frequently talk about. Even within wards considerable contrasts can be found as fieldwork and examination of ED data will confirm.

The government identified deprivation not as an academic exercise but as a means of targeting resources on areas most in need (Figure 8.74). The New Deal for Communities

London Environmental Factfile

- London households each produce an average of 1.3 tonnes of waste each year. The 33 London local authorities collect almost 4 million tonnes of household and commercial waste each year. Industrial waste amounts to about 9 million tonnes each year.

- Although industrial pollution has declined, motor vehicles have been responsible for increased emissions of carbon monoxide, nitric oxide, fine particulates (PM_{10}) and secondary pollutants such as nitrogen oxide and ozone. In 1997 there were 2.7 million vehicles licenced in London.

- Aircraft landing and taking off account for 3% of London's nitrogen oxide emissions. Traffic associated with the airports adds further to this.

- Power stations on the Thames estuary the main sources of sulphur dioxide affecting London.

- The last survey of derelict land shows that there were 1,625ha in London in 1993.

- Water quality in rivers and canals is graded A (best) to E (worst). Most of London's rivers and canals fall into the B to E range of grades.

- Average traffic speeds in the central area between 7.45 and 9.15 am fell from 12.7mph between 1968–70 to 10.0mph between 1996–98. The number of people entering central London between 7–10am in 1997 was 1.059 million.

- Complaints to Environmental Health Officers about noise from domestic premises increased by more than 50% between 1994/5 and 1996/7.

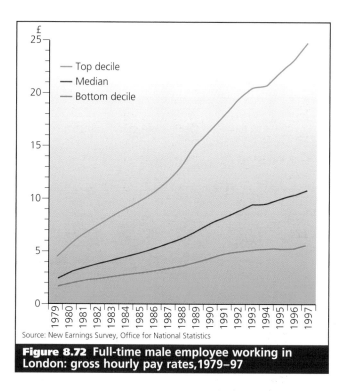

Source: New Earnings Survey, Office for National Statistics

Figure 8.72 Full-time male employee working in London: gross hourly pay rates,1979–97

launched by the government in 1998 aims to 'extend economic opportunity, tackle social exclusion and improve neighbourhood management and quality of life in some of the most rundown areas in the country'.

Detailed analysis down to ED level shows quite clearly that there are small areas of striking affluence in inner London, and to a lesser extent in inner areas of other large cities. There are two main reasons for clusters of high socio-economic status in the inner city:

■ Some areas have always been fashionable for those with money. Areas such as St John's Wood and Chelsea are both only a short journey to the City and West End, and pleasantly laid out with a good measure of open space. The original high quality of housing has been maintained to a very good standard.

■ Other areas have become fashionable in recent decades through the process of gentrification, which in the areas where it takes place reverses the filter-down process of older property (Figure 8.75). Gentrification is marked by the occupation of more space per person than the original occupants from lower socio-economic groups.

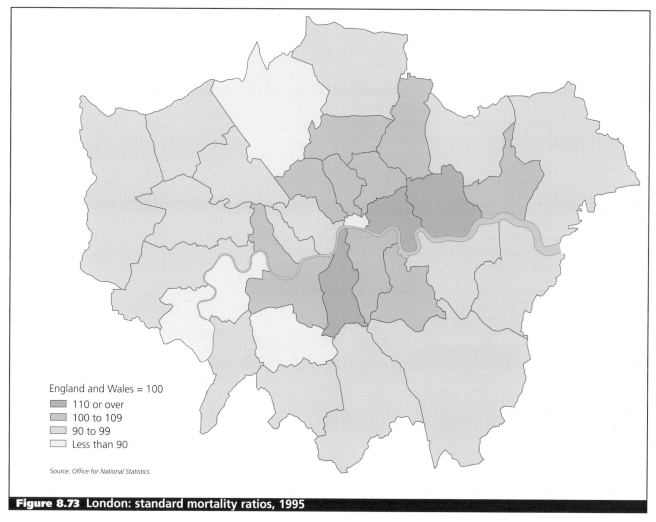

England and Wales = 100

■ 110 or over
■ 100 to 109
■ 90 to 99
□ Less than 90

Source: *Office for National Statistics*

Figure 8.73 London: standard mortality ratios, 1995

Government zones in on deprivation

Multi-pronged attack answers critics of New Labour Policy towards poor. **Simon Buckby** reports

Roy Hattersley, former deputy leader of the Labour party, frequently claims that New Labour does not care sufficiently about the poor. Yet rarely can the socially deprived have received so much attention from a government.

Its work-centred welfare policy is aimed at people living on benefits. The Social Exclusion Unit, due to report next month, has developed plans to tackle homelessness and improve run-down housing estates. And combination punches from a first wave of 12 education action zones, five employment action zones and 11 health action zones are designed to deal a knock-out blow to many of the ill effects of poverty.

The health action zones, the most developed of the three, will receive priority access to capital funding, including money from the private finance initiative, and to National Lottery support.

The zones' priority is to be tough on the causes of poor health. They will 'promote local partnerships to tackle pollution, homelessness, unemployment, and poverty', says Frank Dobson, the health secretary. As further evidence of the government's desire to encourage a multiplicity of agents working to prevent deprivation, the health authorities selected from the 41 applications reflect joint proposals with social services departments, GPs and the voluntary and private sectors.

One health action zone is in the Lambeth, Southwark and Lewisham health authority in south London, where the hospitals, social services departments, police and probation services discovered they were all dealing with problems from the same families.

'Those who complained to the council about poor housing were often the same people who regularly visit their GP or turn up in the accident and emergency department at the hospital, report drug problems, are in conflict with the schools over truancy, or suffer family breakdown,' says Matthew Swindells, general manager for women and children's services at Guys Hospital. 'We are trying to co-ordinate our response to them, rather than treating each of their problems separately.'

The health authority has the highest rate of under-age pregnancies in the country, and 38 per cent of local children are raised in households with no earned income. One of the central projects of the zone is to co-ordinate education campaigns to reduce the number of pregnancies among the young and encourage parental responsibility.

The programme has the support of Simon Hughes, MP for Southwark North and Bermondsey and Liberal Democrat health spokesman.

'An urgent move away from early sex, early children and early single parenting as a result will be one of the best legacies we could leave, both to this generation and the next,' he says.

Beating the blackspots

1. Northumberland: to tackle pockets of severe deprivation in an environment scarred by industrial decline
2. Tyne and Wear: to tackle transport and housing problems
3. N Cumbria: to deal with acute deprivation on west coast
4. Bradford: community-based diabetes service being set up
5. Manchester: to provide employment and training opportunities to people with marital problems
6. S Yorks: to tackle problems of the young in communities ravaged by pit closures
7. Sandwell, West Midlands: to develop community-focused health service
8. Luton: to tackle health needs of Asian women
9. Hackney, Newham and Tower Hamlets: to address greatest concentration of poverty in the country
10. Lambeth, Southwark and Lewisham: area with country's highest rate of teenage pregnancies
11. Plymouth: some of the most deprived neighbourhoods in Britain

Figure 8.74 *Financial Times*, July 1998

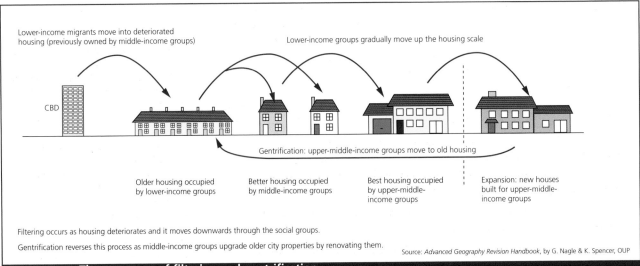

Filtering occurs as housing deteriorates and it moves downwards through the social groups.

Gentrification reverses this process as middle-income groups upgrade older city properties by renovating them.

Source: *Advanced Geography Revision Handbook*, by G. Nagle & K. Spencer, OUP

Figure 8.75 The processes of filtering and gentrification

Figure 8.76 The residential mosaic in the London Borough of Hammersmith and Fulham

The inner-city problem: a sequence of explanations

The nature of and linkage between inner-city problems is well-illustrated by the web of decline, deprivation and despair (Figure 8.77). Inner-city decline is the counterpart of suburbanisation. Over the years a number of different explanations of the inner-city problems have appeared (Figure 8.78). For over two decades after the Second World War inner-city problems were attributed primarily to poor housing and other aspects of the run-down built environment. The solutions were clearance and redevelopment.

From the late 1960s to the mid-1970s the focus of attention shifted to issues of social deprivation. Within this broad area three alternative explanations emerged:

■ A culture of poverty in which families of a certain kind pass on an anti-social life-style from one generation to the next. Social norms in certain areas encouraged vandalism, early school-leaving, early marriage, early child-rearing, crime and a general disrespect for authority.

■ The cycle of deprivation (Figure 8.79) which was seen as preventing the poor from improving the quality of their lives.

■ Institutional malfunctioning characterised by poor links between social and welfare services on the one hand and populations most in need on the other.

Key Definitions ⑤

Deprivation   Defined by the Department of the Environment as when 'an individual's well-being falls below a level generally regarded as a reasonable minimum for Britain today'.

Standard mortality ratio   The ratio expresses the number of deaths in an area as a percentage of the hypothetical number that would have occurred if the area's population had experienced the sex/age specific rates of England and Wales in that year.

Gentrification   A process in which wealthier people move into, renovate and restore run-down housing in an inner city or other neglected area. Such housing was formerly inhabited by low-income groups, the tenure shifting from private-rented to owner-occupation.

Residential mosaic   The complex pattern of different residential areas within a city reflecting variations in socio-economic status which are mainly attributable to income.

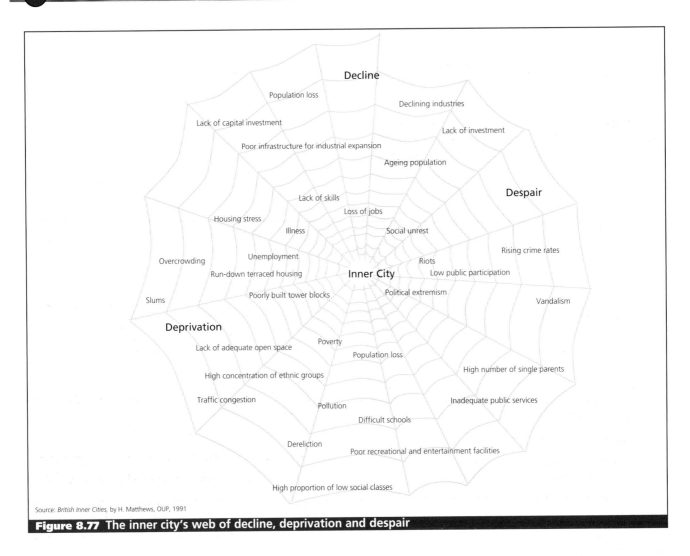

Source: *British Inner Cities*, by H. Matthews, OUP, 1991

Figure 8.77 The inner city's web of decline, deprivation and despair

Figure 8.78 Different explanations of the inner city problem

Perspective	Perceived problem	Goal	Means
Physical decay	Obsolescence	Better built environment	Physical planning
Culture of poverty	Pathology of deviant groups	Better social adjustment	Social education
Cycle of deprivation	Individual inadequacy	Better families	Social work
Institutional malfunction	Planning failure	Better planning	Co-ordinated planning
Resource maldistribution	Inequality	Reallocation of resources	Positive discrimination
Structural conflict	Underdevelopment	Redistribution of power	Political change

From the mid-1970s two additional explanations emerged which focused on the structural (economic) fabric of the inner city:

■ **Resource maldistribution** as a result of the decentralisation of population, industry and commerce to the suburbs and beyond. The massive decline of inner-city manufacturing was seen as the most important deprivation factor.

■ **Structural conflict** whereby large companies benefit from keeping the inner city underdeveloped, viewing it as an area providing a pool of expendable, unskilled labour, and cheap land, which can be taken up or abandoned according to fluctuations in the economy. This explanation sees the need for capitalism to be replaced by a socialist system of production.

Although not all local authority housing estates are in inner cities, a significant proportion are. As a recent study confirmed (*Living with the State*, Institute for Fiscal Studies, 1996) such estates are increasingly becoming the focal points of deprivation. Alarmingly, the report noted that half of all

Figure 8.79 The cycle of deprivation

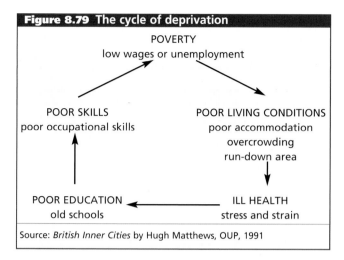

POVERTY
low wages or unemployment

POOR SKILLS
poor occupational skills

POOR LIVING CONDITIONS
poor accommodation
overcrowding
run-down area

POOR EDUCATION
old schools

ILL HEALTH
stress and strain

Source: *British Inner Cities* by Hugh Matthews, OUP, 1991

Figure 8.80 The White City estate, Inner London

men in council housing were unemployed compared with 10% in the 1960s. Half of all council tenants were now in the poorest fifth of the population and those entering such housing were far poorer than those moving out. This is in effect the spatial marginalisation of those who are already socially marginalised. If these trends continue the successful metropolis, according to Siebel (1984) is likely to be divided into three different cities:

- the most visible city is the 'international' city

- hidden behind the international city is the 'normal' city or middle-class city

- in the shadow remains the 'marginalised' city

The processes of reurbanisation and gentrification are likely to accelerate this division. Where gentrification expands the 'international' city, the poor are further pushed out into the worst segment of the housing market.

Urban abandonment

Nearly 1 million homes in England, mainly in northern cities are blighted by low demand, abandonment and dereliction (Figure 8.81). Some of these areas are often surprisingly close to housing where demand is buoyant. Such housing problem areas suffer complex problems that drive residents to move south or to the suburbs. However, within abandoned areas there are often small pockets of stability sustained by extensive family networks. There is also evidence that regeneration schemes and programmes of community support have had positive impacts in helping to sustain local communities.

Suburban decline

A report published by the Civic Trust and Rowntree Foundation in early 1999 concluded that many suburbs built in the middle of the 20th century were approaching a 'crisis of neglect'. Decline is almost always triggered by the closure

of shops which usually happens because of the opening of retail superstores in other suburbs. The high street then becomes dominated by bargain-basement and charity shops. Other characteristics of declining suburbs include:

- public transport cutbacks

- closure of civic services such as libraries, post offices and police stations

- increasing crime, vandalism and graffiti

- a significant increase in the number of rented properties

- the arrival of the massage parlour and the snooker hall.

Examples of suburban decline are Gants Hill in east London and Hayes in west London. However, suburbs account for a third of Britain's housing stock and there are many successful examples.

Questions

1 Study Figure 8.68
   (a) Describe the pattern of deprivation in London.
   (b) Discuss the merits of three of the indicators used to compile the map.
   (c) Outline the merits of two other indicators which might have been used as part of this analysis.
   (d) Why is it important for government and others to be aware of variations in the quality of life?
2 (a) Describe the variation in unemployment by ethnic group in London (Figure 8.71).
   (b) Suggest reasons for such significant differences.
   (c) Analyse Figure 8.69.
   (d) How would you expect such trends to affect variations in the quality of life within London?
   (e) Examine the pattern of mortality within London.
   (f) Explain the main reasons for such large variations in mortality.
3 (a) Explain the processes of filtering and gentrification.

(b) Exemplify the chain of causation from decline through deprivation to despair (Figure 8.77).

(c) Examine the development of explanations for inner-city problems.

(d) With reference to an inner-city area you have studied exemplify the cycle of deprivation.

(e) Discuss the causes and consequences of urban abandonment.

4 **(a)** Study Figure 8.78. Explain the meaning of the terms 'externality' and 'place utility'.

(b) Discuss two positive externalities and two negative externalities that might affect a neighbourhood.

(c) Examine two of the internal factors that might cause a household to examine the possibility of moving house.

(d) Suggest three ways in which place utility might be improved for a household by moving house.

(e) For a neighbourhood or urban area with which you are familiar describe the typical sequence of residential mobility.

Extended Assignment

5 Study Figure 8.79.

(a) Select what you consider to be the five most important indicators of the quality of life. Justify your choices.

(b) On a copy of Figure 8.83 rank the 32 boroughs for each of the five indicators selected. Total the rankings for each borough in the penultimate column and in the last column show the final ranking of each borough (from 1 to 32).

(c) Draw a choropleth map to show your results. Comment on the pattern illustrated by your map.

(d) Select what you believe to be the best two indicators of the quality of life. Use Spearman's Rank Correlation Coefficient to show the extent of the relationship between these two indicators.

(e) Show the relationship between two other indicators through the use of a scattergraph. Draw a line of best fit and comment on the relationship shown.

(f) Use Scamp or another census data computer program to examine the best and worst boroughs according to your ranking on a ward by ward basis. Following the format of Figure 8.79, complete a composite quality of life index for each borough and discuss your findings.

(g) Now focus on the best and worst wards in each of the two boroughs by examining Enumeration District data. To what extent does the quality of life vary at this scale?

Problems of polarisation

Brian Robson on the challenge of tackling the co-existence of local pockets of urban reinvigoration and social polarisation

The abandonment of whole tracts of housing in some of Britain's big northern cities is a new phenomenon. Anne Power and Katharine Mumford have provided a valuable service in focusing on the problem and raising the issue in a Rowntree study report *The Slow Death of Great Cities?*

Power and Mumford draw principally on the experience of four neighbourhoods within Newcastle and Manchester. In the former, between 13 and 20 per cent of council housing is vacant; in the latter, the figure has risen dramatically from 5 per cent two years ago to 15 per cent today.

However, it is not simply council housing that suffers. The authors describe a pattern in which any kind of property (including new, well-built housing association property with gardens) can be abandoned within small tracts that have started upon a seemingly inexorable downward spiral of unpopularity. This is the American experience, from which we had thought British cities had largely escaped. The authors emphasise three factors to explain the trend:

• increasing social polarisation that has left large areas of cities with poor, unemployed ill-qualified people;

• the dominance of social housing within such areas; and

• the competition from low-cost private housing in suburban locations.

Social housing – and particularly council housing – has now become effectively the tenure of last resort. As older stable residents die, allocation policy increasingly becomes a desperate issue of finding anyone to occupy hard-to-let properties.

The areas of abandonment suffer disproportionately from high levels of crime and vandalism. They suffer from falling school rolls as the population disappears (although, interestingly, the authors point to the continuing success of Catholic schools in retaining pupils and maintaining standards through their 'ethos, pastoral role, parental support and teaching methods').

The authors suggest that small triggers can start the process of abandonment – one individual household or person moving into an area or a single family moving out can lead to the start of what they call 'a galloping process'. Demolitions may appear to be a way in which demand and supply can better be balanced, but they argue that demolition (or its threat) can itself exacerbate the problem: 'It appears impossible to sustain social cohesion when large areas have become almost universally poor in a wider context of growing affluence.'

A. Power and K. Mumford: *The Slow Death of Great Cities? Urban Abandonment or Urban Renaissance.* Joseph Rowntree Foundation, York, 1999

Figure 8.81 Source: *Town and Country Planning,* September 1999

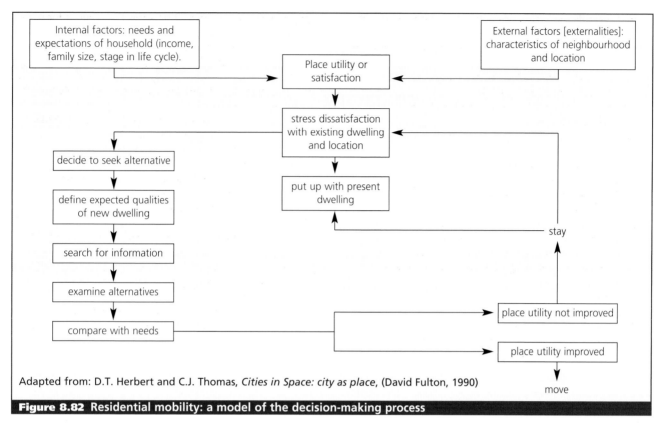

Adapted from: D.T. Herbert and C.J. Thomas, *Cities in Space: city as place*, (David Fulton, 1990)

Figure 8.82 Residential mobility: a model of the decision-making process

Urban planning and management

Although urban planning can be said to be as old as urban settlement the modern era of urban planning did not begin in Britain until the 1940s. What started as a gradual shift towards government involvement in town planning during the first 40 years of the century, climaxed in a radical transformation in official attitudes towards state intervention between the late 1930s and 1945. The passage through parliament of the Green Belt Act in 1944, the New Towns Act in 1946 and the Town and Country Planning Act in 1947 were milestones in spatial organisation, having evolved from a series of reports on major urban areas which appeared a few years earlier (Figure 8.80). There can be little doubt that Britain would be a much worse place without the level of organisation imposed by planning measures, but as Figure 8.80 explains, the urban landscape of today does not reflect the optimistic forecasts of 50 years ago.

The Town and Country Planning Act of 1947

This was the first comprehensive piece of planning legislation in England and Wales. The planning measures of the inter-war period had been piecemeal in nature, attempting to solve particular problems rather than treating the urban system as a whole, and as a result were largely ineffectual. Under the Town and Country Planning Act local authorities were placed under a statutory obligation to prepare development plans which were to state the general use to which each plot of land and building was to be put. Developers had to submit proposals to local authority planning committees which had the power to refuse applications deemed undesirable. Previously a local authority could only do this by purchasing the land on which the development was planned. The general absence of planning controls during the inter-war years had given developers a virtual free hand to take over thousands of acres of countryside and resulted in the huge spread of suburbia. The Town and Country Planning Act has been amended on a number of occasions since 1947 to take account of emerging trends.

Garden cities and new towns

Town planning as we know it today began in the 18th century. Examples are James Craig's New Town in Edinburgh, or the fashionable spa towns of Buxton and Bath in England. The next step in the evolution of town planning came with the Industrial Revolution. In 1817 the utopian socialist Robert Owen created New Lanark around the Clydesdale woollen mills built by his father-in-law. Cadbury (cocoa) and Lever (soap) followed suit and used some of their profits to build the towns of Bourneville and Port Sunlight. The objective of these philanthropists and others was to provide good housing and pleasant environmental conditions for their workers.

Figure 8.83 London: quality of life indicators (1991 Census Data)

Borough	% of HHs with no central heating	% of HHs with more than one person per room	% of HHs with no car	% of economically active population unemployed	% of houses detached and semi-detached	% of HHs social class 1 and social class 2	% HHs owner occupied
Barking & Dagenham	35.66	4.00	42.93	11.96	9.96	10.57	51.76
Barnet	11.38	3.28	30.21	8.49	40.81	34.59	68.94
Bexley	18.14	1.63	26.66	7.80	50.27	22.85	78.79
Brent	16.97	6.69	43.37	13.56	26.93	24.28	57.75
Bromley	12.78	1.33	25.60	7.25	47.01	31.79	77.92
Camden	16.17	4.79	55.82	13.66	3.64	33.92	33.76
Croydon	19.90	2.90	30.48	8.58	34.43	29.98	72.78
Ealing	16.32	5.06	36.58	11.26	21.45	29.50	63.81
Enfield	15.65	3.02	31.83	11.26	24.70	24.57	73.60
Greenwich	23.03	3.35	43.62	13.63	17.96	19.65	47.08
Hackney	22.21	7.48	61.75	22.94	2.24	20.07	26.92
Hammersmith & Fulham	25.91	4.92	52.00	13.04	2.43	31.52	41.93
Haringey	18.84	5.09	49.95	17.48	6.17	27.85	49.76
Harrow	11.41	3.22	26.53	7.63	49.16	32.22	77.89
Havering	13.82	1.72	26.53	7.55	49.68	21.88	78.84
Hillingdon	11.35	3.20	24.44	7.01	48.23	25.79	73.81
Hounslow	18.44	4.42	32.27	9.05	32.09	26.52	61.15
Islington	16.17	4.80	59.95	17.42	1.35	24.65	26.71
Kensington & Chelsea	20.00	5.12	50.49	12.27	2.01	40.47	39.85
Kingston upon Thames	15.22	1.92	26.97	6.21	43.84	35.47	74.72
Lambeth	25.14	5.09	55.41	17.46	7.21	27.02	36.24
Lewisham	27.25	4.07	47.13	14.69	10.33	23.65	37.81
Merton	21.38	3.01	33.38	9.12	16.87	29.74	70.52
Newham	21.19	7.57	53.54	19.57	3.51	14.68	49.80
Redbridge	17.37	2.84	29.94	9.07	25.63	27.05	78.18
Richmond upon Thames	17.25	1.50	28.48	6.71	30.34	42.65	70.03
Southwark	16.56	5.43	57.96	18.63	5.06	20.39	27.17
Sutton	15.27	1.82	26.35	6.80	37.29	28.37	75.25
Tower Hamlets	14.22	11.07	61.55	22.24	1.09	16.98	23.23
Waltham Forest	27.69	4.50	42.92	12.81	12.16	22.40	61.50
Wandsworth	22.47	3.87	44.00	11.77	6.04	33.63	53.62
Westminster	22.23	5.68	57.73	12.33	3.64	32.07	35.08

Of enormous significance was the publication in 1898 of *Tomorrow, a Peaceful Path to Real Reform* by Ebenezer Howard, republished four years later as *Garden Cities of Tomorrow*. Howard was appalled by the squalor of urban Britain and was sure that there was a better way. In 1899 the Garden City Association (GCA) was founded and drew up plans for Letchworth in Hertfordshire which was started in 1903. Not just a visionary writer, Howard had persuaded a number of rich and influential people to invest in real garden cities. In 1918 the GCA published 'New Towns after the War' in which a hundred new towns were proposed. Between 1919 and 1939 four and a half million houses were built, but only one New Town, Welwyn Garden City, begun in 1920, was constructed (Figure 8.85).

Howard's ideas were to provide inspiration for like-minded people in a number of other countries. Howard died in 1928 but his followers continued to campaign and after the

Second World War the government accepted their arguments, largely because of the strong recommendation of the Barlow Report published in 1940. In 1948, Welwyn Garden City became one of London's New Towns and the town thus provides visible continuity between the aims of Howard and the objectives of post-Second World War planners.

When in 1976 it was decided not to undertake any more New Town projects for the time being, 32 such towns had already been designated in the UK (Figure 8.86). Fourteen were designated between the passage of the New Towns Act in 1946 and 1950 with only one, Cumbernauld in Scotland, added in the 1950s. These are regarded as the first generation New Towns and were largely mechanisms for replacing bomb-damaged houses and reducing the extremely high population densities in the inner areas of London and Glasgow. However, a few such as Cwmbran in South Wales were intended to act primarily as small growth poles in

The written plans that ushered in the planning era in the 1940s – Lock on Middlesbrough, Abercrombie on London, Nicholas on Manchester – were full of idealism, certainty and a grand missionary spirit. They captured the flavour of excitement and conviction out of which the 1947 Act emerged.

There was a template for a better life and it was one that could be delivered through the physical redesign of the urban environment. Green belts, new towns, land use zoning and the separation of uses, neighbourhood design, integrated transport; all were important parts of the prescription. Many of those written plans formed the inputs into the rebuilding and redesign of our towns and cities in the 1950s and 1960s and well beyond. They epitomised the dreams of an exhausted but hopeful nation which thought that it could indeed build a new Jerusalem.

But the flame of idealism grew dim as reality returned. So much of what was achieved was so much less than what had been planned: financial constraints, selfish interests, pressures from a changing world economy, misplaced and sometimes corrupt political partisanship were all conducive to a less than ideal outcome.

PERHAPS WE were never bold enough – or the economy was never sufficiently expansive – to wipe away the urban detritus which industrialisation and war had wrought. Perhaps Corbusian ideas were fundamentally implausible as a way of tackling the patching and re-making of an established urban landscape, as opposed to their use in the greenfield contexts of a BrasÍlia.

1 Perhaps the restructuring of the economy since the 1970s – which worked so much to undermine the economic roles of towns and cities – simply exposed the inadequacies of a planning system whose mechanism was based fundamentally on the control, not the attraction of development.
2 Perhaps the unforeseen speed of growth of private cars produced a leviathan which planning ideas could not contain.
3 Perhaps the underlying assumption of a physical determinism was too strongly embedded in planning's ideology.
4 Perhaps planners, as with so many other experts, suffered the inevitable reversal of esteem which came with a more informed and powerful public.
5 Perhaps it was that the bulldozers of the 1950s and 1960s, which had started to clear areas about which few could have had regrets, did not know when to stop.
6 Perhaps the gestation period of implementing grand designs is simply so long that events inevitably leave them looking like tired responses to yesterday's problems.

Whatever the causes, there is no doubt that much of the planning of the post-war years (and the architecture with which planners are inevitably, if unfairly, linked) is now widely condemned.

However, in the present-day climate of infinite regret for the 'little terraces' and the 'neighbourhood' it is so easy to forget that what was needed was better-quality housing fast, and that the lumbering giants of Victorian public buildings were not then as admired as they are today.

Figure 8.84 *Town and Country Planning,* May 1997

commercially depressed development areas. Most were designed for a maximum of 60 000 people, although for some the target was later raised to over 100 000. At the time the proportion of rented housing was, at an average of 80%, above the national norm.

Later New Towns, the so-called second generation, such as Washington on Tyneside and Livingstone in Scotland were seen primarily as new economic growth poles that would bring prosperity to traditional industrial areas. Their projected populations were higher than the earlier New Towns and some were already quite sizeable towns, such as Northampton, Peterborough and Warrington.

Last of all came the 'New Cities', or third generation New Towns of Milton Keynes, Telford and Central Lancashire, which were planned to be larger than their earlier counterparts and act as 'counter magnets' to people who might otherwise have moved to London. All three include within their areas a number of established settlements. Central Lancashire already had a population of 235 000. Milton Keynes was seen as a new growth pole midway between London and Birmingham, while taking overspill from London and other parts of the South East. The M1 motorway forms its north-eastern boundary and it is on the main railway line from London to the Midlands.

The objectives of the New Towns were:

- to strike a balance between the needs of living and working
- to provide good quality housing with all amenities for people across the income range
- to give a green and open quality to the towns – the boundaries of new towns were generously drawn giving architects space to experiment in contrast to the high-rise blocks that were mushrooming in the large cities.

New Towns were designed to be 'self-contained and balanced communities for work and living'. They exhibited many new ideas in urban planning:

- houses grouped in self-contained neighbourhoods, each with its own local service centre – the 'neighbourhood principle'
- through-traffic routed away from housing areas
- pedestrianised shopping centres
- factories grouped into industrial estates
- cycle path networks
- large areas of planned open space.

The New Towns have generally been perceived as successful because:

- they have been a key element in the protection of the countryside from sprawl and spasmodic development
- the quality of housing was far better than most residents could have hoped for within the conurbations

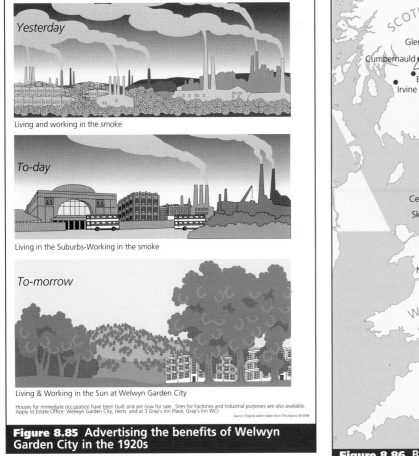

Yesterday

Living and working in the smoke

To-day

Living in the Suburbs-Working in the smoke

To-morrow

Living & Working in the Sun at Welwyn Garden City

Houses for immediate occupation have been built and are now for sale. Sites for Factories and Industrial purposes are also available. Apply to Estate Office: Welwyn Garden City, Herts. and at 3 Gray's Inn Place, Gray's Inn WCl

Source: Original advert taken from *The Express* 9/10/98

Figure 8.85 Advertising the benefits of Welwyn Garden City in the 1920s

Figure 8.86 Britain's New Towns

Figure 8.87 Manor Royal Industrial Estate, Crawley

Figure 8.88 Recent housing development, Crawley

■ they have produced a superior economic performance compared with the national average

■ they have attracted much foreign investment. For example, Telford has the highest concentration of Japanese and Taiwanese manufacturing companies in Britain

■ they have a strong record of spawning small businesses – the small-scale industrial estate with workshops and low-rent 'nursery' units is essentially a new town concept

■ new towns have been pioneers in energy conservation and other environmental matters.

Criticisms of new towns

- People living at or close to areas where New Towns were planned were some of the first to exhibit the now well-known 'not-in-my-back-yard' (NIMBY) syndrome.

- In the early years at least the New Towns were perceived by many as boring and lacking in character. The media frequently ran stories about 'New Town Blues' and the general social problems caused by what was felt to be a 'lack of community' compared with places of origin such as London's East End.

- The London New Towns did not become as self-contained as at first hoped and commuting has been at a higher level than originally envisaged.

- New Towns found it difficult to insulate themselves from the characteristics of the regions in which they were located. High unemployment in the wider region was usually reflected by unemployment above the national average in the New Town.

- The difficulty of attracting the middle class which was necessary to achieve a good social mix.

- The success of the New Towns in creating half a million new homes in 50 years has to be measured against the huge numbers of traditional terraced houses which were structurally sound and simply in need of renovation.

The New Town adventure came to an end when the concept fell out of favour with the two main political parties. To the Conservatives, particularly under the leadership of Margaret Thatcher, they were part of a burgeoning welfare state that needed to be reduced in scope. In contrast, the Labour party was concerned at the way New Towns attracted employment, skilled workers and investment away from the inner cities which were invariably dominated politically by Labour.

The development corporations that set up the new towns have all been wound up. What remains of their publicly owned land is being gradually sold off by the Commission for New Towns (CNT), established in 1961, when it receives what it deems to be appropriate offers. Its job is to manage and dispose of an estate encompassing some 3 million square feet of mainly office buildings and factories, and over 11 000 acres of development land. Through the disposals programme of the CNT the government should eventually recoup the entire historic cost of the New Towns along with a reasonable 'profit'.

Expanded towns

A number of existing well-established settlements also took overspill population from London and some other conurbations as a result of the Town Development Act of 1952. Industrial areas and new housing estates were constructed with both central and local government money.

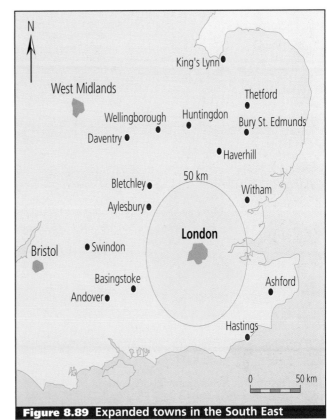

Figure 8.89 Expanded towns in the South East

London entered into house-building agreements with a number of towns such as Andover, Swindon and King's Lynn, which were prepared to take population from London. The expansion schemes had two big advantages:

- **Londoners got better homes and working conditions**

- **new life was brought to declining or static towns. Some expanded towns grew to three or four times their original size and in most services grew correspondingly.**

The expanded towns were in general too far from London for a significant level of commuting and because of their longer history of development, many newcomers saw them as having more character than the New Towns. Figure 8.89 shows the location of expanded towns in the South East.

Green Belts

The idea of the Green Belt dates back, according to some writers, to Ebenezer Howard, although others attribute the concept to earlier thinkers, including Robert Owen. The concept was formalised in 1938 when the then London County Council promoted the Green Belt (London and Home Counties) Act, which set out powers for local councils to acquire land to ensure it could not be built on. One of the wartime planners, Sir Patrick Abercrombie, proposed that restrictions on development could be extended to privately owned land and these powers were given to local councils by the 1947 Town and Country Planning Act.

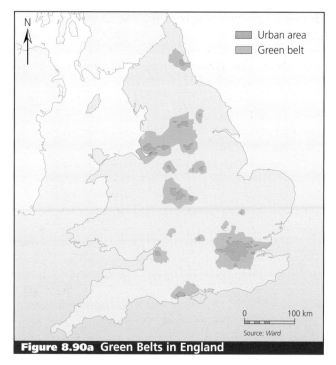

Figure 8.90a Green Belts in England

Figure 8.91 Epsom: new housing on Green Belt land

The first Green Belt was established around London in 1947. Since then Green Belts have been set around a number of other conurbations and cities (Figure 8.90). They have always enjoyed a very high level of public support. Green Belts serve the following purpose:

- limiting urban sprawl
- preventing neighbouring towns from merging
- preserving the special character of historically or architecturally important towns
- protecting farmland from urban development
- providing recreational areas for city dwellers.

Figure 8.90b Green Belts in England

Area	Size (hectares)
Tyne and Wear	200 000
Lancaster and Fylde Coast	5750
York	50 000
South and West Yorkshire	800 000
Greater Manchester, Central Lancs, Merseyside and Wirral	750 000
Stoke on Trent	125 000
Nottingham, Derby	200 000
Burton-Swadlincote	2000
West Midlands	26 500
Cambridge	26 500
Gloucester, Cheltenham	20 000
Oxford	100 000
London	1 200 000
Avon	220 000
Total in England	4 495 300

However, while they contain development within their boundaries there is often intense growth just beyond their outer limits. In order to limit such development London's Green Belt was extended outwards in several areas in the early 1980s. Green Belt restraint has never stopped landowners submitting planning applications for development, and although success in obtaining planning permission is low, the huge financial rewards for those who do succeed has always encouraged land speculation in the rural–urban fringe. The most spectacular breaching of London's Green Belt was the construction of the M25 (Figure 8.92). Despite intense opposition from environmentalists and others, the perceived need to reduce the traffic burden on London was given precedence. Not surprisingly the M25 has increased other pressures on the Green Belt. It has been estimated that

Figure 8.92 The M25 through London's Green Belt

between 5% and 10% of the land statutorily approved as Green Belt in the 1950s has been developed. Land use change has taken place unevenly with considerably more land being developed on the western edge of London than elsewhere, and much of this development has been the direct result of public sector investment. Anyone who wants to build on Green Belt land must demonstrate 'exceptional circumstances' to be allowed to do so.

Examples of development that has been granted include:

- the building of science parks on poorly reclaimed gravel land with employment and environmental gains offsetting the loss of Green Belt
- the construction of housing estates in the extensive grounds of large mental hospitals which have been closed. Development has often been deemed to be more desirable than dereliction (Figure 8.91)
- building on pieces of land cut off by new roads which are too small to be farmed effectively.

Criticisms of the Green Belt

Green Belts have been criticised because they:

- restrict the supply of development land, thus raising its price and that of houses constructed upon it. One consequence of this is higher housing densities
- do not serve the recreational needs of urban areas as a whole, being used largely by local residents
- are usually managed in an unimaginative way with some areas which could be described as being derelict
- result in increased commuting distances and greater traffic-related pollution.

New pressures on the Green Belts and beyond

A report published in 1995 projected that the number of households in Britain would rise by 4.4 million between 1991 and 2016. On that basis counties were told to find space for given numbers of new homes. A number of organisations, however, questioned the line that successive governments have taken on this issue:

- the figure of 4.4 million merely projected past trends and thus should not be seen as a forecast
- no account was taken of income or price, just that every household must have 'access to decent housing' including 'those who would not otherwise be able to afford it'
- making more homes available may itself cause more households to be formed
- the government could act to encourage subdivision of existing large properties. For example, at present, building conversions attract VAT at 17½%, while building new homes is VAT-free.

Britain is currently losing countryside at the rate of an area of new development the size of Bristol each year. Every part of the country is under pressure which is at its most intense in the South East. Between 1945 and 1990 development has eaten up an area of land larger than Greater London.

The favoured option drawn up by Serplan, the regional planning conference of county councils for the South East was a midway figure between the 1 104 000 households it was estimated would be formed in the South East between 1991 and 2016 and the 847 000 homes that planners calculated could be built on land in cities and on already-allocated greenfield sites by 2006. This new strategy threatened dozens of areas previously considered out-of-bounds to house builders. A report published in October 1999 increased the estimated demand for new homes further (Figures 8.93 and 8.94).

Many opposed to such large-scale development argue that local authorities should be allowed to decide for themselves how much should be built in their areas, knowing full well that they would allow very little. But those in favour of significant development say that the high cost of property restricts people's choices and damages their lives. If enough houses are built, homes become more affordable.

Whatever the final outcome, a great deal of new building will take place. To minimise the impact on the region the following might prove expedient:

- development should be directed to those pockets of the South East where it gives most benefit and causes least damage, like the depressed towns of east Kent
- it should make the best use of existing infrastructure
- it has to be built so that it uses land efficiently, creating the least possible environmental impact
- it has to cater for all types of income so that the housing needs of people in the South East are truly met.

Land in South East England which commands around £2500 an acre for farming is worth as much as £500 000 when released for housing around booming towns. Thus, the potential for large profits are there for some, but the impact for most people in the area will be adverse.

Opposition to new building in the South East is mounting in other parts of the country. Politicians and business leaders in the North and Midlands are concerned that such a large increase in the housing stock of the South East will further encourage the migration of skilled workers into the region. This would strip other areas of their most highly prized workers, making it even more difficult to attract inward investment.

Changes on the rural–urban fringe

The rural–urban fringe is the boundary zone where urban and rural land uses meet. It is an area of transition from agriculture and other rural uses to urban use. The pace of

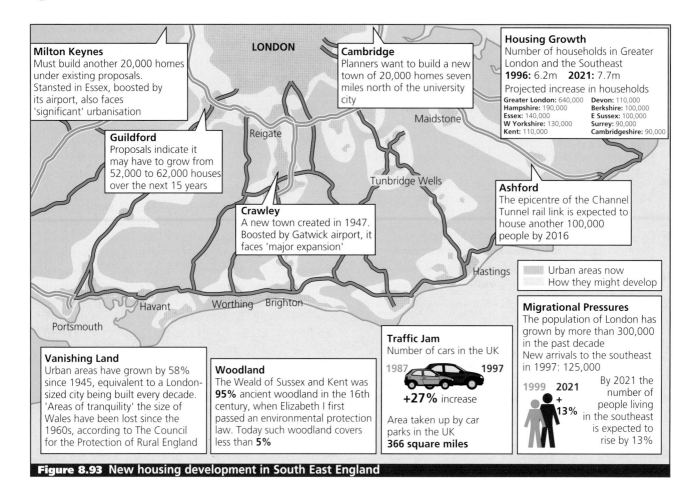

Figure 8.93 New housing development in South East England

Living in the brown belt

This Government cannot tell people where to live; but at the same time it cannot afford to go down in history as the administration which presided over the devastation of the green belt and the concreting over of great swathes of the southern countryside. Precisely this is now being proposed by an independent panel of Government-appointed planning experts, who want to swamp the south-east with 1.3 million new homes built over the course of the next 16 years. This is well above the Government's own planning projections, and 400,000 more new homes than a consortium of south-east local authorities last year concluded were necessary. As Rowan Moore points out on this page, if the proposals were to be accepted, areas of countryside the size of large towns would be swallowed up all around London. The green lung which Londoners need in order to escape urban congestion would begin to choke.

The pressure on the planners is not hard to identify. A powerful lobby of impoverished local authorities, influential developers and greedy farmers and landowners is all too keen to make money from new building on green-field sites. A report earlier this year,

sponsored by the House Builders' Federation, warned that many brown-field sites were 'contaminated', and would be given a wide berth by builders and investors. In charge of environment, Mr Prescott is a doughty fighter, and here is a battle he must fight and win. The Government must lift the planning blight on the King's Cross site, the largest inner-city brown-field site in Europe. It should encourage local authorities to make compulsory purchase orders against landowners who demand unrealistic sums for brown-field land which they are not themselves prepared to develop. Local authorities should be given a firm deadline for producing their figures for the national audit of brown-field sites which John Prescott has demanded. If urbanites wish to move to the country, in the south-east, that is their business. But by the time builders had concreted over 650 square miles of woodland and meadow, in fulfilment of these nightmarish planning forecasts, there would not be much tranquil countryside left. These recommendations must be rejected.

Evening Standard Monday, 11 October 1999

Figure 8.94 Developing brownfield sites

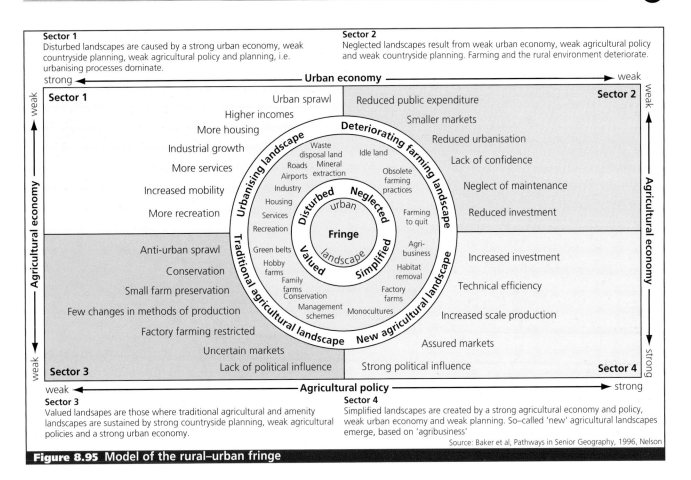

Sector 1
Disturbed landscapes are caused by a strong urban economy, weak countryside planning, weak agricultural policy and planning, i.e. urbanising processes dominate.

Sector 2
Neglected landscapes result from weak urban economy, weak agricultural policy and weak countryside planning. Farming and the rural environment deteriorate.

Sector 3
Valued landsapes are those where traditional agricultural and amenity landscapes are sustained by strong countryside planning, weak agricultural policies and a strong urban economy.

Sector 4
Simplified landscapes are created by a strong agricultural economy and policy, weak urban economy and weak planning. So–called 'new' agricultural landscapes emerge, based on 'agribusiness'

Source: Baker et al, Pathways in Senior Geography, 1996, Nelson

Figure 8.95 Model of the rural–urban fringe

change in the rural urban fringe is determined by a number of factors which are summarised in the Baker Model (Figure 8.95). This model clearly shows that rural–urban fringe areas can vary considerably in character.

Figure 8.97 illustrates a segment of the south-western sector of London's rural–urban fringe drawn from the Ordnance Survey's Landranger No. 187 map (Dorking, Reigate and Crawley). The critical boundaries are those between the continuous built-up area and the fringe, and between the fringe and the rural area proper.

A succession of large open spaces separate the continuous built-up area of London from the fringe settlements. These include Horton Country Park, Epsom Common, Epsom Downs, Banstead Downs and Banstead Wood. The larger fringe towns of Dorking, Reigate and Redhill, linked by the A25, mark the southern boundary of the fringe in this sector. To the south of this line:

- the density of population is considerably lower
- the distance between towns is greater
- more settlement is in the form of villages and hamlets
- agriculture is the dominant land use
- the density of major roads decreases

The rural–urban fringe is characterised by a mixture of land uses all of which require a great deal of space. In the area

under consideration there is a large theme park (Chessington World of Adventures), a horse-racing course (Epsom), golf courses, cemeteries, hospitals, colleges as well as a reasonable number of farms. The demand for these land uses comes mainly from London but they could not be sited within the conurbation because space was not available, except within strictly protected areas (parks, commons, etc.). Thus it was logical for these land uses to locate where the space requirements could be met as close as possible to the continuous built-up area. The fringe uses that might have been able to find a site within London would probably not

Figure 8.96 Epsom race course

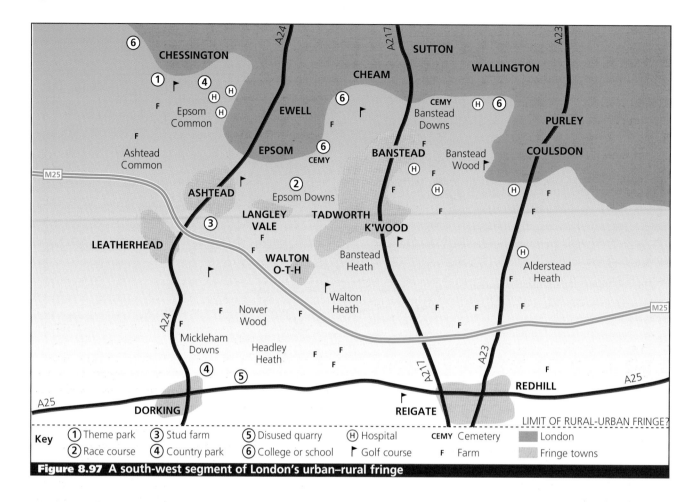

Figure 8.97 A south-west segment of London's urban–rural fringe

Key
① Theme park   ③ Stud farm   ⑤ Disused quarry   Ⓗ Hospital   CEMY Cemetery   ▮ London
② Race course   ④ Country park   ⑥ College or school   ⌐ Golf course   F Farm   ▮ Fringe towns

have been able to afford the high cost of land. Lower prices in the fringe would have provided the answer.

Land use in this part of the fringe changed rapidly until the 1940s when the Green Belt Act imposed strict limitations on future development. New building has taken place when deemed beneficial for the community but developers have found it very difficult to get planning permission for a change in land use. Thus population has not increased in Green Belt areas to the same extent that it has in comparable areas without this protection.

Inner cities

It is reasonable to view the development of inner-city planning in five phases.

Phase 1 (1946–67)

In 1947 the concept of Comprehensive Development Areas was introduced, launching a massive programme of slum clearance. During this period, when physical planning was seen as the solution to the problems of the inner city, one and a half million properties were knocked down. However, apart

Figure 8.98 Recently renovated 1960s tower block

from physical reconstruction inner-city policy lacked further direction, with little thought given to the social consequences of redevelopment. By the end of the period a number of issues were provoking debate:

- in many instances redevelopment did not keep pace with slum clearance, leaving large areas of derelict land
- clearance often divided communities leaving a mood of despondency hanging over those who remained
- the tower blocks into which so many were rehoused soon came to be loathed by the majority of their residents. They proved to be particularly unsuitable environments for the old and the young
- the economic decline and social problems of inner cities were largely ignored
- redevelopment was failing to match the pace of decay.

Phase 2 (1967–77)

During this phase two new strands to inner-city policy were developed: environmental improvement, and social and economic welfare.

Environmental improvement

- The 1969 Housing Act enabled local authorities to set up General Improvement Areas (GIAs) for which grants would be available to improve both housing and the surrounding environment. This was the first move towards renovation as opposed to clearance.
- This trend was strengthened by the 1974 Housing Act which allowed for the establishment of Housing Action Areas (HAAs) in areas of greatest housing and social stress. Successful HAAs could be upgraded into GIAs.

Although the principle of area-based renewal had been firmly established two significant concerns were voiced by critics: (a) the overall scale of improvement was limited by underfunding, (b) in some areas the better-off saw their chance of buying cheap rented accommodation for improvement and owner-occupation – the beginnings of gentrification.

Social and economic welfare

The Urban Programme launched in 1968 was a major effort to assist areas of multiple deprivation, many of which were experiencing high rates of immigration. Local authorities and voluntary agencies bid for grants to fund projects such as pre-school provision, child care and community work. From 1967 Education Priority Areas were established to improve the resourcing of schools in difficult areas. Between 1974 and 1979 Comprehensive Community Programmes were introduced to improve the delivery and coordination of welfare services. However, this initiative was limited in scope and in time because of lack of funding. Two major

investigative projects, under the titles of Community Development Projects and Inner Areas Studies, were undertaken during this phase with the objective of gaining a better understanding of inner-city decline.

Phase 3 (1977–79)

The publication of the White Paper, 'Policy for the Inner Cities', in 1977 marked the beginning of this phase. At last the inner city was viewed as a problem region requiring broad-based action. The Inner Urban Areas Act of 1978 created Partnership and Programme Areas involving action plans to improve social, economic and environmental conditions in seven metropolitan authorities. A key aim of this new urban policy was to retain existing jobs and hopefully attract new jobs.

Phase 4 (1979–97)

The new Conservative government continued most of the initiatives begun by Labour in the late 1970s but added a new emphasis. The objective now was for the public and private sectors to work as partners with economic regeneration being the most important aim with environmental and social issues following. Of particular significance were:

- The Enhanced Urban Programme with assistance extended to cover more areas. Greater emphasis was placed on funding projects with an economic focus.
- Urban Development Corporations (UDCs), introduced in 1981 with the objective of regenerating large tracts of derelict land in inner cities. Thirteen UDCs were established between 1981 and 1993. The first were set up in London's Docklands and on Merseyside.
- Enterprise Zones (EZs) introduced in 1981 to attract growth industries to relatively small (50 to 450 ha) areas of inner-city land. For example, the EZ in London Docklands covered only about one-tenth of the UDC area run by the London Docklands Development Corporation. EZs had a ten-year life span during which time the main benefits were exemption from rates, tax allowances for capital expenditure on industrial and commercial property, and a simplified planning regime. Twenty-three were created, with no more designated after 1983.
- City Action Teams (CATs) set up in nine cities between 1985 and 1987. These teams of civil servants encouraged joint action between regional government departments to encourage new private sector investment.
- The Inner Cities Initiative (ICI) launched in 1986 created Task Forces in unemployment black spots to 'unlock development opportunities'.
- City Grants which replaced and combined a number of earlier schemes to encourage the private sector to develop run-down inner-city areas.

■ Land Registers which identified vacant land held by local authorities so that it could be offered for sale to private developers.

■ City Challenge announced in 1991. Local authorities were challenged to devise imaginative redevelopment projects and bid against each other for the funding available.

Phase 5 (1997–)

With Labour returned to power in 1997 it was almost inevitable that a new inner-city policy would be developed. Regional Development Agencies (for England) commenced operation in April 1999 although the London Development Agency did not start up until a year later. A major role of the RDAs will be to manage the Single Regeneration Budget which replaces the previously fragmented nature of the Urban Programme funding. Local authorities have to bid against each other for SRB project funding. The Social Exclusion Unit is examining all aspects of deprivation and has already set up 12 education action zones, five employment action zones and 11 health action zones.

The Rogers Report 'Towards an Urban Renaissance'

The Urban Task Force chaired by Lord Rogers published its report in June 1999. It acknowledged past efforts at inner-city regeneration but hoped to bring something new by injecting more 'consistency and continuity' into urban policy. Although the remit of the Task Force was not confined to inner-city issues much of its focus was here for obvious reasons. Its key objectives are shown in Figure 8.99. Much discussion is in progress on the report's findings and it will be interesting to see what impact it has on decision-making over the next few years.

Town centres

Town centres have been generally neglected in recent decades as new retail forms have dominated the planning agenda. Five waves of out-of-town retailing have been recognised since the early 1970s (Figure 8.100) as retailers have continued to create new ways to shop in order to boost turnover and increase market share. This has put town centres under intense competition from other locations for the patronage of retailers and developers. Despite central government intervention in the form of PPG 6, it is likely that only the more dynamic town centres will be successful in the future. The worst scenario is that Britain will follow the example of so many American cities to the point where the settlement becomes a 'doughnut' with a hole in the middle.

The centres of many market towns are now in danger of losing the food shopping which brings people into town. The loss of vitality caused by a decline in pedestrian flow leads in time to a loss of viability as retailers fail to reinvest. As a recent article on the subject stated (*Built Environment*, Vol. 24, No. 1) 'The centre is then left to those with time but no money, thus intensifying the pressures towards suburbanisation and spatial segregation that have bedevilled English towns'.

A recent study of shopping centres in inner London found that most serve only a local, limited catchment area with spending power shifting away to some 30 out-of-town centres. In addition, there is competition from superstores and new regional centres such as Lakeside and Bluewater. The use of a car for shopping trips in London rose from 23% in 1981 to 41% in 1991. In comparison to the newer peripheral centres shoppers found it difficult to park and unpleasant to walk in traditional centres.

---

1 All urban neighbourhoods will be managed according to principles of sustainable development. The main environmental indicators – air pollution, ground contamination, energy use, water recycling and waste disposal – will show significant improvement.
2 A more balanced national economy will allow for a more even distribution of economic opportunities and income within cities, between cities and between regions. Key social indicators, such as educational achievement, health, crime and poverty, will have improved.
3 There will be a substantial increase in recycling of previously developed 'brownfield' land in line with local needs and projected demand for housing. A parallel decrease in demand for greenfield sites will have taken place.
4 There will be urban repopulation with year-on-year growth in the number of people living in towns and cities. Movement will have taken place from the outskirts to inner areas, and distinctions between market and social housing will have become blurred.

5 There will be increased quality of life, with at least five major English cities in the European 'top 50'. None will be in the bottom third.
6 Attitude surveys will show that people and investors take a positive view of urban areas, enjoy living in towns and cities, and regard them as safe and attractive places in which to raise children.
7 England will enjoy a world-wide reputation for innovation in sustainable and high-quality urban design.
8 Public services such as health, education and social services will include a clear urban dimension that specifically addresses the needs and aspirations of urban communities.
9 All urban areas will be managed according to standards agreed by the local community.
10 England will have become the leading international location to acquire urban development skills.

Figure 8.99 The Urban Task Force's ten key objectives for urban policy up to 2021

Period	Retail	Store Type	Centre Type
Early 1970s	Food and household goods	Hypermarket	District centre
Mid-1970s	Food	Superstore	District centre
Late 1970s	Household goods	Retail warehouse	
Mid-1980s	Personal goods and fashion	Retail warehouse	Retail park
Late 1980s	All shopping		Regional shopping centre
Early 1990s	Low prices: (food)	Limited line discounter	
	(mixed goods)	Club warehouse	
	(personal and fashion)	Factory outlet	Factory outlet centre

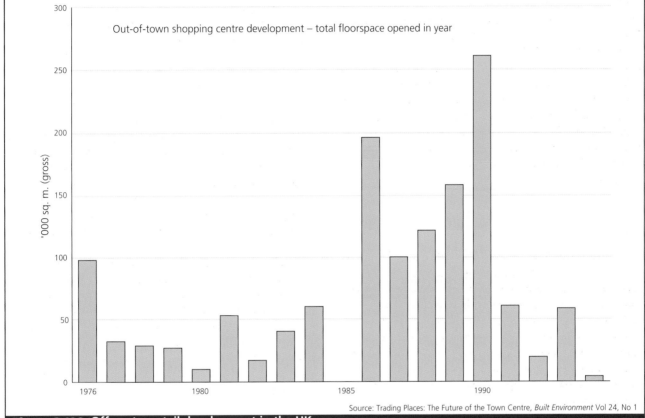

Figure 8.100 Off-centre retail development in the UK

Recently, government policy in the form of PPG 6 (Department of Environment, 1996) has tightened up planning control over off-centre developments, although government views on out-of-town developments had begun to harden from 1993. It seems that central government now regards the availability of a town centre site as sufficient reason for refusing off-centre retail park proposals. Local authorities are increasingly imposing conditions on planning consents for retail parks, usually to restrict retailing to bulky household and electrical goods.

In 1997 the Urban and Environmental Development Group (URBED) published a Department of the Environment sponsored report entitled 'Town Centre Partnerships: Their Organisation and Resourcing'. The report recommended the following:

■ town improvement zones in larger centres loosely modelled on North American business improvement districts

■ investment priority areas using incentives such as rate relief to reduce the cost of occupying property in areas of high vacancy, incentives for refurbishment, and penalties on allowing buildings to decay

■ a restructuring of the business rate to allow local authorities more flexibility. Planners argue that the current operation of the business rate penalises small businesses and increases the likelihood of properties remaining vacant

■ the government should ensure that all the different spending regimes should have a town centre dimension.

Many local authorities have undertaken so-called 'health checks' on town centres and have appointed town centre managers. It is now generally acknowledged that the design of urban space is vital to the success of a town centre. The good practice guide *Managing Urban Spaces in Town Centres* (DoE/ATCM, 1997) highlights all the elements of good urban design including functional, social, perceptual, spatial, contextual, visual and morphological aspects.

Figure 8.101 Declining town centre

1 **(a)** Describe the spatial distribution of Britain's New Towns.
  (b) Explain the reasons for such a distribution.
  (c) Discuss three planned characteristics of New Towns.
  (d) Examine one problem that many New Towns have faced.
  (e) Assess the role of Expanded Towns in the South East.
2 **(a)** Define the term 'Green Belt'.
  (b) Describe and explain the location of Green Belts in England.
  (c) Explain the existence of settlement in some Green Belt areas.
  (d) Discuss the main arguments for and against maintaining a strong Green Belt policy.
3 **(a)** Define the terms (i) greenfield site, (ii) brownfield site.
  (b) Why will so much new housing be required in the South East in the next two decades?
  (c) What impact will this have on the region?
  (d) Which interested parties and why, would prefer the bulk of the new housing requirement to be built on greenfield sites?
  (e) Discuss the reasons for concentrating as much development as possible on brownfield sites.

The 24-hour city concept

Until recently the principal focus of the planning and management of city centres has been on the traditional pattern of daytime retail trading and office hours. However, a growing number of cities are looking to embrace the '24-hour city' concept.

Although the origins of this concept is debateable, Leeds is usually credited as starting the process when the City Council launched a 24-hour city initiative in 1993. The essentials of this scheme, and others which followed such as in Bradford, Reading, Liverpool and Manchester, are:

■ 'stretching' the city centre's 'working life' and encouraging more people to make use of it in the evenings and on Sundays

■ developing a safe and lively city centre which is attractive to all age groups by day and by night

■ building a positive image of the city centre and attracting new investment

■ encouraging environmental regeneration through the redevelopment of redundant and empty buildings and parcels of vacant and derelict land

■ developing new leisure and entertainment facilities

■ encouraging more people to live within city centres.

In Leeds one of the most visible examples of city centre change has been the dramatic growth in the number and variety of cafes, bars and high-quality restaurants, along with the enhancement of retail functions. As a recent article in the journal *Town and Country Planning* states: 'The move towards a "24-hour city" centre seems to provide a variety of opportunities and challenges for the public, businesses, property owners, local authorities, the police, transport operators and city centre residents. All these stakeholders will want to maintain both a formal and an informal watching brief on developments, as the concept of the "24-hour city" continues to take shape and form in a growing number of towns and cities.'

Spatial Focus: São Paulo – World City

Background: urbanisation in Brazil

Brazil, like most other countries in South America, is highly urbanised. Figure 8.102 shows the speed at which Brazil has changed from a rural to an urban society. Between 1970 and 1995 some 30 million Brazilians moved to urban areas, either 'pushed' from their rural environments by adverse factors or 'pulled' by aspirations of a better life in the cities.

The main push factors responsible for rural to urban migration have been:

- the mechanisation of agriculture which has reduced the demand for farm labour in most parts of the country

- the mergers of farms and estates, particularly by agricultural production companies

- the generally poor conditions of rural employment. Employers often ignore laws relating to minimum wages and other employee rights

- desertification in the Northeast and deforestation in the North

- unemployment and underemployment

- poor social conditions, particularly in terms of housing, health and education.

People were attracted to the urban areas because they perceived life there would provide at least some of the following: a greater variety of employment opportunities; higher wages; a higher standard of accommodation; a better education for their children; improved medical facilities; the conditions of infrastructure often lacking in rural areas, and a wider range of consumer services. The diffusion of information from previous migrants was usually such that most potential new migrants realised fully that there was no guarantee of achieving all or even most of the above but most rationalised that their quality of life would hardly decrease overall. Employment was the key. The most fortunate found jobs in the formal sector. A regular wage then gave some access to the other advantages of urban life. However, because the demand for jobs greatly outstripped supply, many could do no better than the uncertainty of the informal sector.

Location and early development

São Paulo was established as a mission station by Jesuit priests in 1554 near the confluence of the Rio Tietê and a southern tributary, the Tamanduatei. It was located 70 km inland at an altitude of 730 m, on the undulating plateau beyond the Serra do Mar. The cool, healthy climate attracted settlers from the coast

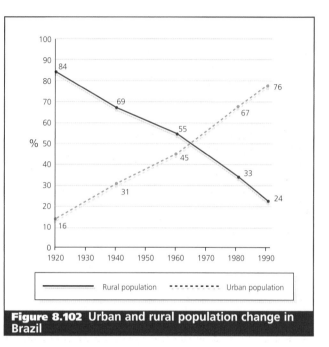

Figure 8.102 Urban and rural population change in Brazil

while the Paraná river system facilitated movement into the interior.

In 1681, São Paulo, as the settlement became known, became a seat of regional government and in 1711 it was constituted as a municipality. Coffee was the catalyst for the rapid growth in the latter part of the 19th century that transformed the city into a bustling regional centre. It became the focus of roads and railways and its prosperity was assured by a rail link with the port of Santos, completed in 1867. This was the only major routeway scaling the great escarpment of the Serra do Mar. The profits from coffee were invested in industry and by the end of the century São Paulo had become the financial and industrial centre of Brazil. The wealth of the coffee barons was lavished on sumptuous town houses, and prestigious public buildings mushroomed in the business district, the Triângulo.

20th-century growth

São Paulo's population growth was relatively slow until the late 19th century. In 1874 its population was only about 25 000. However, the rapidly increasing demand for labour encouraged immigration and the city's population had soared to almost 70 000 by 1890 and reached 239 000 in 1910. The city reached 'millionaire' status in 1934. By 1950 the population had grown to 2.2 million and São Paulo had clearly established its dominant role in the urbanisation of Brazil. Thereafter the population of both the city and the metropolitan area grew rapidly (Figure 8.103) with the latter reaching 16.5 million in 1995, making it the fifth largest city in the world. The annual growth in population is currently 200 000 in the city and 340 000 in the metropolitan area (Figure 8.104). However, a marked change occurred between 1980 and 1991 (Figure 8.105) with negative net migration recorded for both the City of São Paulo and Greater São Paulo. This appears to be the start of the process of decentralisation that has affected most developed countries during the past half century.

Original mono-nuclear form

In the early part of the century São Paulo retained its mono-nuclear structure around the Triângulo, its CBD. From here emanated radial highways which were connected by a concentric sub-system which often acted as 'barriers' between different socio-economic areas. In general, residential areas became progressively poorer with distance from the centre, reflecting limited intra-urban mobility and economic concentration in the central area.

Metropolitanisation and industrialisation

The 'metropolitanisation' of the city occurred between 1915 and 1940 due to high rates of both natural increase and immigration. During this period industrialisation had a marked impact on urban structure. The first industrial districts had been located on the south bank floodplain and terraces of the Tietê, but in the 1930s new industrial satellites emerged in the south-east. These areas were to grow rapidly in the following decades and now form the most important industrial region in metropolitan São Paulo, known as the 'ABCD' complex after its constituent districts of Santo André, São Bernardo, São Caetano and Diadema. This is the focal point of Brazil's motor vehicle industry and also the location of a wide variety of other industries.

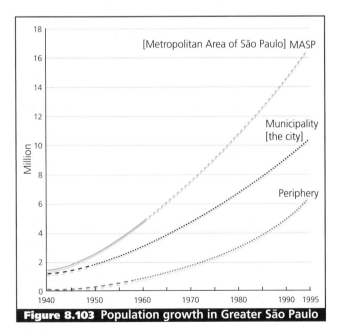

Figure 8.103 Population growth in Greater São Paulo

Until the late 1960s the city grew largely unplanned. The Triângulo developed all the characteristics which make it indistinguishable from any other Central Business District, and the 1970s, in particular, there was a huge increase in the construction of high-rise apartments for the growing middle class in the inner area. The ravines of the River Tietê became sites for numerous *favelas* (many now removed to make way for inner-city parklands), and some of the old mansions near the centre had deteriorated into run-down multi-family dwellings.

Low-grade apartment blocks were built for the workers in the industrial suburbs, but demand far exceeded supply and the authorities were obliged to resort to site-and-service schemes. These too, were insufficient, and large areas of unplanned *favelas* developed. Meanwhile, the very rich had begun to move out to estates within commuting distance via the new motorways of the Triângulo. Here they could live in luxurious villas set in extensive gardens, often with swimming pools and staff quarters. Security men stand guard at the entrance.

Property speculation has resulted in large undeveloped areas within the conurbation which could accommodate a two-thirds increase in population without any further expansion if the undeveloped land were put to use. According to official data, 25% of the city's 1500 sq km remain unoccupied. However, the forested water catchments of the uplands to the south and to the north are now protected by the 1975 Water Resources Protection Law. The two main rivers are said to be dead – most residential sewage and industrial effluent flows directly into rivers and reservoirs; only 10% of the solid waste is collected and treated.

The rapid and generally unplanned growth of São Paulo means that the functional differentiation of different parts of the city is less clearly defined than in most developed world cities. In some cases, squatter settlements are large swathes of urban territory housing hundreds of thousands of people. But more often they are islands of shanty housing, located on vacant land along railway tracks, under flyovers, and in other areas unattractive to the more affluent.

From horizontal to vertical development

In the last 30 years São Paulo has lost its characteristic of being a predominantly horizontal city. As competition for land increased it became more important to take full advantage of inner locations and São Paulo embarked on a period of intense vertical development, a phenomenon in Brazil that first appeared in Rio de Janeiro, a city squeezed between the mountains and the sea. A contributory factor to the high-rise boom has been the increasing concern over crime which reduced single-family residential security to the point where many affluent homeowners began to opt for well-protected apartments.

The establishment of the metropolitan region

It was not until 1968 that the first attempts to formulate a comprehensive planning strategy were made, resulting in the establishment of the metropolitan region in 1973. The fundamental objective was to coordinate administration and

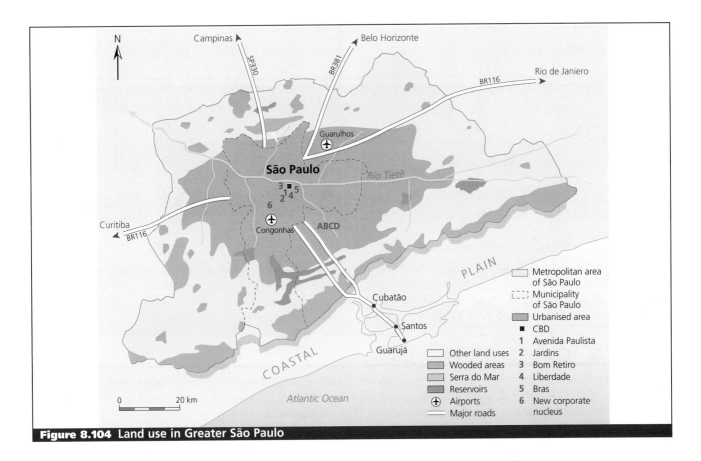

Figure 8.104 Land use in Greater São Paulo

planning at the various levels of government and in 1974 the Metropolitan Planning and Administration System to develop the necessary operational tools. The original mono-nuclear city had developed into a poly-nuclear conurbation, a fundamental fact recognised by the Metropolitan Planning Corporation (EMPLASA) when it set out its objectives.

Redevelopment of the central area

The Anhangabaú Valley is the historic heart of the city and for decades was the centre of the economic and cultural activities of São Paulo. However, the old nucleus gradually lost its significance with increasing suburbanisation and the huge rise in traffic volume which saturated the valley with over 12 000 vehicles per hour. But in December 1991, the old centre recovered its vitality. After a five-year reconstruction programme, the 77 000 sq m area (one-seventh green space) was reopened to the 1.5 million people who daily cross the centre of the city. Traffic now runs underground in 570 m-long tunnels that connect the northern and southern sections of the city. Much of the area is now pedestrianised and the most important historic buildings have been renovated.

Figure 8.105 Population change: natural change and migration, 1970–91

	Greater São Paulo	City of São Paulo	State of São Paulo
Population			
1970	8 139 730	5 924 615	17 771 948
1980	12 588 725	8 493 226	25 040 712
1991	15 416 416	9 626 894	31 546 473
Absolute increase 1970/80	4 448 995	2 568 611	7 268 764
1970/80 Contribution to change			
Natural change (%)	+26.45	+24.05	+6.28
Net migration (%)	+28.20	+19.31	+17.35
Absolute increase 1980/91	2 827 691	1 133 668	6 505 761
1980/91			
Natural change (%)	+24.64	+22.25	+23.64
Net migration (%)	−2.18	−8.90	+2.34

Figure 8.106 Redevelopment in central São Paulo

The Jardins – inner-city affluence

No more than a few kilometres south-west of the central area are the most elegant residential areas, comparable to European and North American suburbs. In this area, known as the Jardins (Figure 8.108), there are 50 m² of greenery per capita; houses average 60 m² per dweller, and the standard of living is that of the upper middle class in most developed countries. The area was laid out in 1915, following the British idea of a garden suburb. It is now dominated by expensive apartment buildings protected by high railings and comprehensive security systems. These exclusive residential neighbourhoods have long since taken over from the city centre as the location of most of the city's best restaurants and shopping streets.

Avenida Paulista – São Paulo's financial centre

To the south-west of the centre and separating it from the Jardins is the 3 km-long Avenida Paulista, the most important financial

Figure 8.108 The Jardins

centre in the city. It is the address of many of the largest banks in the country, two business federations and hundreds of corporations. Mansions once lined this fashionable routeway, but most were replaced in the late 1960s and 1970s by skyscrapers. Real estate along the avenue alone is worth $7 billion. The Museu de Arte de São Paulo (MASP) is also located on the avenue.

The south-west – the new corporate nucleus

The corporate addresses in large cities like São Paulo are often related to major road foci. First it was the Anhangabaú Valley, then Avenida Paulista and most recently the road alongside the Pinheiros River. A privileged rectangular area served by four of the most important avenues in the city: Juscelino Kubistchek, Morumbi, Luis Carlos Berrini and Nacoes Unidas has become the favoured location for corporate headquarters. It is not only easy access that favours this location. Modern concepts of integrated management require large areas – often over 1000 m² per floor, impossible to find in other already crowded commercial areas of the city.

From 1986 to 1991, nearly 100 large corporations moved their head offices to modern buildings in this area. Residents include Philips, Dow Chemical, Johnson and Johnson, Hoescht, Autolatina, Fuji and Nestlé. Finance conglomerates such as Chase

Figure 8.107 Small-scale favela close to CBD

Figure 8.109 Avenista Paulista

Manhatten and Deutsche Bank also have their offices there. The impressive World Trade Centre complex which opened in 1995 acts as the focal point of this office region. Between 1990 and 1993 the area accounted for almost 70% of all new office building in the city.

Shopping malls

Service industry occupies 54.4 million m³ of built area, while manufacturing facilities take up 29.5 million m³. The city has 55 000 shops and 11 shopping malls. The largest shopping mall, Centre Norte, is located on the north bank of the River Tietê, about 4 km north of the CBD. It contains 483 shops on one floor and the parking area can accommodate 17 000 cars.

Transportation

It has been estimated that a quarter of all vehicles in Brazil circulate in São Paulo. Car ownership in the city is rising fast and much has been spent on roads to accommodate this trend. Over 35% of households in the municipality own a car. The noise of traffic on the main roads into and out of the city is incessant 24 hours a day. The answer to at least part of the problem is to invest more in public transport which in 1991 was used by 3.4 million passengers daily.

In 1998 the World Bank approved a $45 million loan to help finance the $95 million São Paulo Integrated Urban Transport Project that will join 270 km of suburban main line track. The main objective is to decrease road traffic which would improve air and noise pollution in the project area.

The environment

São Paulo has done much to improve the air quality of the region in recent years with regard to sulphur dioxide and lead. However, levels of other pollutants such as ozone, carbon monoxide and suspended particulate matter are still of concern causing thousands of premature deaths, high medical costs and lost productivity. Traffic is the greatest emission source in São Paulo.

The City of São Paulo spends $1 million a day on rubbish collection. The cost has risen sharply over the last decade because of (a) a lack of strategic planning, (b) a growing population, and (c) the rising amount of rubbish per person because of increased consumption. Cost is only one aspect of this problem. The other is physical disposal; at present the city has only two landfills for rubbish. In an effort to resolve the latter, two enormous waste incinerators, burning 7500 tonnes a day, are expected to begin operation in São Paulo in 1999.

São Paulo is now involved in a major project financed by the Inter-American Development Bank and the World Bank to clean up the Tietê river, which has for long been a repository for industrial and domestic waste. The Tietê's level of contamination is four, the highest on the scale. A poll conducted in 1998 revealed that 91% of adults in São Paulo believe that the pollution of the river is a serious problem.

'Sub-normal' housing

Brazil's large urban areas could not cope with the large influx of rural migrants. In the early 1970s around 150 migrants arrived in São Paulo every hour. With no prospect of accommodation in the city itself they put up makeshift shelters (barracos) on the outskirts of the city. With such a high rate of in-migration these makeshift settlements or favelas rapidly expanded in size and number. Favelas are found both within the City of São Paulo, which is now basically the inner part of the wider urban area, and of course at the periphery of the wider urban area itself. In the city of São Paulo in 1991 it is estimated that 8.9% of the population lived in favelas. However, as many again were classed as living in corticos, overcrowded and decaying buildings in the city itself. Figure 8.110 shows the number of people living in 'sub-normal' or inadequate housing in the Municipality of São Paulo, according to the city authorities. These figures include those living in both favelas and corticos. Some favelas are found in close proximity to very affluent areas, a situation known as urban dualism.

The corticos are essentially 19th-century mansions in former affluent areas that have been subdivided into a number of one-room dwellings where four or more people live and sleep. Many people live in corticos to be close to their place of work and thus avoid spending money on transportation. For those living at the margin this can make all the difference.

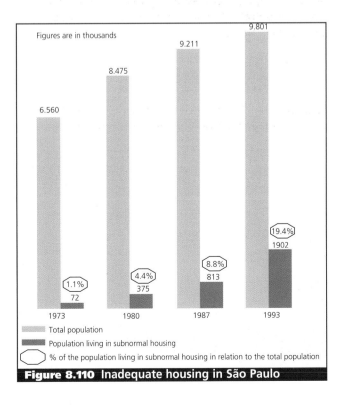

Figure 8.110 Inadequate housing in São Paulo

One solution that São Paulo has employed to improve housing in *favelas* has been to establish self-help schemes. Here residents are supplied with basic building materials to construct their homes to a standard plan. The authorities also supply basic services such as water, sewage and electricity. However in spite of the obvious problems that exist in many residential areas in São Paulo, overall conditions are much better than in most cities in the developing world.

Figure 8.111 Incidence of household environmental problems in Accra, Jakarta and São Paulo 1991/92

Environmental indicator	Incidence of problem (% of all households surveyed)		
	Accra	Jakarta	São Paulo
Water			
No water source at residence	46	13	5
No drinking-water source at residence	46	33	5
Sanitation			
Toilets shared with more than 10 households	48	14–20	<3
Solid waste			
No home garbage collection	89	37	5
Waste stored indoors in open container	40	27	14
Indoor air quality			
Wood or charcoal is main cooking fuel	76	2	0
Mosquito coils used	45	28	8
Pests			
Flies observed in kitchen	82	38	17
Rats/mice often seen in the home	61	82	25

Source: McGranahan and Songsore (1994)

The most obvious explanations for the diffferences between Accra, Jakarta and São Paulo is the relative wealth of the three cities.

The Cingapura Project

This recently announced project has been named after Singapore's huge slum clearance programme. It aims to replace hundreds of *favelas* and derelict areas with low-rise blocks of flats. The inhabitants – the *favelados* – are removed to army-like barracks while construction proceeds. Residents will pay for their new homes with low-interest 20 years mortgages. The Cingapura project is massive: the initial target is to resettle 92 000 families (about 500 000 people) from 243 *favelas*. In targeted areas the project provides:

- new housing
- water supply
- modern sewerage systems
- electricity supply for housing and public areas
- building of new streets and ways of access
- garbage collection

Questions

1 (a) Describe the changing balance between rural and urban population in Brazil between 1920 and 1990.

(b) Discuss the main reasons for rural to urban migration.

(c) What are the possible advantages and disadvantages of such migration to rural areas?

(d) Examine the main problems created in urban areas by a high rate of in-migration.

(e) Discuss two possible benefits that in-migration brings to urban areas.

2 (a) Summarise the data presented in Figures 8.103 and 8.105.

(b) How important has industrialisation been to the growth of São Paulo?

(c) To what extent and why have large commercial organisations decentralised from the historic heart of the city?

(d) Discuss the problems of, and possible solutions for, inadequate housing in São Paulo.

(e) Desribe and attempt to explain the environmental differences between São Paulo, Accra and Jakarta illustrated by Figure 8.111.

- construction of essential community and social facilities

- extensive support for families through its own Social Division.

The fundamental aim of the Cingapura Project is to keep those families who are the main target of the project residing in the same geographic area instead of relocating them to other areas. A section of the project is being developed for senior citizens. This is part of a pilot project the Department of Housing and Urban Development is implementing. Aimed initially at the *favela* problem, the City authorities are now looking at the Cingapura Project as a model for replacing the derelict *corticos*. Because it is so essential for people in such dwellings to live within walking distance of employment in the central area, building peripheral estates is clearly not the answer.

Classifying rural settlement

At the beginning of this chapter reference was made to classification by size. However, rural settlement is also classified by its general pattern over the landscape and by the morphology of individual settlements.

Settlement patterns

The settlement pattern in most areas is the result of the interaction of physical, economic and social factors over a long period of time (Figure 8.112). The basic types of settlement pattern recognised by geographers are (a) nucleated and (b) dispersed.

A dispersed settlement pattern occurs where isolated farms and houses are scattered widely throughout the countryside. Such a pattern is common in sparsely populated areas such as central Wales and the west of Ireland. In such a landscape the clustering of dwellings is very limited and is characterised by hamlets and small villages rather than by larger settlements.

Dispersed settlement has tended to develop where:

- **Natural resources such as fertile land were limited at any one place and sufficient only to support a very small number of people.**

- **The 'agricultural revolution' of the 18th century resulted in the enclosure of land previously farmed under the traditional open field system which had the village as its nucleus. With enclosure many farmers left villages to build isolated farmhouses on their new blocks of land.**

- **The system of land tenure resulted in farms being divided upon inheritance.**

- **Land settled relatively recently was developed under a planned pattern of dispersion. Examples are the Dutch polders and the Canadian prairies.**

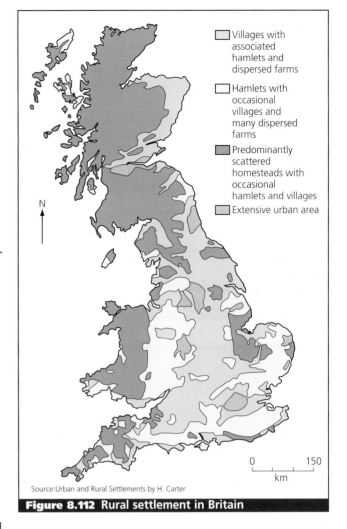

Source:Urban and Rural Settlements by H. Carter

Figure 8.112 Rural settlement in Britain

Villages with associated hamlets and dispersed farms

Hamlets with occasional villages and many dispersed farms

Predominantly scattered homesteads with occasional hamlets and villages

Extensive urban area

0   150
km

A nucleated settlement pattern occurs where most dwellings are clustered around a central feature such as a village green, a crossroads or church. There are few dwellings in the surrounding fields. Thus, hamlets and villages are much larger than in areas with a dispersed pattern of settlement.

A nucleated pattern of settlement has tended to develop where:

- **Cooperative styles of farming operate, such as the open field system discussed above.**

- **Defence has been a concern and appropriate sites (within a meander, on a hilltop) have been available.**

- **Significant supplies of water have been available, such as along spring lines.**

- **Islands of dry ground were found within poorly drained areas.**

- **Very fertile land has led to intensive farming. Here a large number of rural service centres would develop to serve a high rural population density.**

Shape

The morphology (form) of a settlement refers to its shape or pattern. The main types of village morphology recognised are:

- Nucleated: exhibiting a strong concentration of buildings around a focal point.

- Linear: (ribbon) where buildings are spread out in a line because of physical constraints or because development has taken place over time along a road.

- Cruciform: where a village has developed at a road junction and expanded outwards in all directions.

- Fragmented: (loose-knit) where building density is low with no original nucleus to the village.

- Green nucleated: around a central green or common which may include a church and/or pond.

- Planned: where the shape of the village was determined from the beginning rather than evolving over time.

Changing rural environments

Rural areas are dynamic spatial entities. They constantly change in response to a range of economic, social, political and environmental factors. In recent years the pace of change has been more rapid than ever before.

The economy of rural areas is no longer dominated by farmers and landowners. As agricultural jobs have been lost new employers have actively sought to locate in the countryside. Manufacturing, high technology and the service sector have led this trend. Most of these firms are classed as SMEs – small and medium-sized enterprises. In fact in recent decades employment has been growing faster in rural than urban areas. Other significant new users of rural space are recreation, tourism and environmental conservation. The

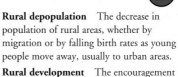

rural landscape has evolved into a complex multiple-use resource.

These economic changes have fuelled social change in the countryside with the in-migration of particular groups of people. To quote Brian Ilbery, a leading authority on rural geography, 'The countryside has been repopulated, especially by middle-class groups ... who took advantage of relatively cheap housing in the 1960s and 1970s to colonize the countryside.' Once they are significant in number the affluent newcomers exert a strong influence over the social and physical nature of rural space. In many areas newcomers have dominated the housing market to the detriment of the established population in the locality. Increased demand has pushed up house prices to a level beyond the means of many original families who then have no option but to move elsewhere.

Gentrification is every bit as evident in the countryside as it is in selected inner-city areas. However, the increasing mobility of people, goods and information has eroded local communities. A transformation that has been good for newcomers has been deeply resented by much of the established population.

In the post-war period the government has attempted to contain expansion into the countryside by creating Green Belts and by the allocation of housing to urban areas or to large key villages. Rural England has witnessed rising owner-occupation and low levels of local authority housing. The low level of new housing development in smaller rural communities has been reflected in higher house prices and greater social exclusivity.

Such social and economic changes have increased the pressure on rural resources so that government has had to re-evaluate policies for the countryside. Regulation has become an important element in some areas, notably in relation to sustainability and environmental conservation.

Murdoch and Marsden (1994) recognise four current types of rural area:

- The preserved countryside which is highly accessible from urban areas. The middle-class population, many of whom are newcomers, lobby strongly against most new development proposals.

- The contested countryside which lies outside the main commuter zones. Here farmers and development interests remain dominant and thus a high proportion of

Figure 8.113 Brockham, Surrey: church and green

development proposals are pushed through. However, newcomers are increasing in number and adopting attitudes similar to those in the preserved countryside.

■ The paternalistic countryside dominated by large private estates and farms. Here development is controlled by established landowners who take a long-term management view of their property. Development is largely related to the economic diversification needed to raise incomes.

■ The clientist countryside of remote rural areas where agriculture is dominant but dependent on government subsidy.

Changing agriculture

The countryside in Britain and other developed nations has been affected by major structural changes in agricultural production. Although agricultural land forms 73% of the total land area of the UK only 2.1% of the total workforce were employed in agriculture in 1994. This was down from 6.1% in 1950 and 2.9% in 1970. In absolute terms nearly a million people were employed in farm work in 1938. By 1998 it was less than 500 000. Even in the most rural of areas, agriculture and related industries rarely account for more than 15% of the employed population.

At the same time the size of farms has steadily increased: 33% of farms in the UK were over 50 ha in 1995, up from 27% in 1970. Such changes have resulted in a significant loss of hedgerows which provide important ecological networks.

Figure 8.114 Arable farmland in mid-Surrey

In 1996 an average farm-worker earned £91 a week less than the average for manufacturing work. As a result farm-workers are among the poorest of the working poor. A quarter of the 11 million people who live in rural England are on or below the margin of poverty. This fact, however, has not been reflected in government funding. In 1996–97, £346 million was allocated to alleviate inner-city deprivation while just over £20 million was set aside to help the rural poor. This disparity is puzzling considering the overlap of urban and rural deprivation (Figure 8.115).

As many farmers have struggled to make a living from traditional agricultural activities a growing number have

---

PERIPHERAL   RURAL

Economic stagnation

Restricted job opportunities

Low wages

High unemployment

Inaccessibility to jobs and services

Social isolation

Absence of basic amenities

Environmental decay

Social and ethnic conflict

Overcrowding and social pathology

Decline in community spirit

Depopulation

Weakened tax base

Residue of ageing and increasingly indigent population

Disinvestment and decline of services (public and private)

High cost and restricted choice of goods and services

INNER   URBAN

Source: Knox and Cottam (1981a)

Figure 8.115 The overlap of urban and rural deprivation

TOURIST AND RECREATION

Tourism
Self-catering
Serviced accommodation
Activity holidays

Recreation
Farm visitor centre
Farm museum
Restaurant/tea room

UNCONVENTIONAL PRODUCTS

Livestock
Sheep for milk
Goats
Snails

Crops
Borage
Evening primrose
Organic crops

Source: Slee, 1987.

VALUE-ADDED

By marketing
Pick your own
Home delivered products
Farm gate sales

By processing
Meat products – patés, etc.
Horticultural products to jam
Farmhouse cider
Farmhouse cheese

ANCILLARY RESOURCES

Buildings
For craft units
For homes
For tourist accommodation

Woodlands
For timber
For game

Wetlands
For lakes
For game

Figure 8.116 Areas of potential farm diversification

sought to diversify both within and outside agriculture (Figure 8.116). However, while diversification may initially halt job losses, if too many farmers in an area opt for the same type of diversification oversupply can result in a further round of rural decline.

Counterurbanisation and the rural landscape

In recent decades counterurbanisation has replaced urbanisation as the dominant force shaping settlement patterns. It is a complex and multifaceted process which has resulted in a 'rural population turnaround' in many areas where depopulation had been in progress. Green Belt restrictions have limited the impact of counterurbanisation in many areas adjacent to cities. But, not surprisingly, the greatest impact of counterurbanisation has been just beyond the Green Belts where commuting is clearly viable. Here rural settlements have grown substantially and been altered in character considerably.

Figure 8.117 shows the changing morphology of metropolitan villages identified by Hudson (1977). Stage 1 is characterised by the conversion of working buildings into houses with new building mainly in the form of in-fill. However, some new building might occur at the edge of the village. The major morphological change in Stage 2 is ribbon development along roads leading out of the village. Stage 3 of the model shows planned additions on a much larger scale of either council or private housing estates at the edge of villages. As time passes the early morphology of the village becomes less apparent. Clearly, not all metropolitan villages will have evolved in the same way as the model, particularly those where Green Belt restrictions are in place. Nevertheless, the model provides a useful framework for reference.

Studies of metropolitan villages in the post-1950 period have charted the gradual disappearance in such settlements of the former integrated community structure and its replacement by a system of two co-existing groups; long-established residents, and newcomers. Figure 8.118 shows at its most extreme the differences that may exist between the two groups. However, in the last decade or so there is evidence of a reduction in socio-economic differences between long-established residents and newcomers. This has been due to the movement of relatively high-paid manual workers to rural areas because of:

■ **the urban–rural shift of manufacturing industry**

■ **the broadening appeal of rural life across the socio-economic groups.**

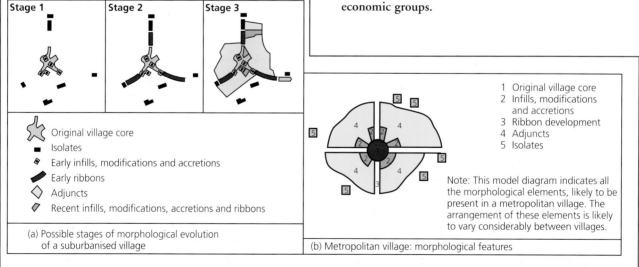

(a) Possible stages of morphological evolution of a suburbanised village

Stage 1 Stage 2 Stage 3

Original village core
Isolates
Early infills, modifications and accretions
Early ribbons
Adjuncts
Recent infills, modifications, accretions and ribbons

(b) Metropolitan village: morphological features

1 Original village core
2 Infills, modifications and accretions
3 Ribbon development
4 Adjuncts
5 Isolates

Note: This model diagram indicates all the morphological elements, likely to be present in a metropolitan village. The arrangement of these elements is likely to vary considerably between villages.

Figure 8.117 Morphology of metropolitan villages

Figure 8.118 Comparison of long-established residents and newcomers to metropolitan villages

Long-established residents	Newcomers
■ Born in village or local area.	■ Born elsewhere, often in another part of the country.
■ Live in council property and tied property.	■ Live in owner-occupied property.
■ Predominantly in socio-economic groups 9, 10, 11, 15.*a	■ Predominantly in socio-economic groups 1, 2, 3, 4.*b
■ Work locally.	■ Work in nearby towns and cities.
■ Travel to work on foot, by bike, by bus.	■ Travel to work by car and train.
■ Earnings below the national average.	■ Earnings above the national average.
■ Average age higher than that of total village population.	■ Average age lower than that of total village population.*c
■ Composite families not uncommon.	■ Simple nuclear families.
■ Relatives live in village and/or surrounding rural area.	■ Relatives unlikely to be found in village or surrounding area.
■ Much use made of village shops and community organisations.	■ Shopping and recreation trips over wide area, especially to surrounding towns and cities.

*a Skilled manual workers, semi-skilled manual workers, unskilled manual workers and agricultural workers respectively. There have always been exceptions to this generalisation; for example, farmers, doctors, parsons and school teachers are examples of long-established village residents who belong to socio-economic groups 1 to 4.

*b Various kinds of employers, managers and professional workers.

*c Some settlements, especially in coastal areas, contain large numbers of retired newcomers.

In addition, a growing number of long-established residents are commuting to work in urban areas because of the decline in rural employment opportunities. The term 'inertia commuters' has been applied to this group of travellers to distinguish them from commuting newcomers, the so-called 'voluntary commuters'.

Areas of tranquillity in the English countryside have shrunk by more than a fifth since 1965. Studies for the Protection of Rural England and the Countryside Commission show that nearly 19 000 km² of tranquil countryside have become blighted by noise and other adverse human impact. These analyses concluded that only three large areas of England are now left untouched by urbanisation and industry: The Marches of Shropshire and Herefordshire, the north Pennines and north Devon. Areas of tranquillity, defined as peaceful and unspoilt places typically between 1 and 3 km from roads, 4 km from a power station, beyond large settlements and the noise of military or industrial activity, remain in all of England's counties. However, they are smaller and more fragmented than they were three decades ago, with large swathes under pressure from development. The greatest change has occurred in the South East. In the 1960s, 58% of the region was considered tranquil. By 1995 this was down to 38%. In contrast the North East was the least changed since the 1960s, losing only 9% of its tranquil area.

Rural depopulation

In more remote rural areas the pattern of population decline that had begun in the 19th century continued into the second half of the 20th century, with the 1971 census showing that almost a third of all rural districts were still losing population. However, because of the geographical spread of counterurbanisation since the 1960s the areas affected by rural depopulation have diminished. Depopulation is now generally confined to the most isolated areas of the country,

but exceptions can be found in other areas where economic conditions are particularly dire. Figure 8.119 is a simple model of the depopulation process.

The issue of rural services

The 1997 Survey of Rural Services was the third in a unique series, providing information about the availability of services in the 9677 rural parishes in England. Earlier surveys were held in 1991 and 1994 (Figure 8.120). Rural parishes are defined as those with a population under 10 000. It should be noted that this is not a village survey, since parishes may contain more than one village. The 10 000 population threshold means that the survey also covered larger rural settlements, including small market towns.

The most recent survey showed that the proportion of rural parishes without key services remained high although there had been no significant decline in most services over the period 1991 to 1997. The sharpest decline was in bus services. However, some services had increased, including childcare and village halls. The main findings of the 1997 survey were:

■ **75% of parishes had no daily bus service**

■ **49% of parishes had no school**

■ **43% of parishes had no post office**

■ **42% of parishes had no permanent shop of any kind.**

The survey found a strong relationship between population size and service provision. For example, 59% of parishes with a population of under 1000 had no permanent shop of any kind, compared with 1% for parishes with a population between 3000 and 9999. Similarly, 67% of parishes with a population of under 1000 had no school, compared with 1% for parishes with a population between 3000 and 9999. Service decline can have a huge impact on rural populations (Figure 8.123).

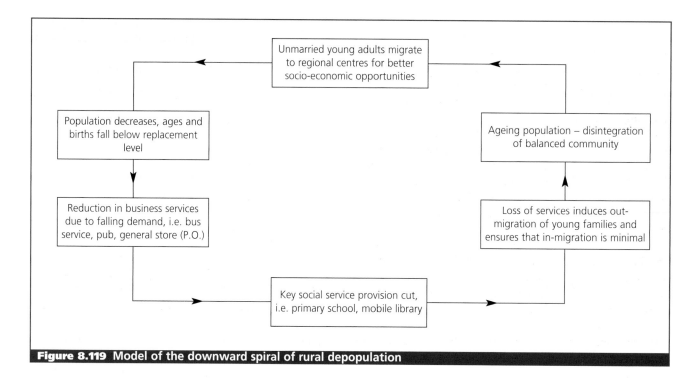

Figure 8.119 Model of the downward spiral of rural depopulation

Figure 8.120 Rural services in England

Percentage of rural parishes without key services

	1991	1994	1997
permanent shop (of any kind)	41	42	42
general store	71	72	70
post office	42	43	43
village hall/community centre	30	29	28
public house	n/a	30	29
daily bus service	72	71	75
school (for any age)	50*	52*	49
school (for 6 year olds)	53*	52*	50

Note: * there may have been a slight undercount of schools in these surveys.

Percentage of rural parishes without other important services

	1991	1994	1997
petrol station or garage	n/a	58	56
bank or building society	89	90	91
public nursery	96	95	93
private nursery	93	90	86
daycare group for elderly	93	92	91
GP (based in the parish)	84	83	83
dentist	91	91	91
pharmacy (of any kind)	81	81	79
Benefit Agency office	n/a	n/a	99
Job Centre	n/a	n/a	99
library (permanent or mobile)	12	16#	12
community minibus or social car scheme	82	79	79
police station	89	91	92

Note: # this is probably a rogue figure and is considered unreliable.

Source: *The 1997 Survey of Rural Services*

Figure 8.121 Rural services in Leigh, Surrey

Figure 8.122 Village pub in Leigh, Surrey

Key villages

Between the 1950s and 1970s the concept of key settlements was central to rural settlement policy in many parts of Britain, particularly where depopulation was occurring. The concept relates to central place theory and assumes that focusing services, facilities and employment in one selected settlement will satisfy the essential needs of the surrounding villages and hamlets (Figure 8.118). The argument was that with falling demand dispersed services would decline rapidly in vulnerable areas. The only way to maintain a reasonable level of service provision in such an area was to focus on those locations with the greatest accessibility and the best combination of other advantages. In this way threshold populations could be assured and hopefully the downward spiral of service decline would be halted. Thus in remote areas the main role of key settlements was to act as 'stabilising centres' in a sea of decline whereas in pressurised areas they acted as 'control centres' to which economic and population overspill could be channelled.

Devon introduced a key settlement policy in 1964 to counter the impact of:

■ rural depopulation

■ the changing function of the village in relation to urban centres

■ the decline in agricultural employment

■ the contraction of public transport.

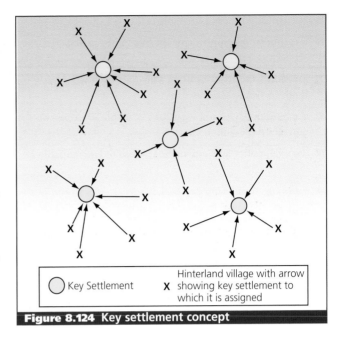

| ⬭ Key Settlement | X | Hinterland village with arrow showing key settlement to which it is assigned |

Figure 8.124 Key settlement concept

The selection of key settlements in Devon was part of a wider settlement policy involving sub-regional centres, sub-urban towns and coastal resorts. The criteria used for selecting key settlements were as follows:

■ existing services

■ existing employment other than agriculture in or near the village

Our dying villages

The villagers of Mollington in Oxfordshire have watched with sadness as many of the things that once made it a community, not just a collection of houses, have vanished.

The primary school, a modern building which had only been in use for 20 years, was shut two years ago due to falling numbers.

Mollington's village shop had already gone the same way, taking the sub-post office with it.

Elderly residents must now struggle into Banbury, five miles away, to collect their pensions, and the sprawling out-of-town Tesco store is now the favoured shopping destination.

At the supermarket check-outs, of course, no village news or gossip is exchanged, and cars are the only practical way to get there.

Sheila Bywaters, 57, president of Mollington's WI branch, said: 'We do have a small pub left in the village, which is thriving, but there's almost nothing else going on, now. My husband hardly knows any men in the village.

'When the school was open there were always open days, barn dances and concerts and everyone got to know each other. When it closed the whole focus of the village disappeared. The Scout group shut four years ago because numbers were down and the Cubs had to combine with another village.'

Mrs Bywaters has personal experience of the common problem of rising house prices, often pushed up by demand from commuters, forcing young people to leave the villages where they grew up.

'My children couldn't afford to buy houses in the village,' she said, 'Even quite ordinary cottages now sell for well over £100,000. As a result, there are virtually no young families in the village. There's not much to attract them anyway. The playgroup is closing this year, and so the village hall won't get used as much and may struggle financially. The only employment in the village is a handful of jobs on farms. Most people commute miles.

'It's becoming just a dormitory village with no vitality or sense of belonging to a community. We've watched all these changes over the years, and it's very sad. The future doesn't seem bright.

'If I could change one thing, it would be to open the school again, but the young families to support it have been forced out.'

Daily Mail, Monday, July 5 1999

Figure 8.123 Our dying villages

- accessibility by road
- location in relation to current bus (and possibly rail) services
- location in relation to other villages which would rely on them for some services
- the availability of public utilities capable of extension for new development
- the availability and agricultural value of land capable of development
- proximity to urban centres (key settlements would not flourish too close to competing urban areas).

Sixty-eight key settlements were selected initially, reduced to 65 in 1970. Although it has been difficult to measure the effectiveness of the policy with precision, depopulation in north and mid-Devon did fall considerably after the introduction of the policy. And in many areas the decline in service provision was slower than the predictions before the policy was implemented. In recent decades the concept has received less attention both in Devon and the country as a whole due to:

- local authority spending cuts resulting in the closure of many public services that local people and councils would have liked to maintain

- the relentless domination of private sector services by large companies based in or just outside urban areas.

In a recent publication by Devon County Council (Devon Structure Plan, First Review 1995–2011, February 1999) the term key settlement is not even used. Policy S2 states:

Particular rural settlements should be identified in Local Plans as Local Centres. These will form the focal points for a modest scale of development, supporting services and the economic well-being of the hinterland. They should therefore:

1 *be accessible to the community they serve and well related to public transport and the highway network*

2 *be defined to ensure that the local needs of all rural areas can be met, taking into account their location relative to other designated Centres, including those in adjoining Districts.*

The 1995 White Paper for Rural England

A White Paper is an official government report which sets out the government's policy on a matter that is or will come before Parliament. The 1995 White Paper for Rural England was the first attempt for over 50 years to address rural issues in a strategy document. Figure 8.125 is a summary of the

---

White Paper plots faster pace of rural lifestyle

Business
Small business development in the countryside is seen as a vital source of job creation. The paper proposes to relax planning controls on the conversion of redundant farm and other rural buildings for business use, with particular emphasis on light industry, exploiting new information technology and 'teleworking'. Approval of such conversions would depend on road traffic forecasts to prevent 'uncontrolled expansion' of such business from damaging the countryside.

Schools
Rural schools are seen as the focus of family life. About 4000 of England's 19 000 primary schools are in rural areas, but 52 per cent of parishes now have no school. The rate of closure is slowing: 350 small rural schools closed over the past 12 years but only 80 since 1990. The paper says one way to stop further closures is to integrate schools more into village life.

Transport
Although 87 per cent of parishes have some kind of bus service, only 29 per cent have a daily service. But the paper sees little hope of reducing reliance on the private car in the countryside. Frequent bus services for all rural communities, no matter how remote or sparsely populated, are not a practical option. The paper would like to see more community bus services, run by volunteers

and subsidised if necessary by the relevant county council. The paper says there should be fewer new trunk roads in the countryside and more spending on improving existing motorways and providing bypasses to relieve villages.

Housing
The paper identifies an acute shortage of cheap rural homes. Increasing the supply of such housing is seen as the key to keeping young people working and living in the countryside. At present only 12 per cent of rural housing is subsidised, compared to 25 per cent in urban areas.

Villages with fewer than 3000 inhabitants will be exempt from a right-to-buy for housing association tenants. That is to prevent such housing disappearing on to the open market and being bought up at prices local people cannot afford.

The Government will also speed the disposal of surplus Ministry of Defence housing. There are an estimated 13 000 empty MoD homes in Britain, many of them in rural areas. Rural households will be encouraged to take in lodgers through the rent-a-room scheme.

More private-sector bodies and charities will be encouraged to bid for funding to provide cheap rural housing.

The Times, 18 October 1995

Figure 8.125 The 1995 White Paper for Rural England

most important aspects of the White Paper. Although a new White Paper is likely as a result of the change of government in 1997, the key issues have not changed.

The rural transport problem

The considerable increase in car ownership in recent decades has had a devastating effect on public transport (Figure 8.126). While this has not disadvantaged rural car owners very much it has considerably increased the isolation of the poor, the elderly and the young. The lack of public transport puts intense pressure on low-income households to own a car, a large additional expense that many could do without. Recent increases in the price of fuel have exacerbated this problem.

There is concern that Britain's rural railway lines are under threat in a repeat of the 'Beeching cuts' of the 1960s (Figure 8.127). The new fears about government intentions towards rural rail closures were first awakened in 1998 when the transport minister said branch lines in sparsely populated areas might be replaced by coaches. It would be possible to convert track beds into guided busways, and then for buses to divert into towns and villages. However, one study of replacing trains with buses found that at most only half of former rail passengers used the bus replacements. With one in five rural households lacking a car and a low level of bus service in many country areas, the train is essential for many.

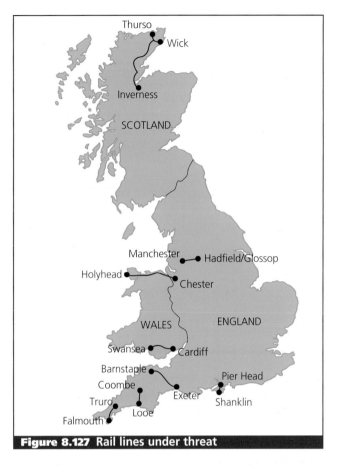

Figure 8.127 Rail lines under threat

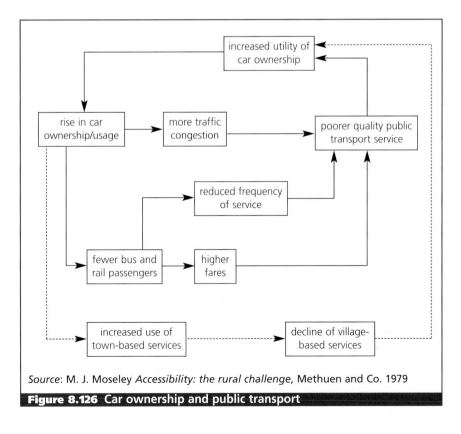

Source: M. J. Moseley *Accessibility: the rural challenge*, Methuen and Co. 1979

Figure 8.126 Car ownership and public transport

The rural housing problem

The lack of affordable housing in village communities has resulted in a large number of young people having to move to market towns or larger urban centres. Only 12% of rural housing is subsidised, compared to 25% in urban areas. The 1995 White Paper on Rural Development sought to improve the rural housing situation by exempting villages with fewer than 3000 inhabitants from the right-to-buy for housing association tenants. This is to prevent such housing disappearing on to the open market and being bought up at prices local people cannot afford. The Government also announced plans to speed up the disposal of Ministry of Defence housing. It estimated that there were 13 000 empty MoD homes in Britain, many of them in rural areas. Rural households would also be encouraged to take in lodgers through the rent-a-room scheme.

The issue of second homes has become increasingly contentious. Figure 8.128, from a book published in the mid-1980s, indicates that some advantages might accrue from second home development. However, recent debate on the issue has centred firmly on the problems created.

Figure 8.128(a) Positive aspects of second home development

Advantages

1. Bring new employment opportunities to areas previously dependent upon a contracting agricultural economy (e.g. building trade, gardening and domestic staff).
2. Local restaurants, shops and garages derive new business and additional profits (which may be essential to year-round economic survival).
3. Specialized shops opened to cater for second home owners also benefit local residents.
4. Property taxes imposed on second homes increase the finances of the local community.
5. Second home owners make fewer demands on local services since education and other community facilities are not required.
6. Renovation of old buildings improves the appearance of the rural area.
7. Rural residents have the opportunity to sell off surplus land and buildings at a high price.
8. Contacts with urban-based second home owners can benefit local residents by exposing them to national values and information, broadening outlooks or stimulating self-advancement via migration.

Figure 8.128(b) Negative aspects of second home development

Disadvantages

1. Concentrations of second homes may require installation of costly sewerage schemes, extension of water and electricity lines to meet peak season demand, and more frequent maintenance of rural roads, with the costs being partly borne by locals.
2. Demand for second homes by urbanites pushes up house prices to the disadvantage of locals.
3. Future schemes for farm enlargement or agricultural restructuring may be hindered by inflated land prices.
4. Fragmentation of agricultural land.
5. Destruction of the 'natural' environment (e.g. soil erosion and stream pollution).
6. Visual degradation may result from poorly constructed or inappropriately located second homes.
7. Second home construction may distract the local workforce from ordinary house building and maintenance.
8. The different values and attitudes of second home families disrupt local community life.

Figure 8.129 New infill housing, Brockham, Surrey

SPATIAL FOCUS

The Village of Urchfont, Wiltshire

A recent article in the *Guardian* newspaper used the village of Urchfont (Figure 8.131) in Wiltshire to exemplify the problems facing rural communities. The village has a population of 1130. The opening paragraph of the article read as follows: 'Rural life is in crisis. Farmers are crying ruin, shops are closing, and wages are low. To top it all, city folk escaping from the urban sprawl are causing house prices to soar'.

The village bus: After deregulation, Urchfont looked likely to lose most of its bus services. Now dozens of villagers depend on the community bus, run by a team of volunteers, and the Wiggly Bus, which is subsidised by the district council and comes out when you call it. In 1997 75% of rural parishes had no daily bus. Traffic on rural roads is forecast to increase by 50% over next 25 years.

The new development: High-speed trains from Pewsey station, a 20-minute drive away, have made it possible for commuters to move into the village, bringing with them London prices (the train to the capital takes 70 minutes). An executive home in a recent development is for sale at £235,000. Each week, 1,700 people leave our major cities.

The new arrivals' home: A couple moved to Urchfont six years ago because they couldn't afford the prices around Newbury. They bought a house for £135,000 and it has just been revalued at £225,000. They came to the village for the quality of life and sense of community.

Manor Farm House: The farmhouse is rented out for £1,800 a month to a couple who have just left London. They find prices comparatively cheap, and have found a period house to buy in the area for £400,000. For farmers, income from residential property can be as important as that from the land. Another local has rented out her farmhouse for £1,200 a month and moved into a house in Urchfont for £390 a month.

Knights Leaze Farm: Once predominantly a dairy farm, its owners, the Bodmans, say they can no longer live on what they get for milk. Now, 90% of their income comes from contract work – farm construction, fencing, hedging, straw-baling. They employ 13 people, but have met some opposition in the village to their heavy lorries and machinery.

The last shop: Until it closed recently, the butcher's was Urchfont's last shop; it also sold a few groceries. It has now been converted into a house and is on the market for £152,000. It has no garden. Seventy per cent of rural parishes are now without a general store; 49% of villages have no school.

The weekenders' cottage: About 7% of houses in Urchfont are weekend homes. Small two-bedroomed cottages would have been suitable for first-time buyers five years ago; now they fetch as much as £150,000. Local people can no longer afford the villages and have to look in nearby towns, according to estate agents. 'A one-bedroomed house will be over £50,000 and £10,000 might be a typical wage.'

Manor Farm: Situated in the centre of the village it's site is unsuitable for the size of today's agricultural machinery, according to its owner, John Snook. The noise and mess of a farm is also thought unacceptable in a conservation area. One use for it would be as a site for new homes, but planning permission is unlikely on a 'zone for employment' and many villagers oppose new housing.

Rookery Farm: Farmer John Snook has decided that he will have to sell his dairy herd, although there have been cows here for over a century and his herdsman will lose his job. 'We've stopped investing. We are lucky we don't have any borrowing – we can get by – but small farms will be packing up in droves.' Farm incomes have collapsed by more than two-thirds in two years.

The council estate: Some houses on the Foxley Fields estate are rented out by a housing association, others have been sold with a covenant on them enabling the council to nominate future buyers. The aim is to ensure the village retains some affordable housing. Of 15 people on the district council's housing waiting list, eight have been on for more than four years.

The *Guardian* 18 October 1999

Figure 8.130 Key issues in the village of Urchfont, Wiltshire

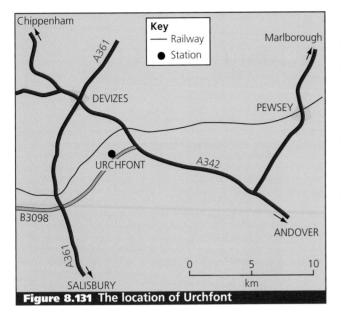

Figure 8.131 The location of Urchfont

The future of rural areas

The Rural Development Commission is the government agency responsible for the well-being of rural England. The stated objectives of the RDC are that the English countryside should be a place where:

■ People both live and work and villages and small towns provide for the varied needs of people in a wide range of circumstances.

■ The economy of all rural areas provides a broad range of job opportunities and makes the most effective contribution to the national economy.

■ Residents are not unduly disadvantaged as a result of living in rural areas and rural communities have reasonable and affordable access to services.

■ Development respects and where possible enhances the environment.

To achieve these objectives the RDC has designated parts of rural England as RDAs. In these rural areas there is clear evidence of relative economic and social disadvantage. Funding is available for the following types of projects: social, community, environment, transport, housing, training, economic, tourism.

RDC funding, with a maximum of 50% of total cost, is normally directed at providing an initial capital contribution to launch a particular project. Projects should aim for self-sufficiency within three years at the most.

Horndean, Hampshire: from village to town?

The settlement of Horndean is situated 15 km north of Portsmouth (Figure 8.132). Horndean developed as a linear village along the London–Portsmouth route. The 1891 census recorded a population of just over 300 with most of the population employed either as agricultural labourers or brewery workers.

A major event in the development of the village was the opening of the Horndean Light Railway in 1903. The railway linked Horndean with Cowplain, Waterlooville, Purbrook and Cosham (on the outskirts of Portsmouth). As a result of this link and the A3 trunk road that passed through the village, Horndean became a growing commuter settlement with new estates established at the edge of the village. By 1931 Horndean had a population of over 2900.

However, the light railway closed in 1935 because of falling demand caused by the steady increase in car ownership. As personal mobility within the region increased further so did the population of Horndean and other villages within easy reach of

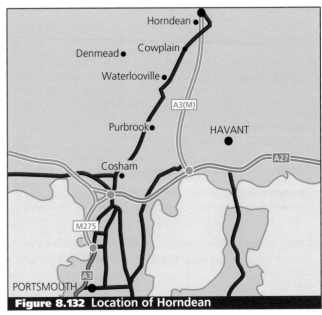

Figure 8.132 Location of Horndean

Portsmouth and Southampton. By 1961 Horndean's population had reached 5500.

The 1962 Ordnance Survey map shows that the settlement still maintained a linear shape. Hazleton Wood, a large forested area, still separated Horndean from Cowplain to the south. By the early 1970s, however, most of Hazleton Wood had disappeared due to the development of a large housing estate. In-filling had also occurred, along with the building of a comprehensive school. The population had now risen to almost 9000.

The next major transport development to influence Horndean was the opening of the A3(M) in 1978 between Horndean and Portsmouth, when the settlement was effectively bypassed. This encouraged further commuting as it was now possible to reach the centre of Portsmouth in 20 minutes. The location of new high technology industries close to the new motorway network around Portsmouth increased the demand for housing in nearby commuter settlements. This led to a new surge of house-building in Horndean in the late 1970s and early 1980s on farmland at the edge of the settlement. The boundaries of Horndean were expanding west towards Denmead and south towards Cowplain. In 1981 the population reached 10 181.

While there was concern about the impact of growth, the benefits of economic expansion could not be ignored and Horndean was identified by the South Hampshire Structure Plan as being one of the settlements in the area best able to support new development. Land for up to 1840 new houses was to be made available between 1983 and 1996 along with a further 5 ha for industrial development. The location of new construction was governed by the fact that some of the countryside in the area is under various forms of protection.

Although the number of jobs provided in and around Horndean has increased considerably, a large proportion of the economically active population still commute to Portsmouth and Southampton.

In 1999 it was estimated that 4902 households with a total population of 12 795 lived in 5140 dwellings. Less than 2% of the working population are employed in agriculture. In early 1999 the unemployment rate in Horndean was under 3%, a strong indication of the affluence of the area.

A more detailed analysis of the historical development of Horndean appears in the Ordnance Survey's 'Mapping News', Issue 8, Autumn 1995.

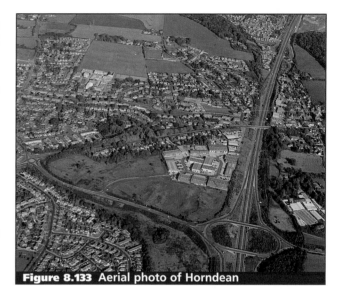

Figure 8.133 Aerial photo of Horndean

Norfolk's Rural Development Area

Norfolk is the third largest Rural Development Area (RDA) in the country (Figure 8.134). It receives about £1 million a year in funding from the Rural Development Commission. Each year an Operating Plan is published showing indicative projects which could be brought forward for financial support in the RDA to combat identified needs.

The Norfolk RDA has a population of 221 000 (mid-1995), 29% of the county's total population. The main areas of population are in the small market towns of Cromer, Fakenham, Hunstanton, North Walsham, Sheringham and Swaffham. The region is predominantly rural and agricultural in nature resulting in a scattered population with the following attendant problems:

■ lack of transport

■ limited access to services

■ social isolation

■ deprivation.

The population of the RDA is older than that for Norfolk as a whole (Figure 8.135). Agricultural employment has declined steadily (Figure 8.136) with a 6% decrease between 1991 and 1995. This decline is forecast to continue in response to structural changes brought about by the Common Agricultural Policy. In addition, the ever-increasing efficiencies of farm management means that farm labourers are not being replaced on retirement and job opportunities for the young are lost. The RDA has a higher concentration of workers in less skilled occupations than the county average. The majority of employers in the RDA are small businesses. Only eight out of the Norfolk's 100 largest companies are found within the RDA.

Levels of unemployment related to the closure of a small village business, while not significant in statistical terms, can have a dramatic impact on that community especially when the negative multiplier effect on other businesses is taken into account. Low wages are also an area of concern. In April 1996, the gross weekly earnings for full-time male employees on adult rates were £340 in Norfolk compared with the national average of £391. Many jobs in the Norfolk RDA are seasonal and part-time due to the nature of employment along the North Norfolk coast. Thus employment prospects are limited resulting in low household incomes and the out-migration of young people.

Among the other problems noted in a recent RDA publication are:

■ village shops have found increasing problems in complying with new hygiene regulations and, as many are also post offices, this puts another vital rural service in jeopardy

■ rural garages are declining as petrol sales are lost to superstores

■ public houses continue to close. The property is often worth more than the business and the large brewery chains frequently sell rural pubs to raise capital

■ there has been a considerable fall in the number of affordable housing developments because of cuts in funding for new projects

■ parts of North Norfolk have become increasingly popular for retirement. Subsequently retirees may lose their personal mobility or a spouse, leaving them isolated in an area, which suffers from poor public transport and limited care provision

■ the suicide rate in the RDA, particularly among young men, is significantly above the national average.

The changing profile of the economy of the region has led to a significant minority of people, particularly in rural communities

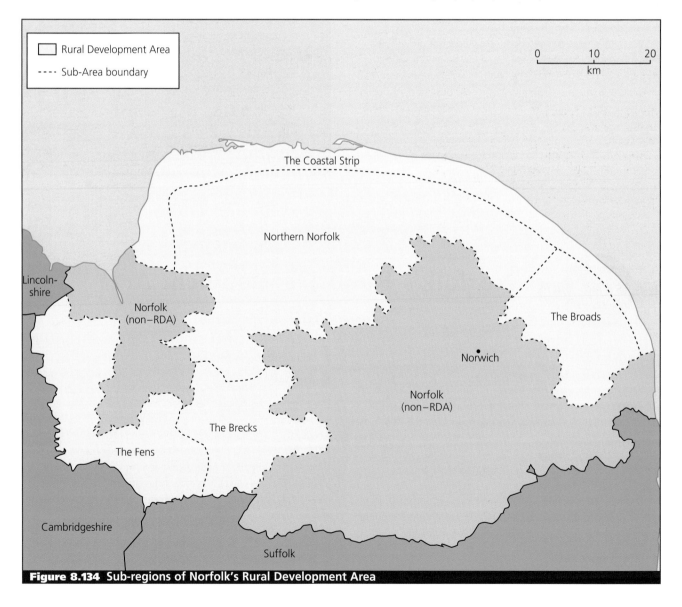

Figure 8.134 Sub-regions of Norfolk's Rural Development Area

under 1000 in population, being socially excluded in various ways. Social exclusion can be identified as 'a lack of access to life opportunities which are open to the mainstream of society'.

The RDA has been sub-divided into five zones. The Rural Development Programme has identified different types of project as suitable for each area (Figure 8.137). Norfolk is concerned not only with rural problems in the RDA but also in its other rural areas. In the recently published Norfolk Rural Economic Strategy a number of strategic priorities were identified of which a selection are listed in Figure 8.138.

Parts of the RDA overlap with that part of Norfolk designated by the EU as eligible for Objective 5(b) funding, providing the potential for joint projects.

Figure 8.135 Age Structure of Norfolk's RDA population in 1995

Age Groups	Estimated Number	Population %	Population 1991 to 1995 Number	Change %
0–4	11 400	5	300	3
5–14	24 900	11	1600	7
15–44	76 800	35	−1700	−2
45–Pensionable Age	50 200	23	5400	12
Pensionable Age	57 800	26	2600	5
Total	221 100	100	8200	4

Figure 8.136 Employment Change in the RDA, 1991 to 1995

Broad Industrial Groups	1991 Number	1995 Number	% change
Agriculture and Fishing	6236	5862	−6
Energy and Water	314	259	−18
Manufacturing	6839	8197	20
Construction	2033	1868	−8
Distribution, Hotels, Restaurants	13 056	12 141	−7
Transport and Communication	1640	2433	48
Banking, Finance and Insurance	4472	4798	7
Public Admin, Education and Health	10 981	10 242	−7
Other Services	1924	2048	6

Figure 8.137 Norfolk Rural Development Area

RDA sub-area	Area characteristics	Types of project considered appropriate
The Coastal Strip:	small seaside holiday resorts in a sensitive environmental area, holiday homes with a high concentration of elderly population and a relatively isolated area.	environmental projects and visitor management schemes, social and community projects, transport schemes and fishing initiatives.
The Broads:	A very sensitive environmental area, popular with visitors but including agricultural and other industries and small local towns.	environmental projects and visitor promotion and management schemes, support for indigenous industries such as boat building, social and economic projects in the small towns.
The Brecks:	a sparsely populated and forested area with some tourism potential.	countryside promotion and tourist facilities.
The Fens:	a high quality farming area with food processing and related businesses but sparsely populated and comparatively isolated.	agricultural diversification projects and initiatives associated with the derived agricultural industries and services, social, community and transport projects.
Northern Norfolk:	a typical rural area with a number of market towns which are important local service and industrial centres. Some low key tourism activity and two disused former RAF Stations.	a programme of market town regeneration throughout the area, support for the survival and expansion of businesses, the development of new tourist attractions and the re-use of former military bases.

Figure 8.138 Norfolk Rural Economic Strategy: a sample of strategic priorities

Context/situation	Problem	Proposed action
Loss of village based services leads to reliance on more distant services creating a need for transport.	Cash circulates less within the local economy and profit 'leaks' out so more goods and services are 'imported', weakening the capacity of new enterprise and community efforts. The potential for communities to profit and reinvest in their local economy is untapped in this part of the country.	Commission and publish a pilot scheme for community-led surveys of a village economy. Promote awareness of the benefits, and support the development, of community enterprises with regard to current best practice. Develop transport brokerage schemes providing Dial-a-Ride and group hire facilities.
Outside the main centres of population, opportunities for development are very limited. However, the role of market towns as generators of economic activity is key to rural development activities.	A balanced portfolio of sites available for employment is a key requisite in attracting new investment/assisting the expansion of locally established businesses.	Extend the 'Strategic Sites' exercise to investigate the smaller scale sites and produce a database of development opportunities. Develop an inward investment strategy that promotes the opportunities of market town locations. Maintain and increase the stock of businesses to encourage quality start-ups and improve survival rate.
Training and job opportunities tend to be concentrated in towns and the city.	Access to opportunities for people in smaller rural communities is more difficult.	Support the development of flexible feeders to existing transport networks i.e. 'Kickstart', car sharing and other alternative transport schemes. Investigate the development of decentralised foyer schemes.
Public sector funding restrictions and inadequate supply of land limit the scope for developing affordable social housing.	Young people/families can be forced to move out of rural communities in order to find affordable accommodation.	Encourage the spread of self-build housing as joint training/housing/employability projects.
Credit is more accessible and affordable to people with a secure/earned income. Interest rates on borrowing are high and the profit from this goes generally to shareholders with no local concern.	Lack of access to mainstream credit sources is often due to low pay and no regular income. This can leave people with high interest rate loans or 'loan sharks' as their only options. People with savings often wish to see their money put to good use for local benefit.	Support the development of Credit Unions, creating access to cheaper credit linked to local savings.

Questions

1 **(a)** Study the OS map extract (Figure 8.140). How large is the land area covered by the map extract?
   (b) With the aid of an atlas place this location in its regional context.
   (c) Describe the pattern of settlement in the area.
   (d) Suggest two reasons for the low density of settlement south of the A64(T).
   (e) Account for the lack of settlement of any significant size in the central area of the map extract.
   (f) What is the evidence that settlement has evolved over a long period of time?
   (g) Compare the site and situation of Sherburn and Snainton.
   (h) To what extent is a hierarchy of settlement evident?
   (i) What clues are available on the OS map about the economy of the area?
   (j) Describe and explain the pattern of communications.

2 **(a)** Define the terms (i) dispersed settlement pattern, (ii) nucleated settlement pattern.
   (b) Describe the distribution of rural settlement patterns in Britain (Figure 8.112).
   (c) Discuss the main reasons for the development of a dispersed pattern of settlement.
   (d) Under what circumstances does a nucleated settlement pattern tend to develop?

3 **(a)** Study Figure 8.117. Suggest a significant reason for the site of the original village core.

(b) Why were isolated dwellings often located within easy of such villages?

(c) Explain the changes shown in Stage 2.

(d) Account for the most complex development of the village shown in Stage 3.

(e) Compare the morphology of a village you have studied with Figure 8.117(b).

4 (a) How might socio-economic opportunities for young adults be better in regional centres than in remote rural areas?

(b) Why do private sector services generally leave rural areas before public sector services?

(c) How might such a trend be reversed?

(d) Explain the demographic changes that generally occur with rural depopulation.

5 (a) Comment on the level of rural service provision in 1997.

(b) To what extent had the provision of services changed since 1991?

(c) Why is the development of small businesses viewed as vital to the stability of many rural areas?

(d) Why are rural schools seen as 'the focus of family life'?

(e) How important is housing in maintaining the viability of rural communities?

6 (a) Describe and explain the location of Norfolk's Rural Development Area.

(b) Why has the RDA been divided into sub-areas?

(c) Comment on the RDA's (i) age structure, (ii) employment structure.

(d) Examine some of the key issues which the RDA proposes to tackle.

7 (a) What is the name given to policies designed to concentrate services in a small number of settlements?

(b) Summarise the changes proposed in Figure 8.139.

(c) Why would a county council consider making such changes?

(d) Give two reasons for selecting Wellby as an 'extended village'.

(e) Why might the inhabitants of Boxford voice opposition to the proposals?

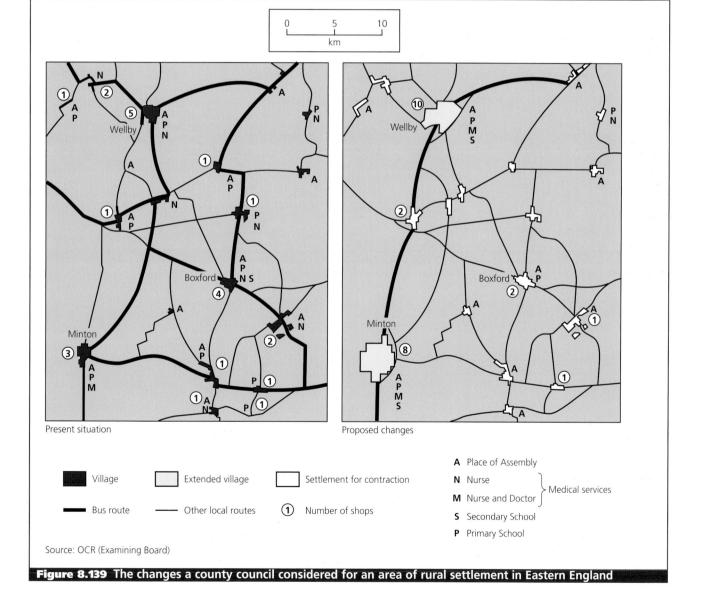

Present situation

Proposed changes

Village   Extended village   Settlement for contraction

Bus route   Other local routes   ① Number of shops

A Place of Assembly
N Nurse
M Nurse and Doctor } Medical services
S Secondary School
P Primary School

Source: OCR (Examining Board)

Figure 8.139 The changes a county council considered for an area of rural settlement in Eastern England

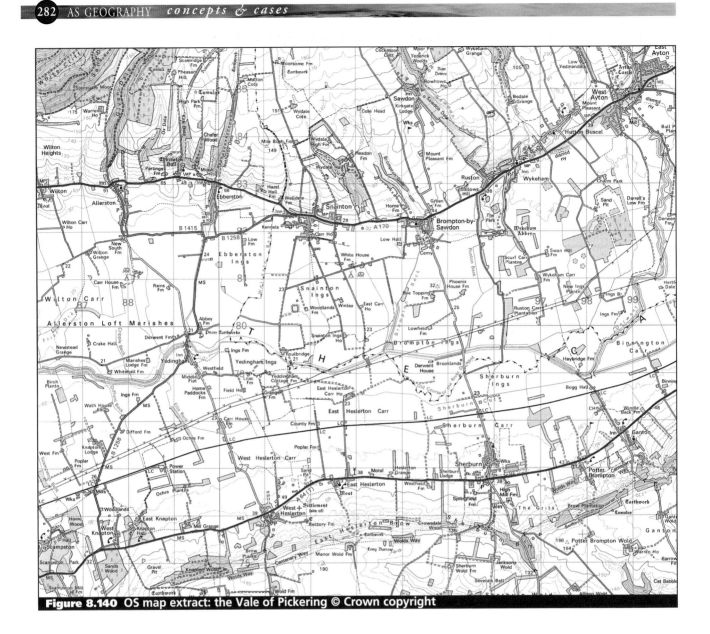

Figure 8.140 OS map extract: the Vale of Pickering © Crown copyright

9 economic *activity*

Sectors and systems

Sectors of employment

In all modern economies of a significant size people do hundreds, and in some cases thousands, of different jobs, all of which can be placed into four broad economic sectors:

■ The **primary sector** exploits raw materials from land, water and air. Farming, fishing, forestry, mining and quarrying make up most of the jobs in this sector.

■ The **secondary sector** manufactures primary materials into finished products. Activities in this sector include the production of processed food, furniture and motor vehicles.

■ The **tertiary sector** provides services to businesses and to people. Retail employees, drivers, architects and nurses are examples of occupations in this sector.

■ The **quaternary sector** uses high technology to provide information and expertise. Research and development is an important part of this sector. Quaternary industries have only been recognised as a separate group since the late 1960s.

As an economy advances the proportion of people employed in each sector changes (Figure 9.2). Countries such as the

Figure 9.1 The primary sector: Fishing fleet in Castletown Bearhaven, County Cork, Ireland

USA, Japan and the UK are 'post-industrial societies' where the majority of people are employed in the tertiary sector with the quaternary sector growing rapidly. In 1900, by contrast, 40% of employment in the USA was in the primary sector. However, the mechanisation of farming, mining, forestry and fishing drastically reduced the demand for labour in these industries. As these jobs disappeared people moved to

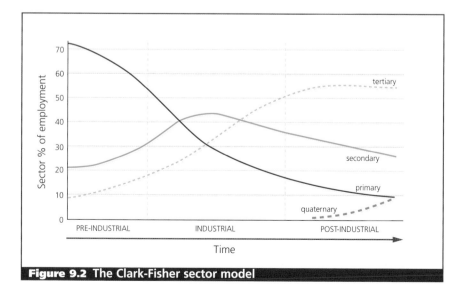

Figure 9.2 The Clark-Fisher sector model

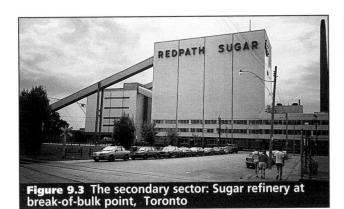

Figure 9.3 The secondary sector: Sugar refinery at break-of-bulk point, Toronto

Figure 9.4 The tertiary sector: Nou Camp stadium, Barcelona

urban areas where most secondary, tertiary and quaternary employment is located. Less than 4% of employment in the USA is now in the primary sector.

Human labour is steadily being replaced in manufacturing too. In more and more factories, robots and other advanced machinery handle assembly line jobs that once employed large numbers of people. In 1950, the same number of Americans were employed in manufacturing as in services. By 1980, two-thirds were working in services.

The tertiary and quaternary sectors are also changing. In banking, insurance and many other types of business, computer networks have reduced the number of people required. But elsewhere service employment is rising such as in health, education and tourism.

The employment structure of a country or region is an important influence on spatial patterns and processes. For example, the geography of South Wales today is very different to that during the era of heavy industry at the height of the Industrial Revolution.

Economic systems

The economic system of a country describes the relationship that exists between consumers, producers and the state. Economists recognise the following types of system:

- **Traditional** – a non-monetary system based largely on subsistence with limited exchange conducted through bartering. Only the most isolated societies in the world remain in this category.

- **Free market** – a system run entirely by private enterprise with no public ownership. This is a theoretical concept as no national economy has ever reached this extreme, but the USA is generally recognised as the country closest to being a free market economy.

- **Centrally planned** – a system where the state decides what is to be produced, where production will be located, and what can be consumed. Such a system, denying private ownership, is associated with communist states. Although another theoretical extreme, many countries, and the former Soviet Union in particular before the fall of communism, came very close to it.

- **Mixed** – where some sectors of an economy are in public ownership and others in private. The nations of western Europe are classic examples of this type of system.

Economic systems can change over time. Traditional systems, which are largely rural based, were once more prevalent than they are today. Since the early 1980s, many countries which had a reasonable balance between public and private ownership – the classic mixed economies – have privatised most of their publicly owned industries on the premise that private ownership is more efficient. This has had a major impact on spatial decision making. For example, the privatisation of electricity generation in Britain hastened the decline of coal as a source of power, resulting in a significant drop in the proportion of coal-fired power stations and closure of many coal mines. The demise of communism is having an important influence on geographical patterns in all the countries it is embracing. In the former East Germany it has resulted in high unemployment and a range of other problems (Figure 9.5). The large-scale rural to urban migration in China at present is a direct result of the shift from a centrally planned economy towards the free market approach.

The Global economy

The major economies

Figure 9.6 shows the relative size of the top ten global economies according to the traditional measure, gross domestic product (GDP). Eight of the top ten are in the north or developed world (Figure 9.7). Because of the rapid growth rates experienced by some of the largest developing

Germany's 'easties' sink in jobs famine

by Peter Conradi and Michael Woodhead, Zwickau

Among the people of Zwickau, an industrial town deep in the former East Germany, the communist era is known merely as the 'old times' – neither good nor bad. It is almost as if the locals have still to be convinced that reunification with the west is better than the 40 years they spent under totalitarian rule.

There was yet more reason for despair last week in the crumbling apartment blocks and decaying industrial architecture of a city best known as the birthplace of the Trabant, the chugging symbol of the communist state's defunct car industry.

With new figures showing unemployment in Germany edging close to a post-war record of 5m, signs are growing that the gulf between the relatively prosperous west and impoverished east is widening. Despite annual subsidies totalling £45–50 billion – equal to 3%–4% of German GDP – the so-called 'new federal states' are sinking deeper into economic misery, fuelling popular disenchantment and boosting support for neo-Nazi groups.

It is now clear that the government seriously underestimated the unemployment that would be caused by the closure of un-productive factories and overestimated how much the private sector would take up the slack. A report, just published by Munich's prestigious IFO-Institut, a think tank, is damning: the economic power of the east is now about 65% of the European Union average – on a par with that of Greece.

Along with nostalgia for the stability of the Communist east, the economic problems have also spawned a revival of neo-Nazi activities. Some of the worst violence has been in Brandenburg, near Berlin, where far-right extremists are trying to establish so-called 'liberated' zones free of immigrants. 'The worst aspect is the sheer level of brutality and the fact that the perpetrators are increasingly younger these days,' said Manfred Füger, a spokesman for the interior ministry in Potsdam.

Zwickau's fate is typical of many one-company towns in the former communist east. The Trabant factory was the biggest employer; when it closed overnight in 1991, its 12,000 workers were out on the street.

Source: *Sunday Times* 8 February 1998

Figure 9.5 Rising unemployment as factories in the former East Germany close

---

countries, economists predict that they could raise their share of world output from around one-sixth to almost a third by 2020.

World trade: from GATT to the WTO

Trade is the most vital element in the growth of the global economy. In 1948, a group of 23 nations agreed to reduce tariffs on each other's exports under the General Agreement on Tariffs and Trade (GATT). This was the first multilateral accord to lower trade barriers since Napoleonic times. Since the GATT was

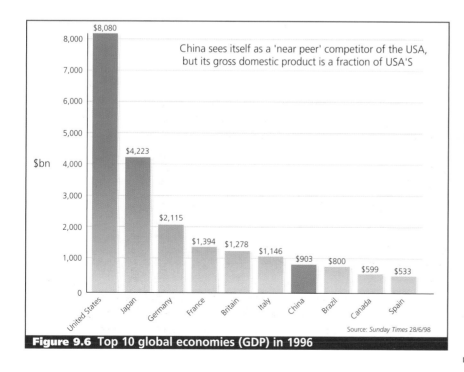

China sees itself as a 'near peer' competitor of the USA, but its gross domestic product is a fraction of USA'S

Source: *Sunday Times* 28/6/98

Figure 9.6 Top 10 global economies (GDP) in 1996

Questions

1 Study Figure 9.2
  (a) Define the terms (i) primary, (ii) secondary, (iii) tertiary, (iv) quaternary.
  (b) Describe the changes which occur in the three stages of the model.
  (c) Explain the sharp decline in agricultural employment over time.
  (d) Why does secondary employment reach a peak in the industrial stage and then decline?
  (e) Account for the changes in the relative importance of tertiary and quaternary sectors over time.

2 (a) What do you understand by the term 'economic system'?
  (b) Describe the characteristics of a traditional economic system. Where in the world and why do such systems still exist?
  (c) Briefly explain the characteristics of the following types of economic system: (i) free market, (ii) centrally planned, (iii) mixed.
  (d) 'The demise of communism is having an important influence on geographical patterns in all the countries it is embracing'. Elaborate.

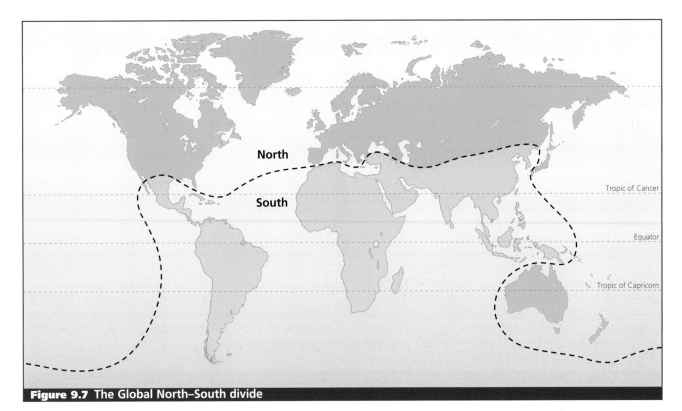

Figure 9.7 The Global North–South divide

established there have been eight 'rounds' of global trade talks (Figure 9.8). The most important recent development has been the replacement of GATT by the World Trade Organisation (WTO). Unlike the loosely organised GATT, the WTO was set up as a permanent organisation with far greater powers to arbitrate trade disputes.

Figure 9.8 A GATT/WTO chronology

1947	Birth of the GATT, signed by 23 countries on 30 October at the Palais des Nations in Geneva.
1948	The GATT comes into force. First meeting of its members in Havana, Cuba.
1949	Second round of talks at Annecy, France. Some 5000 tariff cuts agreed to; ten new countries admitted.
1950–51	Third round at Torquay, England. Members exchange 8700 trade concessions and welcome four new countries.
1956	Fourth round at Geneva. Tariff cuts worth $1.2 trillion at today's prices.
1960–62	The Dillon round, named after US Under Secretary of State Douglas Dillon, who proposed the talks. A further 4400 tariff cuts.
1964–67	The Kennedy round. Many industrial tariffs halved. Signed by 50 countries. Code on dumping agreed to separately.
1973–79	The Tokyo round, involving 99 countries. First serious discussion of non-tariff trade barriers, such as subsidies and licensing requirements. Average tariff on manufactured goods in the nine biggest markets cut from 7% to 4.7%.
1986–93	The Uruguay round. Further cuts in industrial tariffs, export subsidies, licensing and customs valuation. First agreements on trade in services and intellectual property.
1995	Formation of World Trade Organisation with power to settle disputes between members.
1997	Agreements concluded on telecommunications services, information technology and financial services.
1998	The WTO now has 132 members. More than 30 others are waiting to join.

Key Definitions

Gross Domestic Product (GDP) The total value of goods and services produced by a country in a given time period, usually a year.

General Agreement on Tariffs and Trade (GATT) An international agreement to reduce barriers to trade such as tariffs, quotas and subsidies. The GATT was replaced in 1995 by the World Trade Organisation (WTO).

Globalisation The increasing integration of national economies through international flows of trade, investment and financial capital.

Trans-national Corporation A company that manufactures goods or provides services in several countries while directing operations from a headquarters based in one of the countries.

Foreign Direct Investment (FDI) Investment in a country by a foreign company.

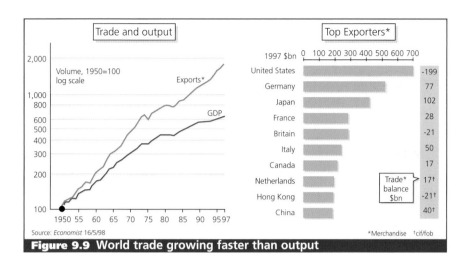

Figure 9.9 World trade growing faster than output

Today average tariffs are only a tenth of what they were when GATT came into force and world trade has been increasing at a much faster rate than GDP (Figure 9.9). In 1996, Europe accounted for 47.9% of world merchandise exports, followed by Asia (25.6%), North America (16.2%), Latin America (4.9%), the Middle East (3.2%) and Africa (2.3%). In terms of the composition of merchandise exports, agricultural products accounted for 11.4%, mining products for 11.2%, and manufactures for 73.3%. In total, global merchandise exports were valued at $5,127 billion in 1996. In the same year global exports of commercial services totalled $1,257 billion.

The activities of the WTO have aroused increasing opposition from groups concerned about the domination of world trade by the rich countries, the environment, workers rights, animal rights, and other issues (Figure 9.10). Figure 9.12 shows the main trading issues confronting the Seattle delegates in the most recent WTO talks as seen by *Time* magazine.

Protesters defy riot police to halt WTO

By Charles Clover in Seattle

The World Trade Organisation was forced to cancel the opening ceremony of its meeting in Seattle last night after protesters choked the city, preventing many delegates from leaving their hotels.

Thousands of 'green' demonstrators and workers' rights marchers blocked the streets around the convention centre where WTO delegates from around the world were due to meet.

Riot police fired rubber bullets and tear-gas in their efforts to move the protesters but many picked up the canisters and hurled them back at the police.

The idea of low-key policing is foreign to Seattle city. By 7am squads of baton-wielding policemen, some carrying machineguns and CS gas, were on the streets round the convention centre.

Two hours later a cordon of environmental protesters with arms linked had surrounded the Sheraton Hotel where the ministers are staying to stop delegates, press and non-governmental organisations getting through.

A gathering of 5,000 'green' protesters massed on the waterfront at Victor Steinbrueck Park and began marching through the city.

Led by Carl Pope of the Sierra Club they intended ultimately to merge with the official demonstration numbering tens of thousands mounted by the trades union group, the American Federation of Labour-Congress of Industrial Organisations.

Protesters carrying effigies of beef cattle skeletons and whales, many dressed in masks, chanted: 'WTO has got to go.'

There were placards calling for 'China out of Tibet', 'Steel workers for labour protection' and 'Vegans' from Philadelphia, Seattle and Oregon. Joe Haptas, carrying a vegan banner, explained that he was protesting against the WTO's frustration of rights for animals around the world.

Masked gangs of black-clad anarchists overturned street furniture and broke windows at multi-national companies such as Starbucks Coffee, Bank of America and Banana Republic. One woman, who would not give her name, said she was against corporate power that trampled the rights of the individual.

Another protester, named Becky, from Colorado, said she was against the American proposal for a 'free logging agreement' that would speed up the destruction of forests.

Source: *Daily Telegraph* 1 December 1999

Figure 9.10 Anti-WTO protests in Seattle

Figure 9.11 Protests in Seattle

The causes of growth

There has been much debate about the cause of economic growth. The Harvard Institute for International Development (HIID), analysing global patterns of growth during 1965–90, concluded that variations between countries were due to:

■ Initial conditions: if other factors are equal poorer countries tend to grow faster than richer ones as the potential for growth from a low base point is much greater.

■ Physical geography: (a) landlocked countries grew more slowly than coastal ones; (b) tropical countries grew more slowly than those in temperate latitudes reflecting the cost of poor health and unproductive farming. However, richer non-agricultural tropical countries such as Singapore do not suffer a geographical deficit of this kind; (c) a generous allocation of natural resources spurred economic growth.

■ Economic policies: (a) open economies grew faster per year than closed economies; (b) fast-growing countries tend to have high rates of saving and low spending relative to GDP; (c) institutional quality in terms of law and order, efficiency of public administration, lack of corruption, etc. delivers a high rate of growth.

■ Demography: progress through demographic transition was a significant factor with the highest rates of growth experienced by those nations where the birth rate had fallen the most.

The stages of economic growth: W. W. Rostow

The American economist Rostow in *The Stages of Economic Growth: A Non-Communist Manifesto* (1960), recognised five

---

THE PLAYERS: Europe and Japan vs. the U.S. and the Third World
THE ISSUE: European and Japanese farmers are swaddled in subsidies. The U.S. and the Third World want access to those markets. Europe and Japan say the U.S. also aids farmers; they oppose the U.S.'s genetically modified food.
THE LIKELY OUTCOME: Look for subsidies to subside but healthy-food fears to grow.

THE PLAYERS: The U.S. and the European Union vs. the Third World
THE ISSUE: Low-cost Third World labor, which threatens jobs in the U.S. and the E.U. The U.S. proposes a WTO study group on labor issues. Some E.U. countries are supportive. Developing nations don't want outside interference in their labor markets.
THE LIKELY OUTCOME: Don't expect much progress here.

THE PLAYERS: The U.S. and E.U. vs. the Third World
THE ISSUE: The U.S. and E.U. want to enforce environmental pacts – such as a treaty restricting endangered-species trade – without wto challenge. But the U.S. seeks to slash wood tariffs, thus increasing deforestation. Third World opposes 'enviro' restrictions.
THE LIKELY OUTCOME: Progress may come on less controversial issues, such as fishing subsidies.

THE PLAYERS: The U.S. vs. everyone else
THE ISSUE: The U.S. wants to slash barriers to several key industries, including health care, banking, education, insurance and e-commerce. But a huge battle looms.
THE LIKELY OUTCOME: wto members have already agreed to negotiate over services. The U.S. will drive hard to get some concessions from the global trade community.

THE PLAYERS: U.S. vs. Japan and the Third World
THE ISSUE: U.S. laws block countries from 'dumping' subsidized products – steel, semiconductors, textiles – on the American market. Third World nations say the laws are protectionist and U.S. should import more.
THE LIKELY OUTCOME: With an election year coming and so many jobs at stake, the U.S. will not give ground.

Source: *Time* 6 December 1999

Figure 9.12 Major issues at the WTO negotiations in Seattle

stages of economic development from the traditional society, characterised by limited technology and a static and hierarchical social structure, to an age of high mass consumption (Figure 9.13). Although Rostow's model was proposed in terms of the national economic unit, he concluded that the development gap was explained by the fact that countries were at different stages of the model.

The crucial part of Rostow's model is the third or 'take-off' stage, the decade or two when economy and society are transformed in such a way that thereafter a steady rate of growth can be sustained. Take-off is launched by an initial stimulus and characterised by a rise in the rate of productive investment to over 10% of national income, the development of one or more substantial manufacturing sectors with a high rate of growth and the emergence of administrative systems which encourage development. After take-off follows the 'drive to maturity' when the impact of growth is transmitted to all parts of the economy with the transition to the age of high mass consumption following in a relatively short time. The model was based on the economic history of over a dozen European countries, all of which are now firmly in the final stage.

Rostow suggested that countries further down the development path would learn from the experience of more developed nations so that the number of years taken for countries to progress through the stages would decrease over time. Thus the 'learning curve' of present-day Brazil would be much shorter than that of the UK in the 18th and 19th centuries.

Rostow noted the US reaching take-off, maturity and high mass consumption in 1860, 1910 and the early 1920s

respectively (Figure 9.14). He saw Canada and Australia as anomalies, with Canada showing the characteristics of high mass consumption by the mid-1920s after take-off a short time before, with maturity following later around 1950. The inversion of the last two stages was made possible by the profitable export of staple commodities in demand on the world market.

The main criticisms of the Rostow model are:

■ there was a low level of explanation concerning the specific mechanisms which link the different stages

■ the model was analogy-based – it was too simplistic to expect that developing nations could easily follow the economic history of the developed world

■ not all countries, which according to Rostow were ready for take-off, could manage the final jump

■ the perspective was endogenic, concentrating on internal factors within individual countries and largely ignoring the importance of external economic relationships

■ some writers argued that Rostow's approach was compromised by his political views (very conservative and strongly anti-communist).

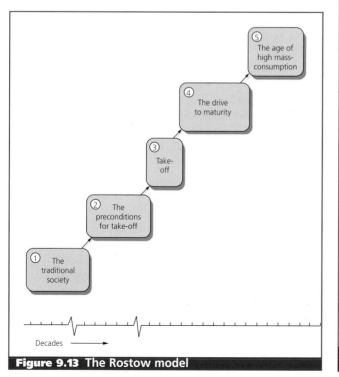

Figure 9.13 The Rostow model

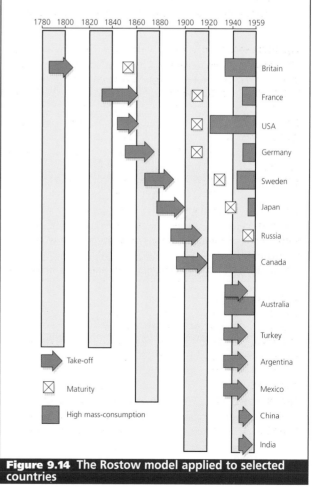

Figure 9.14 The Rostow model applied to selected countries

Alternative theories

As a result of such criticisms alternative theories were developed, the most important of which were:

■ **Dependency Theory (A. G. Frank): an approach which blames the relative underdevelopment of the developing world on exploitation by the developed world. This has been achieved by making the developing world dependent on the developed world, first through colonialism and then by the various elements of neocolonialism.**

■ **World System Theory (I. G. Wallerstein): an approach based on the history of the capitalist world economy since its formation in the 16th century. Countries fall into three economic levels: core, semi-periphery and periphery, and can move from one level to another if their contribution to the world economy changes.**

Figure 9.15 Red Square, Moscow

The economy of the UK

The UK's economy can be seen to comprise of three sectors (Figure 9.16(a)). The goods sector is mainly manufacturing, but it also includes construction, mining, energy and agriculture. Government services include all the things paid for by taxation. Market services are provided (almost entirely) by private sector firms. Wholesaling and retailing are part of the trade, tourism and leisure sector of market services.

The service sector (excluding government) accounts for 45% of UK output, while manufacturing accounts for only 22% (Figure 9.16(b)). The share of market services has been rising steadily in recent decades. By 1997, it was ten percentage points higher than in 1979. Of the large industrial countries the UK now has the second highest share of services in GDP, after the USA.

More than half of the labour force is employed in market services (Figure 9.16(c)), while only 19% work in manufacturing. The highest share of employment in market services is in the South East and London, where it exceeds 65%.

While the workforce as a whole is split almost evenly between men and women, 75% of those working in goods sectors are men, while 70% of those working in government services are women. The sectors which have lost most jobs during the 1990s are dominated by men while the big job-creators are those where women make up the largest share of the workforce. Figure 9.17 shows the distribution of the economically active population.

Globalisation and the communications revolution

The transformation towards a knowledge-based global economy is being driven by a number of forces, at the forefront of which are:

■ **the growing role of information and communication technologies (ICT). The world market for information technology grew twice as fast as GDP between 1987 and 1995**

■ **the increasing importance of services**

■ **the globalisation of markets and societies.**

The term globalisation has been applied to the increased integration of national economies in the 1990s through cross-border flows of trade, investment and financial capital. Globalisation is being driven by plunging communication costs, creating new ways to organise firms at a global level.

The global telephone network has grown nearly tenfold in the past 40 years and reached almost a billion lines by 2000. Teledensity, the number of lines per 100 people, has quadrupled since 1960, although a quarter of the world's nations still have a teledensity of below one.

The new information pipelines linking the players of greatest importance in the global economy are of enormous significance. The transoceanic copper wires that made communications possible in the pre-satellite era are being replaced by arrays of sophisticated fibre optics capable of carrying huge amounts of data.

The cost of computer processing power has been falling by an average of 30% a year in real terms over the past couple of decades. With the costs of communication and computing falling rapidly, the natural barriers of time and space that separate national markets have been tumbling too.

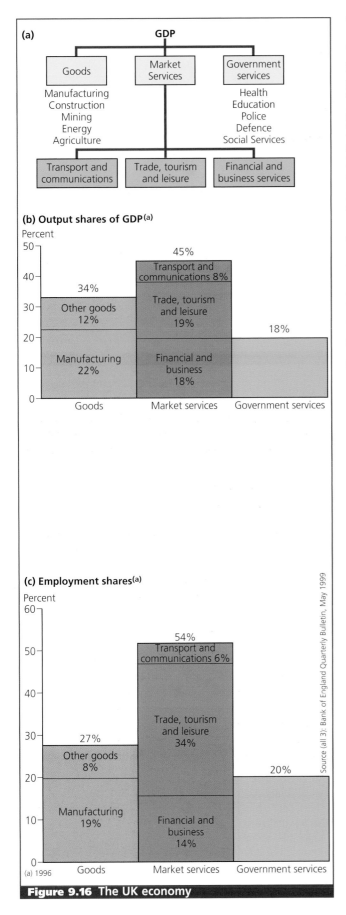

(a) GDP

Goods — Manufacturing, Construction, Mining, Energy, Agriculture

Market Services

Government services — Health, Education, Police, Defence, Social Services

Transport and communications | Trade, tourism and leisure | Financial and business services

(b) Output shares of GDP(a)

Percent

Goods 34%: Other goods 12%, Manufacturing 22%
Market services 45%: Transport and communications 8%, Trade, tourism and leisure 19%, Financial and business 18%
Government services 18%

(c) Employment shares(a)

Percent

Goods 27%: Other goods 8%, Manufacturing 19%
Market services 54%: Transport and communications 6%, Trade, tourism and leisure 34%, Financial and business 14%
Government services 20%

(a) 1996

Source (all 3): Bank of England Quarterly Bulletin, May 1999

Figure 9.16 The UK economy

The 21st century economy

While the 1980s was the 'Japanese decade' according to most indicators, the 1990s turned out to be a decade of unexpected prosperity for the USA – what some American journals called 'the New Economy'. The assertion is that the New Economy is the beginning of a major new wave of innovation. Historically, periods of major innovation have brought substantial increases in living standards. The belief is that the latest wave could make it much easier to address some of the vexing social and environmental problems that affect individual countries and global society in general (Figure 9.18).

Trans-national corporations

Large companies often reach the stage when they want to produce outside of their home country and thus take the decision to become trans-national. The benefits of such a move include:

- **cheaper labour, particularly in developing countries**
- **circumventing trade barriers**
- **tapping market potential in other world regions**
- **avoidance of strict domestic environmental regulations**
- **exchange rate advantages.**

With increased size, greater economies of scale can be achieved, sharpening the company's competitiveness in international markets.

TNCs have a substantial influence on the global economy in general and in the countries in which they choose to locate in particular. They play a major role in world trade in terms of what and where they buy and sell. A sizeable proportion of world trade is actually intra-firm, taking place within TNCs. The organisation of the large motor companies exemplifies intra-firm trade with engines, gearboxes and other key components produced in one country and exported for assembly elsewhere.

Large TNCs often exhibit three organisational levels – headquarters, research and development, and branch plants. The headquarters of a TNC will generally be in the developed world city where the company was established. Research and development will most likely be located here too or in other areas within this country. It is the branch plants that are the first to be located overseas. However, some of the largest and most successful TNCs have divided their industrial empires into world regions, each with its own research and development facilities and a high level of decision making (Figure 9.19).

The world's largest corporations, all of them trans-national, boast annual revenues that are greater than the GDPs of many developing nations (Figure 9.21). Nine of the world's 20 biggest companies are Japanese; six are American; but 12 of the 20 biggest profit makers are American and none

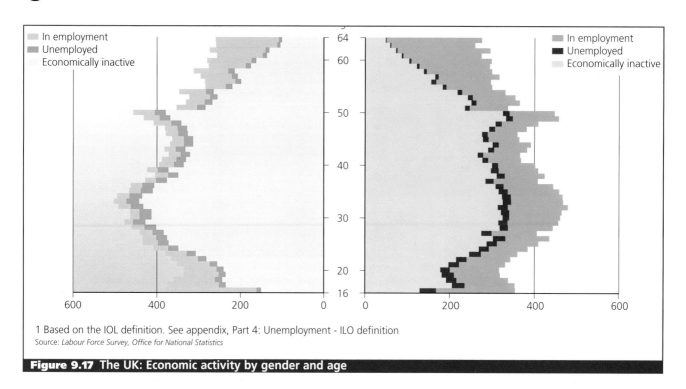

1 Based on the IOL definition. See appendix, Part 4: Unemployment - ILO definition

Source: *Labour Force Survey, Office for National Statistics*

Figure 9.17 The UK: Economic activity by gender and age

is Japanese. The combined stock market capitalisation of the world's top 100 companies is now an enormous $4.5 trillion.

The activities of TNCs have caused considerable debate over a long period of time. The fact that they operate in almost every part of the world indicates that most governments welcome them. This suggests that the advantages of TNCs outweigh the disadvantages to host nations. The advantages of TNCs are:

■ they provide employment and help to raise living standards

■ they may improve the skills of the local workforce

■ the establishment of large firms can create a multiplier effect whereby growth leads to further growth in an upward spiral of development

The 21st Century Economy

Despite Asia's woes, all the ingredients are in place for a surge of innovation that could rival any in history. Over the next decade or so, the New Economy [of the 1990s] – so far propelled mainly by information technology – may turn out to be only the initial stage of a much broader flowering of technological, business, and financial creativity that will sweep across the world.

Call it the 21st Century Economy – an economy that, driven by technological progress, can grow at 3% pace for years to come. The innovation pipeline is fuller than it has been in decades. With the advent of the Internet, the information revolution seems to be spreading and accelerating rather than slowing down. Biotechnology is on the verge of having a major economic impact, and in labs, scientists are testing the frontiers of nanotechnology, with the goal of creating new devices that can transform entire industries.

What's more, the U.S. economy seems to be undergoing a wholesale rejuvenation. Businesses, financial services firms, and universities are reinventing themselves. Even politicians and policymakers are starting to grasp the new technological and economic realities.

To be sure, the path from the New Economy to the 21st Century Economy will likely be a bumpy one. Each innovative surge creates economic and social ills, from recessions to stock-market crashes to widespread job losses – and this one won't be different. But that's the price a nation must pay to achieve the benefits of dynamic change.

Figure 9.18 Source: *Business Week*, 31 August 1998

Age

① Export-led development

② Overseas location of branch plants

③ Shift of R&D and HQ functions

④ Rationalisation

Number of plants / facilities

Increasing globalisation

Time

Activities concentrated in home country where labour and sourcing are established. However, exports may be subject to tariffs and other restrictions

Incentives include cheaper labour, access to markets, and financial assistance from host governments

New locations become semi-autonomous as products are more carefully tailored to new markets

Increasing competition or recession necessitates concentrating activities in the best locations

Figure 9.19 The development of TNCs: locational changes

- new technology may be introduced into the host country
- a widening of the host nation's economic base
- foreign currency is brought into the country which improves the balance of payments situation.

However, critics of TNCs point to the following:

- the employment created may only require low levels of skill with higher level jobs filled by employees from the headquarters country of the TNC
- the bulk of the profits made may be exported to the headquarters country of the TNC
- large TNCs may try to influence government policy to the detriment of the majority of people
- companies may adopt the lowest environmental and health and safety standards they can get away with
- TNCs can move investment from one country to another very quickly creating sudden unemployment when they pull out of a country
- TNCs involved in the exploitation of raw materials usually export them directly so that the benefits of the employment created and the value added in manufacture go to the headquarters country in the developed world.

Foreign direct investment

The phenomenal growth in foreign direct investment (FDI) is

the most obvious sign of the increasing integration of the world's economies (Figure 9.20) and much of this investment is by trans-national corporations. Most FDI is in the developed world but investment in developing economies rose substantially in the 1980s and 1990s. Of the emerging economies, the big five in order of importance, all with an FDI stock of over $50 billion, are China, Brazil, Mexico, Singapore and Indonesia. The UN expects the growth in FDI to continue as more governments liberalise their investment rules to attract FDI in the quest for capital and growth.

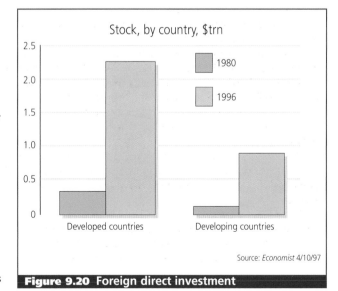

Stock, by country, $trn

1980

1996

2.5

2.0

1.5

1.0

0.5

0

Developed countries

Developing countries

Source: *Economist* 4/10/97

Figure 9.20 Foreign direct investment

Figure 9.21(a) The world's largest corporations: the top twenty in *Fortune's* Global 500

Rank 1997	1996			Revenues $ mil.	Profits $ mil.	Rank	Assets $ mil.	Rank	Stockholders' equity $ mil.	Rank	Employees Number	Rank
1	1	GENERAL MOTORS	U.S.	178 174.0	6698.0	6	228 888.0	46	17 506.0	42	608 000	3
2	2	FORD MOTOR	U.S.	153 627.0	6920.0	5	279 097.0	34	30 734.0	12	363 892	5
3	3	MITSUI	JAPAN	142 688.3	268.7	336	55 070.5	143	5 272.1	252	40 000	292
4	4	MITSUBISHI	JAPAN	128 922.3	388.1	287	71 407.8	123	7 569.4	177	36 000	314
5	6	ROYAL DUTCH/SHELL GROUP	BRIT./NETH.	128 141.7	7758.2	3	113 781.4	87	59 981.8	2	105 000	105
6	5	ITOCHU	JAPAN	126 631.9	(773.9)	484	56 307.9	136	2 956.6	353	66 752	473
7	8	EXXON	U.S.	122 379.0	8460.0	1	96 064.0	102	43 660.0	4	80 000	149
8	11	WALL-MART STORES	U.S.	119 299.0	3526.0	25	45 525.0	162	18 502.0	36	825 000	2
9	7	MARUBENI	JAPAN	111 121.2	140.4	388	55 403.4	140	3 563.9	323	64 000	187
10	9	SUMITOMO	JAPAN	102 395.2	209.8	360	42 866.1	171	4 318.6	296	29 500	346
11	10	TOYOTA MOTOR	JAPAN	95 137.0	3701.3	21	103 893.8	93	45 158.2	3	159 035	48
12	12	GENERAL ELECTRIC	U.S.	90 840.0	8203.0	2	304 012.0	29	34 438.0	10	276 000	15
13	13	NISSHO IWAI	JAPAN	81 893.8	24.7	443	40 799.3	180	2 019.6	403	18 158	408
14	15	INTL. BUSINESS MACHINES	U.S.	78 508.0	6093.0	8	81 499.0	117	19 816.0	29	369 465	17
15	14	NIPPON TELEGRAPH & TELEPHONE	JAPAN	76 983.7	2361.3	56	113 409.5	88	35 989.7	6	226 000	27
16	78	AXA	FRANCE	76 874.4	1357.0	109	401 206.0	13	13 075.0	69	80 613	146
17	20	DAIMLER-BENZ	GERMANY	71 561.4	4639.2	12	76 190.7	121	19 510.8	32	300 068	9
18	24	DAEWOO	SOUTH KOREA	71 525.8	526.9	253	44 860.6	165	6 325.3	212	265 044	20
19	18	NIPPON LIFE INSURANCE	JAPAN	71 388.2	2118.3	62	316 530.4	24	5 575.3	241	75 851	163
20	21	BRITISH PETROLEUM	BRITAIN	71 193.5	4 046.2	16	54 009.1	144	23 221.3	19	56 450	222

Source: *Fortune*, 3 August 1998

Figure 9.21(b) Industry totals: the top twenty in *Fortune's* Global 500

Rank		Number of companies	1997 revenues $ millions	Profits $ millions	Rank	Assets $ millions	Rank	Stockholders' equity $ millions	Rank	Employees Number	Rank
1	BANKS: COMMERCIAL AND SAVINGS	68	1 243 155	49 353	2	17 372 718	1	686 105	1	2 879 083	3
2	MOTOR VEHICLES AND PARTS	25	1 150 812	33 564	4	1 257 833	5	234 318	5	3 659 169	1
3	TRADING	19	1 013 106	2 714	34	486 545	12	51 074	18	580 090	18
4	PETROLEUM REFINING	31	945 174	54 928	1	864 312	9	363 941	2	1 077 912	9
5	ELECTRONICS, ELECTRICAL EQUIPMENT	25	782 434	21 458	6	969 969	7	234 695	4	3 656 120	2
6	TELECOMMUNICATIONS	22	534 222	40 371	3	869 711	8	291 179	3	2 384 667	5
7	FOOD AND DRUG STORES	28	486 405	8 809	17	233 172	15	72 208	14	2 782 912	4
8	INSURANCE: LIFE, HEALTH (STOCK)	19	425 851	13 132	10	2 375 456	2	134 868	8	570 553	19
9	INSURANCE: LIFE, HEALTH (MUTUAL)	17	410 825	11 933	12	1 782 715	3	36 739	24	415 298	26
10	GENERAL MERCHANDISERS	13	373 322	9 696	13	227 227	18	68 128	15	2 380 587	6
11	INSURANCE: P&C (STOCK)	16	351 062	19 241	7	1 476 487	4	171 081	6	518 891	23
12	UTILITIES, GAS AND ELECTRIC	16	307 208	8 967	16	674 152	11	150 078	7	645 450	16
13	CHEMICALS	16	294 472	11 981	11	329 295	13	99 120	9	968 841	10
14	COMPUTERS, OFFICE EQUIPMENT	9	264 385	14 877	8	242 453	15	76 830	12	913 196	11
15	FOOD	13	258 779	13 354	9	204 938	20	75 959	13	1 146 006	8
16	METALS	13	168 947	3 144	31	191 216	23	46 970	20	566 384	20
17	MAIL, PACKAGE, AND FREIGHT DELIVERY	8	168 743	3 953	29	197 248	21	49 344	19	2 287 173	7
18	PHARMACEUTICALS	10	161 822	24 463	5	211 391	19	90 166	11	608 893	17
19	AEROSPACE	8	154 422	4 836	25	150 708	26	35 727	25	859 673	12
20	ENGINEERING, CONSTRUCTION	10	150 331	(1 099)	45	174 001	24	29 166	31	460 675	25

Source: *Fortune*, 3 August 1998

FDI in Britain

In 1997, the stock of FDI in Britain stood at $345 billion (Figure 9.22), the highest for any economy outside the USA. The inflow has been high in recent years, reaching £16 billion in 1996 and, taking the total for the five year period 1992–96, to more than £53 billion. The figure for 1997 was even higher at £23 billion. There are now in excess of 2400 foreign-owned manufacturing firms in the UK, representing 26% of net output and 17% of the manufacturing workforce. The attractions of a UK location as cited by foreign companies are:

- relatively low levels of corporate and personal taxation
- labour flexibility
- access to the EU market
- a less oppressive regulatory climate than in many other countries
- a stable economy with low inflation
- welcoming national, regional and local agencies involved in economic regeneration
- an attractive quality of life
- the English language, the second language for so many foreign executives.

Figure 9.23 FDI in Crawley, Sussex

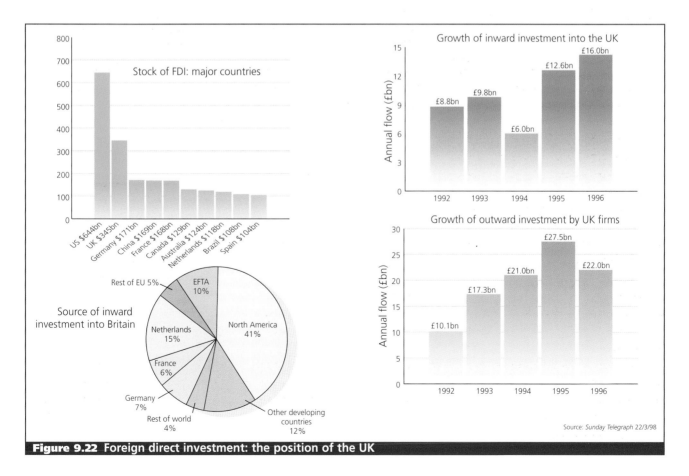

Figure 9.22 Foreign direct investment: the position of the UK

Source: *Sunday Telegraph* 22/3/98

The nature of FDI in Britain has changed considerably over the years. Initially branch plants carrying out routine tasks dominated foreign investment but increasingly research and development has located here.

Questions

1 (a) What is meant by the North-South divide at the global scale?
   (b) Which countries in Figure 9.6 are considered to be in the south?
   (c) define the term 'gross domestic product'.
   (d) Compare the GDP of the world's three largest economies.
   (e) For a named country illustrated in Figure 9.6 give three reasons to explain its high GDP.
   (f) Apart from GDP discuss three other factors which affect the quality of life in a country.

2 (a) Define the terms 'exports' and 'trade balance'.
   (b) Describe the changes in world exports and GDP from 1950 to 1997 (Figure 9.9).
   (c) Suggest reasons for these trends.
   (d) Explain why it is important for countries to export at least some of the products they produce.
   (e) Despite the efforts of GATT and the WTO why do many countries still take measures to limit imports from other countries?

3 (a) What do you understand by the term 'model'?
   (b) Briefly outline the characteristics of each of the five stages in Figure 9.13.
   (c) Why was Britain the first country to achieve 'take-off'?
   (d) Explain why the sequence of economic development in Canada and Australia was different from the other developed nations shown in Figure 9.14.
   (e) Outline the main criticisms of the model.

4 (a) Why do many large companies seek to become trans-national in character?
   (b) Discuss the locational changes illustrated in Figure 9.18.
   (c) Describe and explain the location of the world's 20 largest corporations shown in Figure 9.21.
   (d) Comment on the balance between the secondary and tertiary sectors in terms of the 20 largest industries in the world.
   (e) Discuss the advantages and disadvantages of TNCs to developing countries.

5 (a) What do you understand by the term 'foreign direct investment' (FDI)?
   (b) Study Figure 9.20. By how much did the stock of foreign direct investment change in developed countries and developing countries between 1980 and 1996?
   (c) Explain the pattern of national ownership of FDI (i) globally, (ii) in Britain.
   (d) Why has FDI in Britain been running at such a high level over the past decade or so?
   (e) What are the advantages to Britain of a high level of inward investment?
   (f) British firms invest heavily overseas. What are the advantages of such investment to Britain?

Britain is the world's second largest outward investor with a total outflow of almost £98 billion in the period 1992–96. Such outward investment strengthens the global reach of UK companies, bringing in substantial flows of money from profits made overseas.

National variations in economic development

Economic development varies not only between countries but also within countries. In Britain the contrast between the north and the south is often referred to in the media. London and the South East is the undoubted core of the national economy. In France, the Paris Basin dominates the country's economy; in Brazil it is the Southeast; in Italy the north; in Ireland the Dublin region, and so on. Thus it not surprising that geographers and economists have tried to develop a body of theory to explain the fact that one region dominates the national economy in most countries.

Myrdal and Hirschman

The Swedish economist Gunnar Myrdal originally framed his cumulative causation theory (1957) in the context of developing countries but it can also be applied to more advanced nations. According to the theory, a three-stage sequence can be recognised:

■ the pre-industrial stage when regional differences are minimal

■ a period of rapid economic growth characterised by increasing regional economic divergence

■ a stage of regional economic convergence when the significant wealth generated in the most affluent region(s) spreads to other parts of the country.

Figure 9.24 Reykjavik: dominating the economy of Iceland

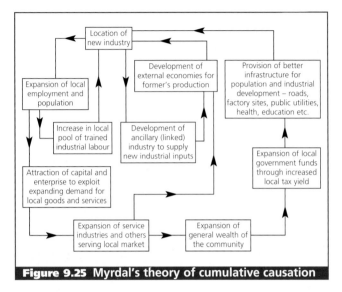

Figure 9.25 Myrdal's theory of cumulative causation

In Myrdal's model, economic growth begins with the location of new manufacturing industry in a region with a combination of advantages greater than elsewhere in the country. Once growth has been initiated in a dominant region spatial flows of labour, capital and raw materials develop to support it and the growth region undergoes further expansion by the cumulative causation process (Figure 9.25). A detrimental 'backwash effect' is transmitted to the less developed regions as skilled labour and locally generated capital is attracted away. Manufactured goods and services produced and operating under the scale economies of the economic 'heartland' flood the market of the relatively underdeveloped 'hinterland' undercutting smaller scale enterprises in such areas.

Even so, increasing demand for raw materials from resource-rich parts of the hinterland may stimulate growth in other sectors of the economies of such regions. If the impact is strong enough to overcome local backwash effects a process of cumulative causation may begin leading to the development of new centres of self-sustained economic growth. Such 'spread effects' are spatially selective and will only benefit those parts of the hinterland with valuable raw materials or other significant advantages.

The American economist Hirschman (1958) produced similar conclusions to Myrdal although he adopted a different terminology. Hirschman labelled the growth of the 'core' (heartland) as 'polarisation', which benefited from 'virtuous circles' or upward spirals of development whereas peripheral areas (the hinterland] were impeded by 'vicious circles' or downward spirals. Figure 9.26 gives examples of such downward spirals. The term 'trickle-down' was used to describe the spread of growth from core to periphery. The major difference between Myrdal and Hirschman was that the latter stressed to a far greater extent the effect of counterbalancing forces overcoming polarisation (backwash), eventually leading to the establishment of economic equilibrium. The subsequent literature has favoured the terms core and periphery rather than Myrdal's alternatives.

Figure 9.27 is an application of core-periphery theory to Brazil. With the onset of industrialisation the Southeast benefited from spatial flows of raw materials, capital and labour. The latter two came from abroad as well as internal sources. The region grew rapidly through the process of cumulative causation. This had a backwash effect on the periphery, resulting in widening regional disparity. Figure 9.28 summarises the physical and human advantages of the Southeast.

More recently, however, some parts of the periphery, with a combination of advantages greater than the periphery as a whole, have benefited from spread effects (trickle down)

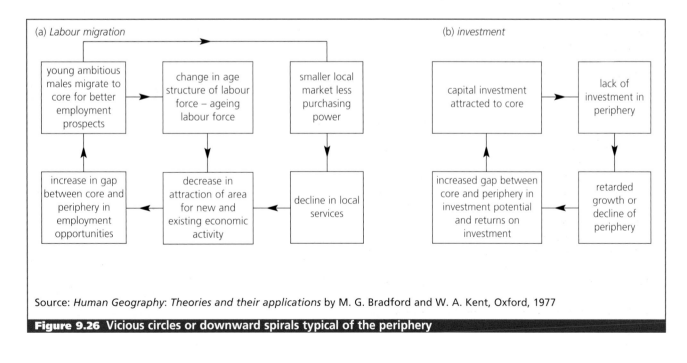

Source: *Human Geography: Theories and their applications* by M. G. Bradford and W. A. Kent, Oxford, 1977

Figure 9.26 Vicious circles or downward spirals typical of the periphery

emanating from the core. The South has been the most important recipient of spread effects but parts of the Centre-West have benefited as well. In contrast, the Northeast and North have found substantial economic development much more elusive, as Figure 9.29 indicates.

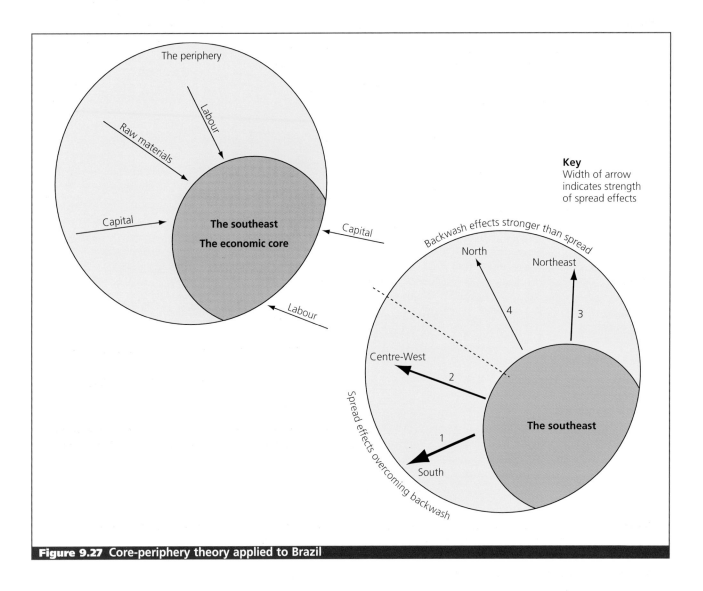

Figure 9.27 Core-periphery theory applied to Brazil

Friedmann

J. R. Friedmann (1964) related the work of Myrdal and Hirschman to a general theory of urbanisation (Figure 9.30) and linked regional income differences to the stage of development of city systems. According to Friedmann the process of convergence is only initiated when urbanisation has progressed enough to generate self-sustained growth.

Friedmann identified three types of region within the periphery:

■ Upward Transition Areas – where inward investment is increasing, the industrial base is widening and growth rates are above the national average.

■ Downward Transition Areas – where industry mix is dominated by stagnant or declining industries, unemployment is high, and new investment difficult to come by.

■ Resource Frontiers – areas of low population density and very limited economic activity apart from the exploitation of one or a number of resources of high value on the world market.

In Brazil, the South and Centre-West can be classed as upward transition areas. The Northeast is the country's problem region illustrating the characteristics of a downward transition area. The North is a classic resource frontier.

Key Definitions 2

Cumulative causation The process whereby impulses of economic growth or economic decline within regions (or countries) are, via the operation of market forces, self-reinforcing. This is a consequence of multiplier and scale economy effects.

Core A region of concentrated economic development with advanced systems of infrastructure, resulting in high average income and low unemployment.

Periphery A region of low or declining economic development characterised by low incomes, high unemployment, selective out-migration and poor infrastructure.

Spread effect The movement of resources from the core to the periphery. Spread effects are spatially selective.

Backwash effect The movement of resources from the periphery to the core. This is to the detriment of the former and the benefit of the latter.

Climate: attractive warm temperature with adequate rainfall for agriculture, industry and settlement.

Minerals: large deposits of iron ore, manganese and bauxite. Gold is still mined.

Energy: the focus of energy supply in Brazil. HEP on plateau rim; nuclear power near Rio; biomass; offshore oil and gas.

Transport: focus of road and rail networks with highest densities. Main airports and seaports. A significant pipeline network.

Government Policy: investment capital and managerial power were centralised in the Southeast, particularly in the 1950s and 60s. More recently the government has tried to spread development to other parts of the country but this has proved to be difficult.

Innovation: the centre of research and development in both the public and private sectors.

Relief and Soils: the Serra [mountain ranges] run parallel to the coast, while the interior is open, tabular upland composed of layers of lava that have been weathered to form rich terra roxa soils.

Agriculture: major world region for coffee. Also important for beef, rice, cacao, sugar cane, and fruit.

Labour: greatest population density. Highest educational and skill levels in Brazil.

Multinational Companies: more located in the Southeast than in the rest of Brazil. The region has grown rapidly through the process of cumulative causation.

Finances: São Paulo is by far the largest financial centre in South America.

Source: *Brazil: Advanced Case Studies* by P. Guinness

Figure 9.28 The combination of advantages explaining the status of the Southeast as Brazil's economic core region

Figure 9.29 Socio-economic indicators for Brazil and its five regions

| Indicator | Year | Brazil | Region | | | | |
			Southeast	South	Centre-West	Northeast	North
Rate of urbanisation	1991	75.6	88.0	74.1	81.3	80.7	59.0
% households with electricity	1996	92.0	98.0	97.0	93.0	78.0	60.0
Hospital beds per 1000 population	1992	–	4.2	4.1	4.3	3.1	2.3
% urban households served by public water supply system	1996	91.1	95.5	94.9	82.7	86.1	69.1
% illiteracy rate of children 10–14 yrs old in urban areas	1996	5.0	1.7	1.7	2.1	13.0	5.9
% children in urban areas not attending school	1996	6.3	4.7	5.3	5.2	9.5	7.9
% urban households having a telephone	1996	30.3	34.9	30.3	33.8	20.0	22.1
% urban households with TV sets	1996	91.3	95.0	93.1	88.8	83.8	84.3
% urban household units served by refuse collection	1996	87.4	92.9	95.6	89.2	72.9	64.7
Annual residential electricity consumption (KWh)	1996	2035.0	2372.0	2014.0	2145.0	1282.0	1916.0
% population under 18 yrs old	1991	41.0	37.0	38.0	42.0	47.0	48.0

Source: *Brazil: Advanced Case Studies* by P. Guinness

Stage 1. Relatively independent local centres; no hierarchy. Typical pre–industrial structure; each city lies at the centre of a small regional enclave.

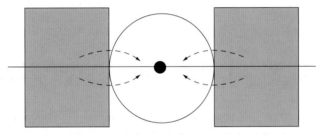

Stage 2. A single strong core. Typical of period of incipient industrialisation; a periphery emerges; potential entrepreneurs and labour move to the core, national economy is virtually reduced to a single metropolitan region.

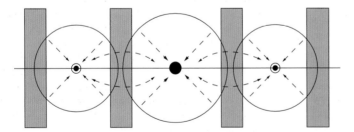

Stage 3. A single national core, strong peripheral sub-cores. During the period of industrial maturity, secondary cores form, thereby reducing the periphery on a national scale to smaller inter-metropolitan peripheries.

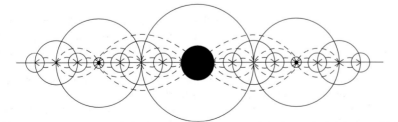

Stage 4. A functional interdependent system of cities. Organised complexity characterised by national integration, efficiency in location, and maximum growth potential.

Figure 9.30 Friedmann's development model

Britain's north–south divide

Social and economic conditions vary significantly by region in Britain (Figures 9.31, 9.32 and 9.33). The South East (including London) is by far the richest region while Northern Ireland lags behind the rest of the country. However, the regional divide is not as great as in some other European countries such as Italy and France.

Socio-economic differences between what may broadly be termed 'north' and 'south' have been a matter of concern in Britain for much of the 20th century. The causes of the north–south divide can be found in Britain's economic history. The northern regions that had experienced industrial growth in the 19th century suffered severely in the depression of the 1930s and during the decades that followed there was a steady decline in the activities on which much of the original prosperity of these regions was based. The major industries central to this decline were coal, iron and steel, shipbuilding, heavy engineering and textiles (Figure 9.34). Unemployment is the trigger for a wide range of social and economic problems (Figure 9.35). It can have a devastating impact at a range of scales from the individual to the national.

Successive governments have committed considerable resources to encourage the development of new opportunities in the traditional industrial areas and in other peripheral areas in Britain. The origins of regional policy in Britain can be traced back to 1928 when the Industrial Transference Board was established to retrain men from declining industries and by the use of grants enable them to move and find employment in expanding industries. Figure 9.36 shows the areas currently eligible for regional assistance. The European Union's Structural Funds have become an increasingly important source of development funding to the UK's disadvantaged regions, contributing £922 million in 1995.

Government yet to bridge the North–South divide

By Roland Watson, Chief Political Correspondent

Tony Blair will admit today that two and a half years of Labour Government have failed to achieve significant inroads in bridging Britain's regional divide.

The Prime Minister will claim that government policies, such as the minimum wage, the New Deal for the unemployed and the working families' tax credit, have succeeded in lifting thousands of people out of poverty, but he will publish a report showing that those living in the North tend to be blighted by higher crime, worse housing, higher unemployment and poorer health than their fellow citizens in the South. He will try to head off criticism from Labour MPs that the Government is doing too little for the deprived regions and calls for changes in policy to correct the imbalance.

Mr Blair will declare that the impression of a North–South divide is far too simplistic and that the chief finding of a 100-page Government report, titled *Sharing the Nation's Prosperity*, is the wide diversity in living standards within regions.

Mr Blair is travelling to Manchester and Liverpool for a two-day trip designed to show that poverty and prosperity are living cheek-by-jowl, rather than divided by a line across the middle of the country. The aim of the exercise, according to Downing Street, is for Mr Blair to challenge some of the 'old assumptions' that problems in the South are largely those of high prices and schools shortages, whereas in the North they are of unemployment and poverty. However, Mr Blair will admit that while the report shows evidence that the Labour Government has begun to make a difference, it is too early to draw definitive lessons.

Instead, adopting the language of One Nation politics and attempting to allay the fears of the middle classes that he is about to abandon their concerns, he will say that he intends to govern for North, South, East and West.

The report, drawn up by the Cabinet Office, shows that Liverpool is the most deprived district, while London has both the greatest affluence and also some of the worst pockets of poverty. The North West and London have most districts falling within the country's 150 most deprived districts. In London, half the capital's 32 districts, minus the City of London, are in the top 50.

The North East, North West and Yorkshire and Humberside share broadly similar profiles and are the three most deprived regions, according to an index of indicators measuring crime, housing, unemployment and health.

Based on that scale, the South West, which boasts the most number of self-employed people, is the least deprived, followed by the South East, the East, East Midlands, West Midlands and London.

But Mr Blair will insist that the real story is within the regions. Five London districts are among the top ten of the country's most deprived: Newham, Hackney, Tower Hamlets, Southwark and Islington.

In Yorkshire, Leeds and West Yorkshire are thriving, but South Yorkshire has fared much worse.

However, the report also shows that Britain's regional imbalances are narrower than other EU countries. The divide between regions in Germany is much wider, as is the case in Belgium.

Source: *Times*, 6 December 1999

Figure 9.31 Britain's north–south divide

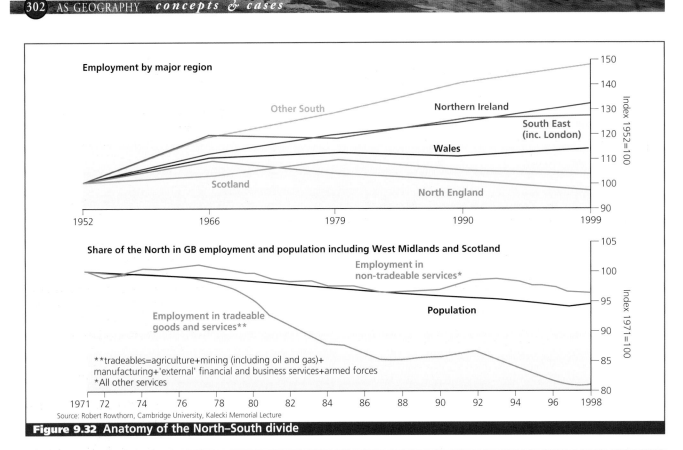

Employment by major region

Other South

Northern Ireland

South East (inc. London)

Wales

Scotland

North England

Index 1952=100

1952   1966   1979   1990   1999

Share of the North in GB employment and population including West Midlands and Scotland

Employment in non-tradeable services*

Employment in tradeable goods and services**

Population

Index 1971=100

**tradeables=agriculture+mining (including oil and gas)+
manufacturing+'external' financial and business services+armed forces
*All other services

1971 72   74   76   78   80   82   84   86   88   90   92   94   96   1998

Source: Robert Rowthorn, Cambridge University, Kalecki Memorial Lecture

Figure 9.32 Anatomy of the North–South divide

Figure 9.33 The UK: Regional variations in the quality of life

	Average gross weekly household income 1995–8 (£)	ILO unemployment rates Spring quarter 1998 (%)	% of households with no car 1997	Infant mortality rate 1997	% achieving 5 or more GCSE grades A*–C 1997	Gross domestic product (£ million) 1996
United Kingdom	408	6.1	30	5.9	46.2	656 184
North East	333	8.2	42	5.8	38.1	23 163
North West	384	6.6	31	6.8	43.7	67 596
Yorkshire and the Humber	360	7.0	34	6.5	39.7	49 334
East Midlands	401	4.9	27	5.7	43.4	43 479
West Midlands	381	6.3	32	7.1	42.2	53 659
East	419	5.0	23	4.8	48.5	58 959
London	491	8.1	39	5.8	44.0	98 292
South East	474	4.3	19	5.0	51.1	101 145
South West	405	4.5	24	5.8	50.1	50 539
Wales	360	6.7	31	5.9	43.7	26 253
Scotland	371	7.4	35	5.3	54.8	54 141
N. Ireland	336	7.3	30	5.6	53.5	14 545

In a recent paper (*Regional Studies*, 1997 Vol. 31.9) Taylor and Wren state the main arguments in favour of a strong and effective regional policy in the UK:

■ Reducing unemployment in depressed areas has direct economic and social benefits. The increased income of the previously unemployed benefits others through the multiplier effect. The national economy also gains through reduced transfer payments and increased tax revenues.

■ Reducing spatial unemployment disparities will lessen inflationary pressure in the national economy. The labour shortages that tend to arise in the South East, pushing up wages when the economy is buoyant, would be less likely to occur.

■ Selective migration from depressed areas can in the longer term be harmful to both the origin and destination regions. The depressed areas lose their most highly skilled workers while more affluent areas may

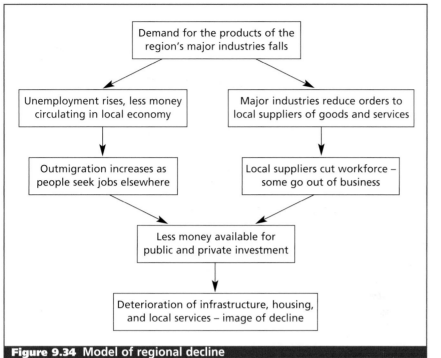

Figure 9.34 Model of regional decline

Figure 9.35 The impact of unemployment

Scale	Impact
Individual and family	Increase in suicide, homicide, alcoholism, admission to mental hospitals, prison entry and family break-up. Racial minorities bear brunt of unemployment.
Local community	Community spirit slowly declines. Housing deteriorates and local shops close because of shortage of money. Total quality of environment degenerates.
City region	As businesses close the tax base falls. Public services suffer as a result. If prospects are better elsewhere people move out. Impact felt most in inner city.
National	(a) loss of output (GNP) (b) loss of tax revenue (c) increasing expenditure on unemployment benefits and other transfer payments.

suffer because of the increased pressure on available resources.

■ Reducing unemployment in areas of high unemployment is politically necessary. The evidence for this assertion is in the sharp differences in regional voting patterns.

Thus, according to Taylor and Wren regional policy is desirable not only for reasons of equity (fairness) but also in terms of efficiency as real benefits will accrue to the national economy as a whole.

There can be little doubt that without a regional policy favourable to the periphery the North–South divide would be much greater than it is. On the other hand, however, regional policy has not been as successful as its supporters originally expected. The facts of economic life are that when market forces encourage spatial concentration it is not easy for governments to counteract such a trend.

Questions

1 (a) Define the term 'cumulative causation'.
   (b) Why is it likely that the location of a large new industry in a region will encourage in-migration?
   (c) Explain the potential benefits of the new industry to the small firms already located in the region.
   (d) What impact will new company investment have on local government finances?
   (e) Why is it frequently the case that once one major company locates in a region others follow?

2 (a) Define the terms (i) core, (ii) periphery.
   (b) Explain the polarisation of growth in the economic core region.
   (c) Summarise the combination of advantages that explain the status of the Southeast as Brazil's economic core region.
   (d) Under what conditions are spread effects likely to occur?

3 (a) Why can the concept of a North–South divide in Britain be seen to be too simplistic?
   (b) Describe and explain the trends illustrated by Figure 9.32.
   (c) Comment on the regional differences shown in Figure 9.33.
   (d) Suggest and justify two other indicators that could be used to exemplify regional contrasts.

4 (a) Under what circumstances might the demand for a region's major products fall on a long-term basis?
   (b) Discuss the sequence of regional decline that may follow.
   (c) Why and how does unemployment impact at a range of scales?
   (d) What is the social and economic justification for using resources to reduce unemployment?
   (e) What measures can a government undertake to reduce regional inequality?

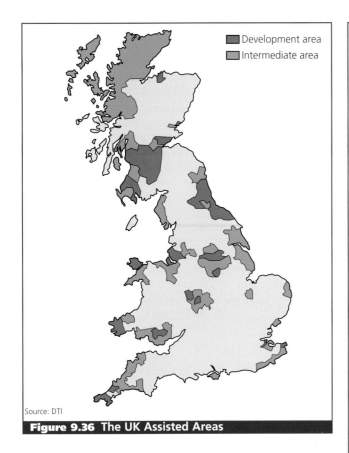

Source: DTI

Figure 9.36 The UK Assisted Areas

Development area
Intermediate area

It seems unlikely that the gap between north and south is going to close in the near future. The Office of National Statistics estimates a population increase of 12.8% in the South East and 9.14% in London between 1996 and 2001. This population shift is mainly due to the fact that the new service industries are predominantly based south of Birmingham.

Changes in the primary sector

In developed countries all components of the primary sector shed jobs heavily in the latter part of the 20th century, a trend which is also now affecting a growing number of developing countries. The contraction of a significant industry always causes considerable distress to both the individuals and the regions affected. The coal industry in Britain is a classic case.

The decline of Britain's coal industry

The peak output for British coal was in 1913 when 290 million tonnes were hewn from coal faces. The tale thereafter was one of lost markets at home and abroad. The decline in employment has been dramatic. The peak year for employment was 1920 when 1.25 million people worked in

(a) Production of energy in the United Kingdom

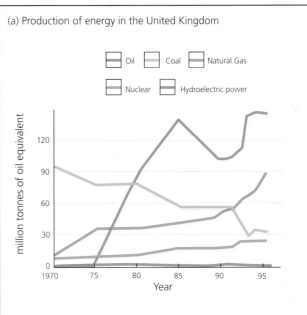

Oil   Coal   Natural Gas
Nuclear   Hydroelectric power

(b) Decline in coal production in the United Kingdom

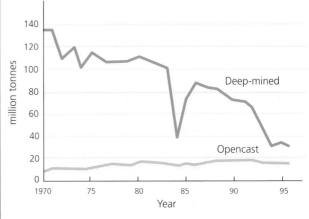

Deep-mined

Opencast

(c) Decline in coal mining employment in the United Kingdom
Thousands, 1983–96

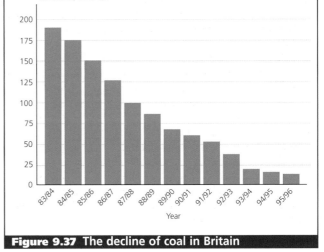

Figure 9.37 The decline of coal in Britain

the mines. This fell to 700 000 in 1955; 235 000 in 1979; and 17 000 in 1997. Although environmentalists have welcomed the reduction in the use of coal, the worst polluter of all the fossil fuels, the impact on many mining communities has been devastating, particularly where alternative employment opportunities have been lacking. Figure 9.37 illustrates the decline of coal in Britain in recent decades. In some of Britain's coalfield areas, for example the Northumberland–Durham, North Wales and Kent coalfields, not a single mine remains open (Figure 9.38). For such areas the economic and social consequences have been particularly hard. A significant proportion of regional development funding in Britain in the post-war period has been directed at mining communities.

Because there is a limited amount of good quality and easily accessible coal at every mine, the eventual closure of a mine is a fact of life. Yet what really hit most mining regions was the speed of closure. On the Durham coalfield 90 000 men were employed in 116 pits in 1947 (Figure 9.39); by 1992 only two mines remained and both of these were soon to close. Some open-cast mining continues but this employs only a small number of people.

The demand for British coal has declined for the following reasons:

■ some industries, such as the railways (now electricity and diesel) and gasworks (now natural gas) in particular, which once used large amounts of coal now use none at all

■ household use has fallen sharply, from 36 million tonnes in 1958 to less than 5 million tonnes by 1995. Very few new houses are built with coal fires and most houses which could burn coal do not because compared to other fuels it is dirty, bulky to store, requires manual handling, and is usually more expensive

■ power stations, traditionally the largest user of coal, are switching to alternative sources. In 1990, coal accounted for 65.3% of electricity generation in Britain. By 1996 this was down to 43%. During the same period gas-fired power plants increased their contribution from 0.7% to 21.5%. Combined cycle gas turbine power plants are both cheaper to build and operate than big coal-fired plants. Nuclear power has also increased its share of electricity generation, from 21.3% in 1990 to 28.9% in 1996

The Pits!

Britain's coal industry is staring into an abyss. The closure, announced last week, of one of the country's largest and most modern pits has sounded the death knell, many believe, for the once mighty industry.

The closure of Asfordby in Leicestershire was blamed on geological problems. But the fact that one of Britain's finest pits could shut, has underlined the threat to coal in the face of growing competition from gas-fired power stations, nuclear power and cheaper imported coal. Many say that, without government help, the British coal industry has no future.

Neil Greatrex, president of the Union of Democratic Mineworkers, said: 'I believe that if nothing is done, then by the year 2005 the coal industry in Britain will be finished.'

The hornet's nest of emotion over Asfordby was stirred up by RJB Mining, the company which bought most of British Coal in 1994.

RJB has so far been protected by long-term contracts with the power generation industry, inherited from British Coal. But that safety net goes when the contracts expire next March.

The coal business is no stranger to controversy. As recently as 1992, the industry erupted when Michael Heseltine, then President of the Board of Trade, announced the closure of 31 pits – some of which were later salvaged by RJB and others – and the loss of 30,000 jobs.

Five years on from Heseltine's bombshell, Asfordby's fate has invoked bitter memories. This latest closure will leave Britain with 23 deep mines compared with more than 1,000 when the industry was nationalised in 1947.

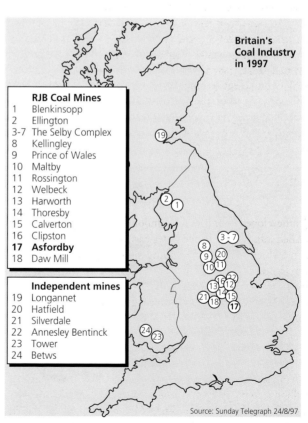

RJB Coal Mines
1   Blenkinsopp
2   Ellington
3-7 The Selby Complex
8   Kellingley
9   Prince of Wales
10 Maltby
11 Rossington
12 Welbeck
13 Harworth
14 Thoresby
15 Calverton
16 Clipston
17 Asfordby
18 Daw Mill

Independent mines
19 Longannet
20 Hatfield
21 Silverdale
22 Annesley Bentinck
23 Tower
24 Betws

Britain's Coal Industry in 1997

Source: Sunday Telegraph 24/8/97

Figure 9.38 Source: *Sunday Telegraph*, 24 August 1997

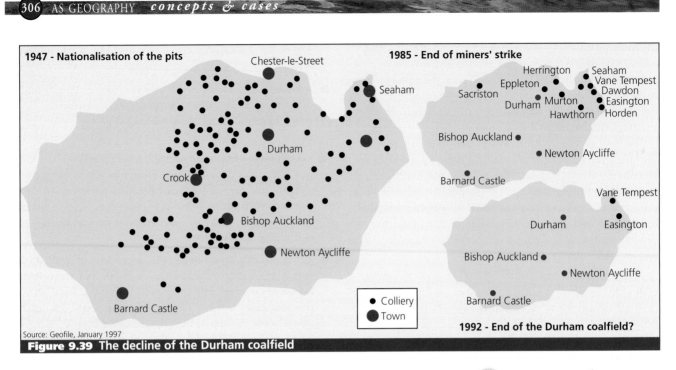

1947 - Nationalisation of the pits

1985 - End of miners' strike

Chester-le-Street

Seaham

Durham

Crook

Bishop Auckland

Newton Aycliffe

Barnard Castle

Sacriston
Eppleton
Herrington
Seaham
Vane Tempest
Durham
Murton
Dawdon
Easington
Hawthorn
Horden

Bishop Auckland

Newton Aycliffe

Barnard Castle

Vane Tempest

Durham
Easington

Bishop Auckland

Newton Aycliffe

Barnard Castle

1992 - End of the Durham coalfield?

● Colliery
● Town

Source: Geofile, January 1997

Figure 9.39 The decline of the Durham coalfield

- in the steel industry technological advance and rationalisation of the industry has significantly reduced the demand for coal to less than 10 million tonnes from 28 million tonnes in 1958

- industries other than those named above have reduced consumption from 34 million tonnes in 1958 to less than 7 million by 1995 for a combination of cost, technological and environmental reasons

- imported open-cast coal is usually significantly cheaper than underground coal mined in Britain.

Longstanding concerns about the extensive use of coal were highlighted by the Kyoto Protocol, the environmental protection agreement negotiated by the UN in 1997. The UK government has given a commitment to reduce by 20% Britain's emissions of carbon dioxide by 2010. This can only happen by using less coal or by introducing new environmental technology. However, clean-coal technology is expensive and is not viable in the short-term without substantial financial support.

1 (a) Describe the decline in coal production and employment in Britain.
   (b) Discuss the main reasons for these trends.
   (c) Suggest reasons for the location of the coal mines that remain open in Britain.
   (d) Describe and attempt to explain the pattern of decline on the Durham coalfield.
   (e) Outline the consequences of pit closures in mining communities.

Figure 9.40 Abandoned colliery

Changes in manufacturing industry

Classification and importance

Manufacturing industry in an advanced economy comprises a wide range of processes and products. Some industries such as textiles and shipbuilding, which were of major importance during the Industrial Revolution, have shed employment consistently over much of the 20th century. Production has generally been lost to competitors benefiting from cheaper

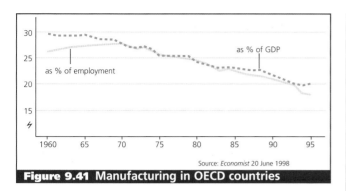

Source: *Economist* 20 June 1998

Figure 9.41 Manufacturing in OECD countries

Figure 9.42 Major industrial zone, Toronto

labour and other advantages. However, even for industries with stable or rising production, employment has generally declined as increasing investment in capital has replaced labour (Figure 9.41).

Manufacturing industry is often described or classified by the use of opposing terms. The most frequently used are:

■ large scale and small scale – depending on the size of plant and machinery, and the numbers employed

■ heavy and light – depending on the nature of processes and products in terms of unit weight

■ market oriented and raw material oriented – where the location of the industry or firm is drawn either towards the market or the inputs required, usually because of transportation costs

■ processing and assembly – the former involving the direct processing of raw materials with the latter putting together parts and components

■ capital intensive and labour intensive – depending on the ratio of investment on plant and machinery to the number of employees

■ Fordist and flexible – Fordist industries, named after the assembly line methods used in the early automobile industry to mass produce on a large scale making standardised products. Flexible industries make a range of specialised products using high technology to respond quickly to changes in demand

■ national and trans-national – many firms in the small to medium size range manufacture in only one country. Trans-nationals, which are usually extremely large companies, produce in at least two countries but may manufacture in dozens of nations.

Location, scale and change

At the global scale, the developed world still retains the lion's share of industrial production although there have been significant changes in the last 30 or 40 years caused by the rapid development of the Newly Industrialised Countries

(NICs). Some of these, led by South Korea and Taiwan, have invested considerable sums of money in manufacturing facilities in the developed world in recent years. The share of manufacturing accounted for by the developing world will take another major step forward as the largest countries in this global sector, led by China and India, extend their industrial capabilities.

Within each country, rich or poor, there are areas where manufacturing is highly concentrated and other regions where it is largely absent. In the USA, the north-east 'manufacturing belt', which covers only one-eighth of the country, has over 40% of all manufacturing jobs (although at the turn of the 20th century the figure was around 70%). Everywhere the most significant locational change has been from traditional manufacturing regions, more often than not on coalfields, to higher quality of life regions offering the infrastructural requirements of modern industry.

Within individual regions manufacturing has historically concentrated in and around the largest urban agglomerations. However, in recent decades there has been a significant shift of industry towards 'greenfield' rural locations. This movement has been so great that it was generally recognised as the most important locational change in the developed world in the latter part of the 20th century.

At the urban scale the relative shift from inner city to suburbs has increased as the century has progressed, impacting clearly on both of these urban regions. Although there has been much debate about the demise of the inner city many would agree that the loss of employment, much of it in manufacturing, was the initiating factor in the cycle of decline.

Industrial location: influential factors

Everyday decisions are made about where to locate industrial premises, ranging from small workshops to huge industrial complexes. For each possible location a wide range of factors can impact on total costs and thus influence the decision-making process.

Raw materials

The processes involved in turning a raw material into a manufactured product usually result in weight loss so that the transport costs incurred in bringing the raw materials to the factory will be greater than the cost of transporting the finished product to market. If weight loss is substantial the location of the factory will be drawn towards that area where it is most costly to transport raw material(s). The clearest examples of this influence are where one raw material only is used. In Britain, sugar beet refineries are centrally located in crop growing areas because there is a 90% weight loss in manufacture. In many processing industries technological advance has reduced the amount of raw material required per finished product and in some cases less bulky and cheaper substitutes have been found.

Tidewater locations are particularly popular with industries using significant quantities of imported raw materials. Examples include flour milling, food processing, chemicals and oil refining. Tidewater locations are break-of-bulk points where cargo is unloaded from bulk carriers and transferred to smaller units of transport for further movement. However, if raw materials are processed at the break-of-bulk point, then significant savings in transport costs can be made.

Energy

The Industrial Revolution in Britain and many other countries was based on the use of coal as a fuel. This was usually much more costly to transport than the raw materials required for processing. It is therefore not surprising that outside of London most of Britain's industrial towns and cities developed on coalfields or at nearby ports. The coalfields became focal points for the developing transport networks: first canals, then rail, and finally road. The investment in infrastructure was massive so that even when new forms of energy were substituted for coal, many industries remained at their coalfield locations, a phenomenon known as **industrial inertia**.

During the 20th century the construction of national electricity grids and gas pipeline systems has made energy virtually a ubiquitous resource in the developed world. As a result, most modern industry is described as **footloose** in that

it is not tied to certain areas because of its energy requirement or other factors.

Transport

Although once a major locational factor the share of industry's total costs accounted for by transportation has fallen steadily over time. For most manufacturing firms in the UK, transportation now accounts for less than 4% of total costs. The main reasons for this reduction are:

■ major advances in all modes of transport

■ great improvements in the efficiency of transport networks

■ technological developments moving industry to the increasing production of higher value/lower bulk goods.

Land

The space requirements of different industries, and also of firms within the same industry, vary enormously. Technological advance has made modern industry much more space-efficient than in the past. On the other hand, modern industry is horizontally structured (on one floor) as

Figure 9.43 Modern industrial estate, London

opposed to, for example, the textile mills of the 19th century with four or five floors. In the modern factory transportation takes up much more space than it used to.

Capital

Capital represents the finance invested to start up a business and to keep it in production. That part of capital invested in plant and machinery is known as fixed capital as it is not mobile compared with working capital (money). Capital is obtained either from shareholders (share capital) or from banks or other lenders (loan capital). Some geographers also use the term 'social capital' which is the investment in housing, schools, hospitals and other amenities valued by the community that may attract a firm to a particular location.

Virtually all industries have over time substituted capital for labour in an attempt to reduce costs and improve quality. Thus, in a competitive environment capital has become a more important factor in industry. In some industries the level of capital required to enter the market with a reasonable chance of success is so high that only a few companies monopolise the market. This has had a major influence on the geography of manufacturing.

Labour

The interlinked attributes of labour that influence locational decision making are cost, quality, availability and reputation.

Although all industries have become more capital intensive over time, labour still accounts for over 20% of total costs in manufacturing industry. The cost of labour can be measured in two ways; as wage rates and as unit costs (Figure 9.44). The former is simply the hourly or weekly amount paid to

employees while the latter is a measure of productivity, relating wage rates to output. Industrialists are mainly influenced by unit costs which explains why industry often clusters where wages are higher rather than in areas where wage rates are low. It is frequently, although not always, the high quality and productivity of labour that pushes up wages in an area. Certain skills sometimes become concentrated in particular areas, a phenomenon known as the **sectoral spatial division of labour**.

Variation in wage rates can be identified at different scales. By far the greatest disparity is at the global scale. The low wages of developing countries with reasonable enough levels of skill to interest foreign companies has been a major reason for trans-national investment in regions such as south-east Asia and Latin America. A filter-down of industry to lower and lower wage economies can be clearly seen, particularly in Asia.

At the continental scale, wage differentials can still be substantial. Within the EU there is a large gap between Germany, the highest wage economy in the world, and Portugal and Greece. At the national scale, wage variation is usually of a lesser magnitude, particularly where trade union membership is strong. In the UK, a history of national pay bargaining has resulted in a narrow regional wage range with much of the absolute difference between regions accounted for by industry mix.

The availability of labour as measured by high rates of unemployment is not an important location factor for most industries. The regions of the UK that have struggled most to attract new industry are the traditional industrial areas which have consistently recorded the highest unemployment rates in the country. In such regions, although there are many people available for work, they frequently lack the skills required by modern industry.

The reputation of a region's labour force can influence inward investment. Regions with militant trade unions and a record of work stoppages are frequently avoided in the locational search.

Markets

Where a firm sells its products can also have a considerable influence on where the factory is located. Where the cost of distributing the finished product is a significant part of total cost, and the greater part of total transport costs, a market location is logical. However, there are other reasons for market locations: industries where fashion and taste are variable need to be able to react quickly to changes demanded by their customers. One of the reasons why the global car giants spread themselves around the world is to ensure that they can produce vehicles which customers will buy in the different world regions.

Agglomeration economies

Agglomeration economies, known as external economies of

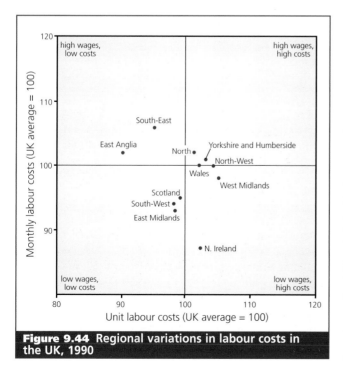

Figure 9.44 Regional variations in labour costs in the UK, 1990

scale by economists, are the benefits that accrue to a firm by locating in an established industrial area. They can be subdivided into:

■ **urbanisation economies**, which are the cost savings resulting from urban location due to factors such as the range of producer services available and the investment in infrastructure already in place

■ **localisation economies**, which occur when a firm locates close to suppliers (backward linkages) or firms which it supplies (forward linkages). This reduces transport costs, allows for faster delivery and facilitates a high level of personal communication between firms.

When an urban-industrial area reaches a certain size, however, urbanisation diseconomies may come into play. High levels of traffic congestion may push up transport costs, and intense competition for land will increase land prices and rents. If the demand for labour exceeds the supply wages will rise. Locating in such a region may then no longer be advantageous with fewer new firms arriving and some existing firms relocating elsewhere.

Government policy

Clearly, in the old-style centrally planned economies of the communist countries the influence of government on industry was absolute. In other countries the significance of government intervention has depended on:

■ the degree of public ownership

■ the strength of regional policy in terms of restrictions and incentives.

Governments influence industrial location for economic, social and political reasons. Regional policy largely developed after the Second World War, although examples of legislation with a regional element can be found before this time. There is a high level of competition both between countries and between regions in the same country to attract foreign direct investment.

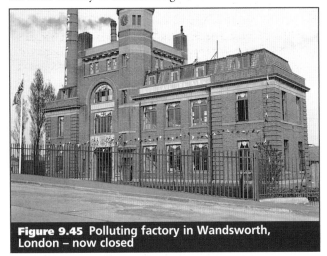
Figure 9.45 Polluting factory in Wandsworth, London – now closed

Currency exchange rates

A rise in the value of the pound compared to other currencies may be good news for Britons going on holiday abroad but it creates problems for British companies that sell to other countries. The price of British products increases in foreign markets and potential customers may well opt for lower priced products from elsewhere. If a North American or Asian trans-national is looking to expand production in Europe,

1 (a) What is a raw material?
(b) Why are tidewater locations popular with industries using bulky raw materials?
(c) Referring to examples, explain industrial inertia.
(d) Why has the share of industry's total costs accounted for by transportation fallen over time?
2 Study Figure 9.44
(a) Which regions of the UK have (i) the highest wage rates, (ii) the highest unit labour costs?
(b) Explain the difference between these two measures of the cost of labour.
(c) Discuss briefly the importance of the following characteristics of a labour force (i) skill level, (ii) availability, (c) reputation.
(d) Consider the influence of markets and agglomeration economics on the location of industry.

selling to the continent as a whole, its decision about where to site may well be influenced by the relative values of European currencies, particularly if there is little difference in other factors.

Deindustrialisation and the filter-down process

The declining importance of manufacturing in the developed world

In the USA and Britain the proportion of workers employed in manufacturing has fallen from around 40% at the beginning of the 20th century to barely half that now. Not a single developed country has bucked this trend (Figure 9.46), known as deindustrialisation, the causal factors of which are:

■ the filter-down of manufacturing industry from developed countries to lower wage economies, such as those of South East Asia;

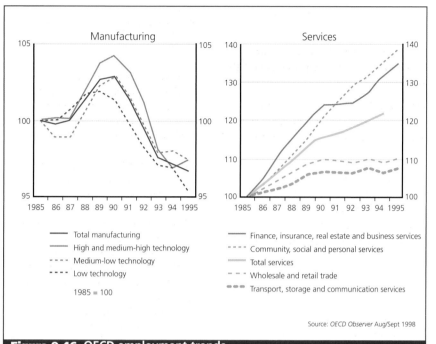

Manufacturing

105 — 105

100 — 100

95 — 95

1985 86 87 88 89 90 91 92 93 94 1995

— Total manufacturing
— High and medium-high technology
- - - Medium-low technology
- - - Low technology

1985 = 100

Services

140 — 140

130 — 130

120 — 120

110 — 110

100 — 100

1985 86 87 88 89 90 91 92 93 94 1995

— Finance, insurance, real estate and business services
- - - Community, social and personal services
— Total services
- - - Wholesale and retail trade
- - - Transport, storage and communication services

Source: *OECD Observer* Aug/Sept 1998

Figure 9.46 OECD employment trends

- the increasing importance of the service sector in the developed economies.

There can be little surprise in the decline of manufacturing employment for it has mirrored the previous decline in employment in agriculture in the developed world. So, if the decline of manufacturing in the developed world is part of an expected cycle, the consequence of technological improvement and rising affluence, why is so much concern expressed about this trend? The main reasons would appear to be:

- the traditional industries of the Industrial Revolution were highly concentrated, thus the impact of manufacturing decline has had severe implications in terms of unemployment and other social pathologies in a number of regions

- the rapid pace of contraction of manufacturing has often made adjustment difficult

- there are defence concerns if the production of some industries falls below a certain level

- some economists argue that over-reliance on services makes an economy unnecessarily vulnerable.

Rather than being a smooth transition, manufacturing decline tends to accelerate during periods of economic recession. As Figure 9.47 indicates, the recession of the early 1980s was particularly severe for the manufacturing sector in the UK. In general high-technology manufacturing, such as computers, aerospace, pharmaceuticals and electronics has been able to hold its share in the economy. However, medium and low technology manufacturing, such as chemicals, food products and textiles, has declined markedly.

From Shell, the oil giant, to Fujitsu, the Japanese chip maker, the fallout from Asia's slump is cutting deep into Britain's industrial heartland, write **John Waples** and **David Parsley**

THERE is now little doubt that the battle to protect Britain's shrinking manufacturing employment base will become one of the dominant issues for the government. One element at stake is the country's reputation as an industrial magnet capable of attracting big multinationals to use Britain as the springboard for their European expansion plans.

Opinion is split as to whether the current spate of foreign-owned plant closures and cutbacks are a series of significant but isolated events or the start of a damaging, widespread trend.

Britain is particularly exposed to the world semi-conductor recession because inward-investment chiefs specifically targeted the sector for development and proved highly successful – at least initially – in winning microchip projects.

Some economists fear that Britain is facing a repetition of the wave of branch factory closures that caused such havoc in Scotland and northern England during the 1980–82 recession. As the pound soared, factories like the Linwood car plant near Glasgow, the Speke factory on Merseyside, the British Steel plant at Consett, and the RCA record plant near Durham were axed in what became a manufacturing jobs massacre.

Figure 9.47 Source: *The Sunday Times,* 20 September 1998

The filter-down process of industrial relocation

The filter-down process is based on the notion that corporate organisations respond to changing critical input requirements by altering their geographical location of production to minimise costs and thereby ensure competitiveness in a tightening market.

The economic core (at a national and global level) has monopolised invention and innovation, and has thus continually benefited from the rapid growth rates characteristic of the early stages of an industry's life cycle (the product life cycle), one of exploitation of a new market. Production is likely to occur where the firm's main plants and corporate headquarters are located. Figure 9.48, illustrating the product life cycle, indicates that in the early phase scientific-engineering skills at a high level and external economies are the prime location factors.

In the growth phase, methods of mass production are gradually introduced and the number of firms involved in production generally expands as product information spreads. In this stage management skills are the critical human inputs. Production technology tends to stabilise in the mature phase. Capital investment remains high and the availability of unskilled and semi-skilled labour becomes a major locating factor. As the industry matures into a replacement market the production process becomes rationalised and often routine. The high wages of the innovating era, quite consistent with the high level skills required in the formative stages of the learning process, become excessive when the skill requirements decline and the industry, or a section of it, 'filters-down' to smaller, less industrially sophisticated areas where cheaper labour is available, but which can now handle the lower skills required in the manufacture of the product.

On a global scale, large trans-national companies have increasingly operated in this way by moving routine operations to the developing world since the 1950s. However, the role of indigenous companies in developing countries should not be ignored. Important examples are the 'chaebols' of South Korea, such as Samsung and Hyundai, and Taiwanese firms such as Acer. Here the process of filter-down has come about by direct competition from the developing world rather than from the corporate strategy of huge North American, European and Japanese trans-nationals.

The revolution in transport and communications has made such a substantial filter-down of manufacturing to the developing world possible. Containerisation and the general increase in scale of shipping have cut the cost of the overseas distribution of goods substantially while advances in telecommunications have made global management a reality.

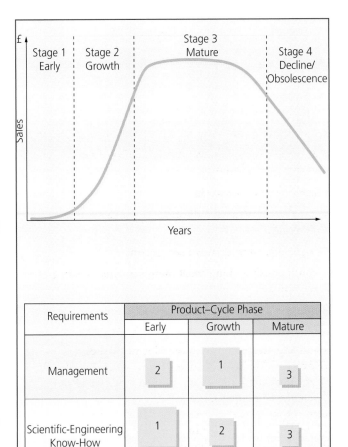

Requirements	Product–Cycle Phase		
	Early	Growth	Mature
Management	2	1	3
Scientific-Engineering Know-How	1	2	3
Unskilled and Semiskilled Labour	3	2	1
External Economies	1	2	3
Capital	3	1a	1a

Source: Based on Oakey 1984 and Erickson & Leinbach 1979

Figure 9.48 The product life-cycle

Key Definitions 4

Deindustrialisation The long-term absolute decline of employment in manufacturing.

Reindustrialisation The establishment of new industries in a country or region which has experienced considerable decline of traditional industries.

Economies of scale A situation in which an increase in the scale at which a business operates will lead to a reduction in unit costs.

Product life cycle The pattern of sales in the life of a product usually divided into four stages: early, growth, maturity and decline.

New techniques and trends

Organisational innovation

'Lean' manufacturing techniques were first developed in the 1950s in Japan, by a Toyoto manager called Taiichi Ohno. Lean production involves:

- **carrying minimal stocks**
- **having parts delivered direct to the assembly line 'just in time' (JIT)**
- **'right-first-time' quality management**
- **continuously seeking small improvements to gain greater efficiency**
- **seeing the factory as part of a supply chain with its suppliers downstream and its customers upstream.**

Lean manufacturing seeks to combine the best of both craftwork and mass production. It seeks to use less of each input and to eliminate defects – if a fault is spotted the production line is halted immediately and remedial action taken. The process eliminates waste by making only as much as is wanted at any given time. Advances in manufacturing software programs have allowed companies to integrate the various aspects of their work to a higher degree than ever before. Figure 9.49 compares the Fordist and Toyotist (flexible) models of manufacturing.

At first JIT transferred the burden of storing parts to the suppliers of assembly plants but at its most advanced it extends all the way along the supply chain. JIT encourages suppliers to concentrate around assembly plants to ensure rapid delivery. In fact, some assembly plants insist on suppliers being no more than a certain distance away. Ideally the number of suppliers should not be too extensive and long-term contracts should be agreed. Increasingly production

plants and suppliers are collaborating on research. As supply interruptions would prove disastrous agreements are often signed between employers and employees to avoid such an occurrence. Single union agreements are an important part of this process.

It was not until the mid-1970s that US firms began to employ lean manufacturing, as Japanese goods were making considerable inroads into US markets. Once it was proved that such techniques could work outside Japan, large European companies followed suit in an effort to bridge the productivity gap.

The adoption of the lean system paved the way for further increases in productivity and manufacturing flexibility through the integration of IT-based AMT (Advanced Manufacturing Techniques). This replaced simple automating devices with numerically controlled tools, industrial robots and flexible-transfer machines, and eventually computer-integrated manufacturing systems. These new techniques have been largely responsible for significant advances in productivity in a number of large mature industries (Figure 9.50).

Robots

Increasing worldwide sales of industrial robots will bring the global robot population to more than one million by the end of 2000 (Figure 9.51), according to forecasts by the UN. More than half of this total will be in Japan, which leads the current world ranking by a wide margin, ahead of the USA, Germany and South Korea.

The car industry is the largest user of robots. In 1997, Japan had over 830 robots for every 10 000 car workers compared to 370 in the USA. As one robot generally performs the tasks of at least two persons it has been estimated that robots in the Japanese car industry correspond to about 20% of the labour force. Unlike other countries, Japan also makes widespread use of robots in the engineering and electrical machinery industries. Compared with wages, robot prices have fallen significantly in the 1990s.

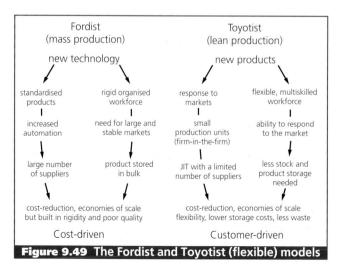

Figure 9.49 The Fordist and Toyotist (flexible) models

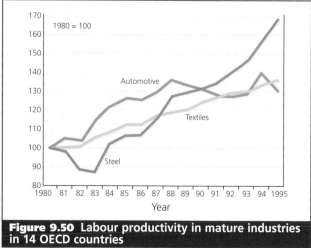

Figure 9.50 Labour productivity in mature industries in 14 OECD countries

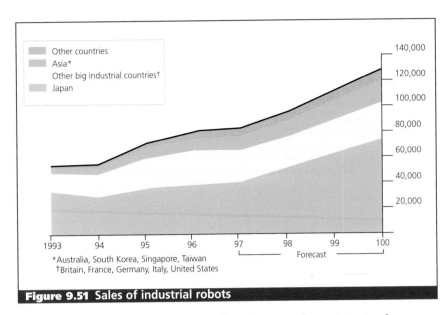

*Australia, South Korea, Singapore, Taiwan
†Britain, France, Germany, Italy, United States

Figure 9.51 Sales of industrial robots

Reindustrialisation, rationalisation and restructuring

The development of new industries has to a limited extent offset the decline of traditional manufacturing. The sector at the forefront of reindustrialisation is high technology. In terms of the process as a whole small firms have led the way.

The increasing level of global competition has driven all industries to improve their productivity. The consequences have generally been:

■ **rationalisation or 'downsizing' of the workforce with the expectation that a smaller number of workers will maintain the same level of production**

■ **closure of inefficient plants**

■ **restructuring by (a) introducing more efficient production methods; (b) merging with another company in the same sector.**

Large TNCs continuously compare all aspects of production in their plants in different countries and can move quickly to rationalise either when the market becomes more crowded as a result of increasing competition or during a recession when demand in general declines.

1 **(a)** Describe the trends in manufacturing employment in OECD countries between 1985 and 1995 (Figure 9.46).
   (b) Suggest reasons for these changes.
   (c) To what extent did total employment in services change during the same time period?
   (d) Why did some services increase in employment much faster than others?

2 **(a)** Explain the meaning of the terms 'product life cycle' and 'external economies'.
   (b) Describe how the volume of sales changes with each stage of the product life cycle.
   (c) Examine the importance of the major production factors at each stage of the product life cycle.
   (d) How might the changing importance of production factors influence industrial location?

Manufacturing Industry in Brazil

Figure 9.52 provides a brief review of the historical development of manufacturing industry in Brazil. While the country clearly recorded industrial growth prior to the 1930s it was not until that decade that industrialisation in the true sense arrived in Brazil. Before this time the range of industrial production was extremely restricted. For example, the 1920 Census showed that 47% of the industrial labour force was employed in just two industries: food processing and textiles. At that time agriculture still employed more than two-thirds of the total labour force.

The shortage of industrial goods during the Second World War gave added impetus to the government's policy of import substitution. With a wider industrial base the volume of industrial production tripled between 1947 and 1961, and in 1958 Brazil displaced Argentina as the leading industrial nation in Latin America. The car industry was promoted as the key industry around which a range of ancillary industries would develop.

Investment from abroad is an essential element of industrial development in Brazil. The USA is by far the largest source of inward investment with a total stock of almost $36 billion in 1995. Germany was in second place with $11 billion. Although not all of this is in manufacturing industry, the latter remains the main focus of foreign investment interest.

Manufacturing Industry in Brazil: A Brief History

1. Very limited industrialisation prior to late 1800s – (a) until independence in 1822 Portuguese mercantilist policies prohibited industry in the colony (b) the free trade policy and heavy reliance on raw material exports during the years of the Brazilian Empire 1822–89 presented unfavourable conditions for industrial development.
2. Late 1800s/early 1900s – attempts to achieve economic independence [after becoming a Republic in 1889] spurred industrial growth. Profits from coffee invested in food processing, textiles, iron and steel, and HEP.
3. Between the early 1930s and late 1960s the dominant process was import-substitution industrialisation [ISI]. In the 1930s the emphasis was on the production of non-durable consumer goods.
4. Early/mid-1940s – the government moved from being just a regulator to a supplier of goods and services, exemplified by the construction of the first integrated steel mill by the state in Volta Redonda.
5. Late 1940s/early 1950s – ISI reached a second stage with the expansion of the durable goods sector.
6. Second half of 1950s – government expenditure and foreign capital helped to raise investment to unprecedented levels. Hugh public investment in energy and transport; the construction of Brasília; the establishment of the car industry.
7. Early 1960s – a range of economic problems restricted industrial development.
8. The change to a military government in 1964 brought about economic stability leading to the 'economic miracle' of 1968–73 when the output of manufacturing industry grew by an average of over 13 per cent per year led by cars, iron and steel, and petro-chemicals. Brazil entered the 1970s with the world's highest growth rate.
9. Mid/late 1970s – after the first oil crisis in 1973/4 growth was achieved in capital goods and basic intermediate goods at the expense of a sharp rise in external debt. Second oil price rise in 1979 created more problems.
10. Recession, inflation and debt in the 1980s and early 1990s provided a very uncertain environment for industry.
11. Mid/late 1990s – industrial expansion led by the car industry with new era of economic stability due to introduction of the Real Plan in 1994. Increasing diversification.

Source: *Brazil: Advanced Case Studies* by P. Guinness

Figure 9.52 Manufacturing industry in Brazil

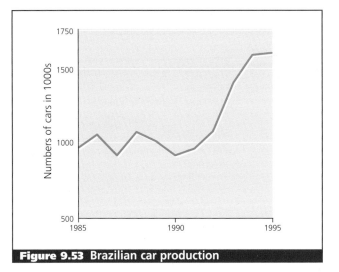

Figure 9.53 Brazilian car production

Brazil's car industry

In 1995, Brazil produced 1.6 million motor vehicles (Figure 9.53), making it the seventh largest car manufacturer in the world. Total vehicle sales in the country accounted for 65% of the South American market. By the turn of the century Brazil is expected to replace Italy as the fifth biggest producer with a capacity of 2.6 million cars a year. Strong domestic demand, the advantages of the regional free trade market, Mercosur and export markets are all contributing to the industry's rapid growth. The industry, with the inclusion of parts, is expected to invest $20 billion in Brazil between 1995 and 2000. Figure 9.54 shows the estimated investment by car manufacturers by region for the same period.

Volkswagen is the biggest car manufacturer in Brazil producing 35% of total output, followed by Fiat with 27%. The Brazilian automotive parts industry is the largest and most advanced in the developing world.

The car industry has been the most important single element of the Brazilian economy since its inception in the 1950s. Much of the investment in Brazil announced by multinational companies goes to the automotive sector. With a ratio of one car for every 11.3 people the growth potential is substantial and much higher than in mature markets like Europe and the USA. Also, Brazil's current vehicle fleet, with an average age of 15 years for buses and 11 years for cars, promises a booming replacement market. The sales volume that can be achieved in Brazil is only possible in countries of continental proportions and a rapidly growing middle class – which rules out China, India and Eastern Europe, at least in the near future. Annual vehicle sales in Brazil almost doubled between 1992 and 1996 making the country one of the world's fastest-growing domestic car markets.

In 1992, the government helped to kick-start the industry's growth by cutting taxes on small-engined cars, which now account for 57% of the Brazilian market. Since then growth has been driven by a combination of increasing competition and low (for Brazil) inflation. The latter has provided the country's lower-middle class with greater purchasing power and access to consumer credit. The removal of barriers to the importation of

Manufacturing industry in Brazil is heavily concentrated in the Southeast which comprises the four states of São Paulo, Rio de Janeiro, Minas Gerais and Espirito Santo. A further level of concentration can also be recognised because the state of São Paulo dominates the industrial scene, accounting in 1995 for 38.5% of the total tax charged on sales of manufactures in Brazil. The state of São Paulo alone is responsible for nearly one-half of Brazil's total GDP.

Figure 9.54 Estimated investment 1995–2000, in the assembly of cars, trucks and jeeps ($ billions)

Southeast

Fiat	Minas Gerais	3.0
Ford	São Paulo	2.5
General Motors	São Paulo	2.4
Volkswagen	São Paulo and Rio de Janeiro	2.3
Mercedes	Minas Gerais and São Paulo	0.8
Toyota	São Paulo	0.6
Honda	São Paulo	0.3
Scania	São Paulo	0.16
BMW	Undecided	0.15

South

General Motors	Rio Grande do Sul and Santa Catarina	1.1
Renault	Paraná	1.0
Audi	Paraná	0.5
Chrysler	Paraná	0.3
Volvo	Paraná	0.15

Northeast

Asia	Bahia	0.72
Inpavel	Pernambuco	0.3
Hyundai	Bahia	0.29
General Motors	Undecided	0.16
Subaru	Ceará	0.15
Skoda	Bahia	0.1
Troller	Ceará	0.016

North

Zam	Acre	0.014
Nanjing	Tocantins	0.009

Centre-West

Mitsubishi	Goiás	0.035

cars and parts has forced Brazil's established car manufacturers – Volkswagen, Fiat, General Motors and Ford – to modernise their outdated plants. In the past these companies had made substantial profits by producing small volumes of old-fashioned cars and selling them at high prices. The objective now is to produce the same models and engines as in their European plants and to achieve similar levels of productivity.

Initial location in the Southeast

The first Brazilian car was a Model 'T' Ford assembled from imported parts in São Paulo in 1919. However, prior to Juscelino Kubitschek becoming President in 1956 only a few cars were assembled in Brazil and there was one government-owned factory manufacturing trucks. Foreign companies were invited by the Kubitschek government to establish branch plants and then pressed to manufacture more and more of the parts for their vehicles in Brazil. Motor vehicles was seen as a key industry that would stimulate the development of other industries because of the great variety of components required for the finished product.

It is unsurprising that foreign car manufacturers first located in and around São Paulo. The largest metropolitan area in South America offered the following advantages:

- a good regional raw material base
- the largest pool of skilled labour in South America
- proximity to Santos, now one of the largest port complexes in the world
- a mature web of industrial linkages
- the largest market by far in the country
- the hub of road, rail and air transport and telecommunications
- welcoming federal and state governments.

Once the industry had become established in the São Paulo region it was only a matter of time before other locations in the Southeast region became attractive, namely Rio de Janeiro and Belo Horizonte. For example, in 1978 Fiat opened a large plant in Belo Horizonte employing over 10 000 people and producing 130 000 cars a year.

Although car makers are now looking beyond the Southeast in terms of location the region is still attracting substantial inward investment. For example, Mercedes-Benz is building a $460 million plant in Minas Gerais state (Figure 9.55). This is only the company's second fully fledged car plant outside Germany, the first being in Alabama, USA. High labour costs in Germany are a major reason for the company's decision to extend manufacturing overseas.

The spread to the South

Among the new manufacturers to locate in Brazil , the most ambitious project belongs to the French company Renault which has invested $1 billion in a huge plant producing the 'Megane'. The site selection process promoted what the press termed a 'fiscal war' between a number of states anxious to benefit from such substantial inward investment. For Renault, locating in Brazil is a matter of survival. Ranked eighth in the world with a production of 1.8 million vehicles per year, the newly privatised company lacks space to grow in Europe where competition is intense. The selection of Curitiba as the location for the new Renault plant marked an important stage in the development of the industry – the beginning of its spread outwards from the Southeast. Major factors in Renault's choice of location were:

- the high quality of life in Curitiba
- improvements in the port of Paranaguá (Curitiba's outport) where handling charges are considerably below those of Santos
- the success of existing multinational companies in Curitiba such as Phillip Morris, Bosch and Volvo (tractors).
- proximity to the large markets in the Southeast
- Paraná borders Argentina and Paraguay, both viewed as expanding markets.

More recently the state of Rio Grande do Sul has been criticised

Newcomers flood auto market

In 1997, Brazil ranked eighth in the world vehicle industry, with an output of 2,067,452 units, excluding motorcycles. In theory, it is one of the most promising markets in the world. The country is huge (8.5 million square kilometers). The population is large (165 million inhabitants). And its economy is the ninth-ranked worldwide (a $760 billion Gross Domestic Product), though the per capita income is relatively low (around $4,600).

It's no wonder that several car makers have disclosed plans to settle here, partially encouraged by heavy fiscal breaks. Last October, Honda began production of its first car in Brazil, the Civic sedan. The Japanese automaker has publicly announced a goal of getting a humble stake of the market in the coming years, near 8%. Volkswagen has been the main producer for decades – its market share nears 33%, followed by Fiat, General Motors and Ford. So far, Toyota has had a modest participation, by producing a single utilitarian vehicle since the 1960s. But this year, it will kick off production of the Corolla sedan at a new plant in the state of São Paulo.

There are many expansion plans. In the second half of this year, Volkswagen will inaugurate a new plant in the southern state of Paraná to produce the Audi A3, Santana and Golf models. In 1999, General Motors should open a plant in Rio Grande do Sul to make the Astra sedan and a subcompact car. Ford is to begin construction of a plant in the same state, and Fiat is upscaling its factory in Minas Gerais to produce light trucks and vans. In the same state, Mercedes Benz – which has locally produced trucks and buses since the 1960s – will manufacture the compact car Classe A as of this year.

Meanwhile, French group Peugeot-Citroen will set up a plant in the state of Rio de Janeiro to assemble compact and mid-sized automobiles by the end of 1999. Renault, for its turn, is slated to open a plant in Paraná to assemble sedan and van models, in the coming year. BMW, in a joint venture with Chrysler, will set up an engine plant in Paraná (the motors will be exported to Great Britain and the United States). Still in Paraná, BMW will produce, in association with its British subsidiary Rover, the utilitarian vehicle Defender. Chrysler alone will assemble the Dakota pick-up model. Korean automakers Asia Motors and Hyundai, reported plans of settling here too, but the Asian financial crisis is likely to postpone their projects.

If all projects reported by local producers and newcomers are put into practice, the aggregate investment in the automotive sector will reach $20 billion by the year 2000, and total output would increase by 40% or 50%.

Figure 9.55 Source: *Gazeta Mercantil,* **9 March 1998**

Figure 9.56 New car plant in Northeast Brazil

1. Foreign trans-nationals assembling components mainly produced in developed countries for the Brazilian market.
2. Foreign trans-nationals assembling components mainly produced in Brazil for the Brazilian market.
3. Foreign trans-nationals assembling components mainly produced in Brazil for the South American market in general.
4. Foreign trans-nationals exporting parts and some cars to developed countries.

Future
5. Foreign trans-nationals exporting cars to developed countries in significant volumes.
6. Brazilian car manufacturer(s) compete for the domestic market with foreign trans-nationals.

Figure 9.57 The Brazilian car industry: stages of development

by competing states for offering an over-generous package of incentives to General Motors. GM was given R$ 253 million before plant construction had even started. Opponents argue that:

- the state is failing to collect taxes from those who can afford them
- the capital intensive nature of the modern car industry brings only limited new employment.

New plants planned in the Northeast

Foreign vehicle manufacturers have also targeted the Northeast in terms of car plant location. This has been the result of government incentives aimed at encouraging the industry to spread out beyond the Southeast and South. Asia Motors should be the first car manufacturer to benefit from the incentives included in the new Provisional Measure (MP). Planning to build either in the state of Bahia or in Ceara an investment of $500 million should realise production of 60 000 cars by 1999.

While labour availability is high in the region, levels of education and industrial skill are low compared to the Southeast and South. Thus low productivity appears to be the major obstacle for the car industry to overcome in this problem region.

Stages in development

Figure 9.57 can reasonably be seen as the life-cycle of the Brazilian car industry. The country has recently moved into stage four, as exemplified by GM's São Jose dos Campos plant near São Paulo which exports engines to the USA. The eventual aim shared by the government and leading manufacturers is that cars produced in Brazil will be shipped in large numbers to the developed world, a purpose for which the Northeast is strategically located. At present such exports are very limited. When stage five is fully achieved Brazil will really have come into her own as a global player in the industry.

Urban and rural manufacturing

The urban–rural shift

The relatively compact nature of towns and cities during the Industrial Revolution of the 19th century resulted in a concentration of manufacturing industry in the inner cities of the 20th century as the era of the motor vehicle allowed cities to sprawl far beyond their previous limits. As time progressed, however, the disadvantages of inner-city locations became more and more obvious. The first reaction to the constraints of inner-city sites was to select new suburban locations but increasingly, from the 1960s in particular, manufacturing industry has been attracted to rural areas. This latter movement has generally been recognised as the most important trend in the location of manufacturing industry in Britain in the second half of the 20th century.

The explanation for the inner-city decline of manufacturing industry lies largely in the constrained location theory which identifies the problems encountered by manufacturing firms in congested cities, particularly in the inner areas:

■ The industrial buildings of the 19th and early 20th century, mostly multi-storey, are generally unsuitable for modern manufacturing which prefers a single-storey layout.

■ The intensive nature of land use usually results in manufacturing sites being hemmed in by other land users thus preventing on-site expansion.

■ The size of most sites is limited by historical choice and frequently deemed to be to small by modern standards, making change of use to housing, recreation or other uses likely. Old sites can rarely accommodate industrial estates, the preferred form of industrial location in most local authority areas.

■ Where larger sites are available the lack of environmental regulations in earlier times has often resulted in high levels of contamination. In such situations reclamation is very costly indeed.

■ The high level of competition for land in urban areas has continuously pushed up prices to prohibitive levels for manufacturing industry in many towns and cities.

Other factors specific to inner cities which have contributed to manufacturing loss are:

Figure 9.58 Example of new rural manufacturing

- Urban planning policies in the form of the huge slum clearance schemes of the 1950s, 1960s and 1970s meant that factories located in slum housing areas were frequently demolished too.

- Regional economic planning also had an impact in some areas, London in particular. The availability of incentives in Development Areas provided the stimulus that some firms needed to abandon the inner city.

- Before the era of decline, important inter-firm linkages had been built up in inner city areas. As these links were steadily broken, the locational '*raison d'etre*' of the remaining inner city firms gradually evaporated.

Rural areas, in contrast have offered plenty of available land at relatively low prices. This has allowed firms to purchase generous allocations of space for single-storey development and for future expansion. With agriculture being the previous use in most cases, pre-construction costs have been minimal.

Other reasons for the urban–rural shift include:

- Residential preference in terms of the high perceived quality of life in rural areas. This tends to be truer for small- to medium-sized enterprises than for large firms. For the latter it is likely that the decision makers will live near company headquarters rather than near the new site in question.

- The cost and turnover of labour: both tend to be lower in rural compared to urban areas.

- The relative lack of traffic congestion may reduce transport costs.

- Green belt policies preventing development at the edges of cities have encouraged 'leapfrogging' into rural areas proper.

The service sector

The service sector is by far the most dominant employer in developed economies and it is also of considerable significance in most developing countries. As an economy becomes more sophisticated the contribution of the service sector increases in terms of employment, GDP and all other standard measures. It can be argued that a service sector can be recognised historically at all levels of development but it was not until the Industrial Revolution that it really came to maturity. Its economic influence increased steadily to attain a prominent role in the second half of the 20th century. As Figure 9.21 shows, in terms of the industry totals for the world's 500 largest corporations, major service industries are very well represented. However, this data does not include smaller companies. For example, insurance is a giant industry with a $2 trillion turnover, several times more than the entire oil industry. It controls £10 000 billion in equities, a third of the value of global stockmarkets.

The service sector encompasses a wide range of industries, occupations and products, and can be classified in a number of different ways (Figure 9.60). Perhaps the most important distinction is between producer services and consumer services. Producer services are those supplied to other firms or organisations, helping them to deliver their product or service to the final consumer. Many producer services are 'high order', such as market research, management consultancy, advertising and legal services. In contrast, many consumer or household services are 'low order' and provided generally on a personal basis. Examples include retailing, refuse collection, hairdressers and dry cleaners. Generally speaking, services are grouped together in 'central places' which are hierarchically arranged in terms of the services offered: places with larger populations offer a wider variety of services than smaller places.

Figure 9.59 Tourism in Iceland – a rapidly growing service industry

The variety of services has increased over time in response to:

- the adoption of new technology
- increases in personal disposable income
- greater leisure time
- demographic changes
- new social values.

It is not difficult to think of a considerable number of services available today which did not exist 30 or 40 years ago.

The aspect of service provision which is of greatest interest to geographers is location. Important locational factors affecting service provision include:

- the distribution and density of population
- variations in purchasing power
- availability of labour with appropriate skills
- proximity to other service activities
- demographic factors such as age and gender.

The distribution of services in a region can be to a considerable extent explained by reference to OS maps at various scales and census data, and by conducting judicious field research.

Figure 9.60 Summary of alternative classifications for service industries

Dichotomous classifications

A Producer (or intermediate) services
   Consumer (or final) services

B Local (population-serving, industry-serving) services
   Non-local services (serving regional, national or international markets)

C Market services (funded from private resources)
   Non-market services (funded from public resources)

D Footloose services   stresses influence of location on
   Tied services   operation

E Office services   based on relative importance of
   Non-office services   office occupations

Other classifications

A Complementary services
   Old services
   New services

B Productive services
   Individual consumption services
   Collective consumption services

C Distributive services
   Producer services
   Social services
   Personal services

Source: *Service Industries: Growth and Location* by P. Daniels, CUP, 1986

In the UK, service industries employ about 17 million people, accounting for two-thirds of GDP. The South East has the highest concentration of service employment with the lowest rates found in the East Midlands and the West Midlands. Britain is now the world's second biggest exporter of services (Figure 9.61).

Business and financial services

The increasing concentration of high level services

As a result of a combination of globalisation and informationalisation, the production of services has become increasingly detached from that of production of goods. Although producer services do depend to a certain extent on production, the financial services sector has no direct relationship to manufacturing. In locational terms contrasting trends have been at work. As manufacturing has dispersed worldwide, high level services have increasingly concentrated and have been doing so in places different from the old centres of manufacturing.

At the top of the urban service hierarchy are the global cities of London, New York and Tokyo. These cities are the major nuclei of global industrial and financial command functions. The economic strength of these cities has become more and more detached from the local economies in which they are located and they have become embedded in a truly global set of economic relations. The prominence of the global cities is now less as centres of corporate headquarters and more as leaders in financial markets and financial innovation. In an era of hypermobility of financial capital the volume of activity conducted in these centres is absolutely crucial to their success as margins on transactions have become increasingly slight as a result of intense competition. However, the knowledge structures and institutions which comprise the growth engines of the global cities are difficult for other cities lower down the hierarchy to capture.

Below the three global cities is a second level of about 20 cities, including Paris, Brussels, Milan, Chicago and Los Angeles, which also have significant global connections. All the cities in the two top rungs of the global hierarchy offer a wide range of highly specialised services. There is a production process in these services which benefits from proximity to other specialised services. For example, the production of a financial instrument requires inputs from accounting, advertising, legal expertise, economic consulting, designers, public relations and printers. Time replaces weight in this process as a force for agglomeration.

CREATIVE INDUSTRIES – OFFICIAL SURVEY FINDS FASHION, FILM, POP MUSIC AND ADVERTISING AMONG FASTEST-GROWING SECTORS

UK is world's second biggest services exporter

By Christopher Adams, Economics Staff

There may be more to Cool Britannia than slick government marketing. Official figures published yesterday showed explosive growth in the creative industries has helped the UK overtake Germany and France to become the world's second biggest exporter of services.

Fashion, film, pop music and advertising are among the fastest-growing sectors of the UK economy. An expanding appetite in other countries for a taste of British style and culture is outpacing overseas demand for core services such as transportation and travel.

In the first survey of its kind on the service sector, the Office for National Statistics reports that the creative industries have helped lift the UK's share of global trade in services to more than 6 per cent.

Rapid growth in the financial services, computing and information technology industries have also boosted exports.

The service sector has grown much more rapidly than the rest of the UK economy, displacing the long-suffering manufacturing industry. Since 1970, average annual growth for services and manufacturing has been 2.6 and 0.7 per cent respectively. Services account for about two-thirds of total economic output.

But accurate and detailed information about the composition of the services sector is lacking. DeAnne Julius, a member of the Bank of England's monetary policy committee, warned recently that failure to understand how the services industry behaved could be undermining the Bank's ability to set interest rates appropriately.

The ONS survey, published yesterday, showed worldwide exports jumped from $82.7bn to $93.9bn in 1997. The US is the biggest exporter of services worldwide.

The biggest components of UK trade in services are the transportation and travel sectors. Export revenue derived from civil aviation and maritime shipping grew 1.6 per cent to £10.2bn during 1997. People travelling to the UK accounted for £13.8bn or 15 per cent of exports.

But growth in both of these sectors appears pedestrian compared with the increase in overseas earnings from Britain's creative industries.

Helped by British-made box office successes such as *Four Weddings and a Funeral* and *Trainspotting*, total revenue from film and television exports increased by an average annual rate of about 7 per cent between 1993 and 1997. Together with music, design, software, publishing, fashion, arts and antiques, these sectors employ roughly 1m people yet account for some 16 per cent of the global market in creative goods and services.

The largest single consumer of British film and television programmes is the US, which accounted for £330m or just over a quarter of the industry's export revenue in 1997. Sweden, Germany and France follow some way behind.

The UK's surplus in services more than doubled from £5.7bn to £11.9bn in the five years to 1997. Until the second half of last year, the buoyancy in services was the main driving force behind economic growth. Services output has since begun to slow.

The US is also the UK's biggest trading partner in all services. Services trade with Europe moved from a deficit of £1.69bn in 1992 to a surplus of £534m in 1997.

Source: *Financial Times*, 12 February 1999.

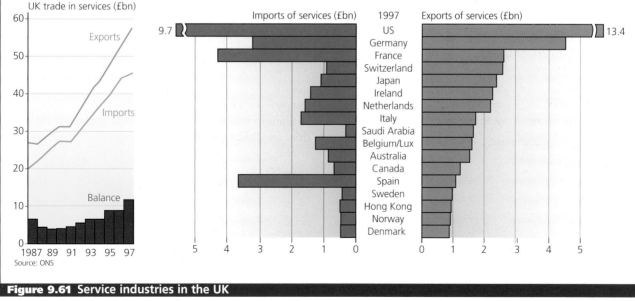

Figure 9.61 Service industries in the UK

The City of London

The City of London is the world's leading financial centre with the largest share of many world markets. With a resident population of less than 3000, its workforce is just under 300 000, 75% of whom work in banking, finance, insurance and business services. Office floor space in the City amounts to 7.2 million square metres. The net overseas earnings of UK financial institutions, predominantly based in the City of London, totalled over £22.7 billion in 1996. More trading in the dollar and the deutschmark takes place in London than in either the USA or in Germany. The following facts help to underline London's importance to the global financial system:

- The London foreign exchange market is the largest in the world. Its daily turnover of $464 billion accounts for 30% of global business.

- London is the world's largest international insurance market, with a net premium income of £14 billion in 1996.

- The London Stock Exchange, with 60% of global turnover in 1996, is the world's largest centre for the trading of foreign equities. More foreign companies are listed on the LSE than on any other exchange.

- There are approximately 570 foreign banks in London, more than in any other city.

- London is the world's second largest fund management centre, after Tokyo, with over $1.8 trillion in equities under management.

- London is the largest centre in the world for maritime services with net overseas earnings of £2 billion.

- Banks based in the City of London invest more capital abroad than those of any other country, amounting to 18% of external bank lending globally.

- London is the global clearing centre for gold forward trading and financing.

- London is second in the world, after Chicago, in exchange traded derivatives.

To remain in such a formidable position continuing investment is vital. Thirty per cent of high-tech City office buildings were constructed since the late 1980s. Information technology expenditure in the City in 1996 amounted to £2.2 billion. The Corporation of London's Economic Development Unit is responsible for maintaining London's prime global position. Using the Corporation's statutory powers to invest in economic development, the Unit's objectives are:

- to ensure that all the leading entities in global finance and commerce are in the City and that they have the professional support to function efficiently

- to enhance the quality of the working and living environment within the City

- to ensure the provision of an efficient infrastructure and a high quality workforce by working closely with the property and training sectors

- to market the attributes of the City and of London as a whole on a world wide basis

- to achieve an orderly property market by using its influence as both planning authority and landowner.

The Unit has specialist teams such as (a) the City Property Advisory Team which helps business occupiers, developers, owners and investors; and (b) a European Office to help promote dialogue between the City and Europe.

A survey of office rents in the world's major commercial centres for June 1998 found the City of London to be the second highest, after Hong Kong, with an average of $1173 a square metre (including service charges and taxes). During the initial phase of a substantial foreign company's presence in London, there is a marked preference for location within the City. However, following an initial period of settling into the London market many foreign firms expand and become more locationally footloose, usually requiring larger premises at more economic rents in areas such as the South Bank and West End.

The latest threat to the position of the City within Europe has come from the launch of the euro (Figure 9.62). However, after a nervous beginning it is clear that the arrival of the euro has allowed London to capitalise on its expertise. Financial services now account for 8% of Britain's GDP. Add in shipping, real estate, accountancy and legal services and the activities cover a quarter of national output.

The decentralisation of 'back office' functions

In the 1980s and 1990s in particular routine 'back office' functions have been moved away from core locations to less expensive sites as companies have sought to reduce costs in order to remain competitive. Back office functions process large volumes of paper, electronic transactions and telephone enquiries. They include international call centres and customer services such as direct banking and computer support. These routine functions have relocated to:

- elsewhere in metropolitan areas
- peripheral locations at the edge of urban areas
- more distant locations in peripheral regions

City wins the euro war

The launch of the euro was supposed to bring misery to the City Jobs would be lost, business would migrate to Frankfurt and Paris and the citizens of Britain would be punished for refusing to embrace Europe's splendid new currency.

Many sensible people believed this story, including Lord Levene, then just starting his year as Lord Mayor of London. The Treasury was sufficiently worried to make the future of the City one of the five economic tests for deciding whether to join the single currency.

Such fears were understandable. After all, the new European Central Bank was to be based in Frankfurt. The French and German governments also made little secret of their determination to use the single currency as a lever against London. There was much anguished talk of the cost of being excluded from strategic discussions.

Last week Lord Levene admitted that it has not worked out like that. This week Eddie George sounded as though he was about to adopt the slogan: 'Britain will do very well out of the euro.' The governor of the Bank of England declared that 'the City is in very good shape'. The advent of the euro had left it 'at least as strong as before'.

George was equally clear why. He cited 'the English language, the European time zone, the sheer momentum of the City's critical mass, supported by the availability of a professionally qualified and technically skilled workforce which gives the City remarkable resilience and flexibility, aspects of our commercial and physical infrastructure, and so on'.

Figure 9.62 Source: *Daily Telegraph*, 11 December 1999

- developing countries with pools of labour able to handle such tasks.

In Britain the recent wave of private sector relocations has mirrored the decentralisation of public sector back offices, such as the Inland Revenue, from London to the regions in the 1960s. The North East in particular has benefited from recent decentralisation. Examples include British Airways which opened its ticket telesales operation on the Newcastle Business Park in 1991, and Abbey National (Teesside), Orange (Darlington) and Ladbrokes (Peterlee).

Although relocation can reduce costs in a number of ways the main savings are in terms of labour and office space costs. Labour accounts for about 70% of total costs in back office functions and thus considerable savings can be made by moving from London to lower wage regions within Britain. Labour availability is also important as turnover rates are high, often exceeding 15%. This reflects the moderate salaries and the routine, unchallenging nature of many of the tasks involved.

Some companies have really sought to slash labour costs by moving back office functions to the developing world. India and Russia, for example, have become major centres for subcontracting computer programming.

Key Definitions ⑤

Global city Major world city supplying financial, business and other significant services to all parts of the world. The world's major stock markets and the headquarters of large TNCs are located in global cities.

Back offices Offices of a company handling high volume communications by telephone, electronic transaction or letter. Such low- to medium level functions are relatively footloose and have been increasingly decentralised to locations where space, labour and other costs are relatively low.

SPATIAL FOCUS Dublin's International Financial Services Centre [IFSC]

Dublin's IFSC was established in 1987 as a spatially discrete 'offshore financial centre'. Overcoming early problems the IFSC emerged as an important financial centre in Europe and in terms of offshore funds it is in competition with Luxembourg and the Channel Islands. An article on the subject (*Area* (1998) 30.2) stated 'the IFSC represents an important attempt to reposition

Ireland in the international divison of labour'. By 1996, the 11 ha site constructed on a disused dockland area:

- contained over 400 companies
- employed over 2300 people
- contributed IR£200 million in corporate taxes.

Figure 9.63 Dublin's International Financial Services Centre

Although many of the companies represented are essentially back offices the Irish government has made every effort to attract more sophisticated financial and business services.

The IFSC has benefited from a low tax regulatory environment set up by the Irish government but sanctioned by the European Commission (Figure 9.64). The justification for EU approval was Ireland's high unemployment rate and to benefit from the incentives on offer firms were required to make firm job commitments. Dublin's position in the EU and the package available proved particularly attractive to insurance and reinsurance back office banking functions and corporate treasury activities. For the latter the major firms represented include IBM, Hewlett-Packard, General Electric and Heinz.

The back offices pay relatively low wages but the IFSC is also about graduate employment opportunities and emerging institutional thickness. The Irish government has been particularly anxious to encourage the latter and upgrade the nature of investment. Dublin is becoming an important centre of US corporate treasury operations and banking activities for the whole

Figure 9.64 Dublin's IFSC tax benefits

10% corporate tax rate on trading income
• Guaranteed to 31 December 2005
• Time limit on approvals 31 December 2000
A range of double taxation treaties
• 24 countries covered, including USA, Canada, Japan, Korea, Germany
No withholding tax on dividends and interest
Availability of tax-based financing
Zero tax on certain fund-management entities owned exclusively by non-Irish residents

of Europe. In 1997, *Fortune* magazine ranked Dublin as the best city in Europe in which to do business. This corresponds very much with the image of Ireland in the 1990s as the 'Celtic Tiger Economy'.

The IFSC has had a significant multiplier effect in the Dublin region. Local financial services outside the boundary of the IFSC, such as tax consultants, accountants and law firms have benefited in particular.

Retailing

The retailing of goods and services is a major source of employment in all towns and cities. It has a significant impact on land use and is a considerable influence on the daily lives of most people. Traditionally, urban retail organisation has been in the form of a hierarchy with the central business district (CBD) at the top with subsequent layers, the number depending on the overall size of the urban area, down to the corner shop. The lowest levels of the retail hierarchy mainly provide convenience or low order goods. With ascent of the hierarchy comparison or high order goods become increasingly dominant, although a full range of convenience goods can usually be found at the top of the hierarchy alongside the goods and services for which consumers want to compare price and quality.

The concept of threshold, explained in the previous chapter, is central to the understanding of this hierarchical organisation. Low order centres that require small threshold populations to survive will be spread throughout the urban area with most customers living within walking distance although they will still probably use their cars to shop. In contrast, the large threshold populations required by the highest order goods ensures that these will only be found in the CBD.

Retailing has been affected by a number of important organisational trends in recent decades:

- **the number of independent traders has declined sharply in the face of severe competition from the multiples which compete among themselves for market share**

- **big retailers such as Marks and Spencer have gone transnational by opening stores in a number of other countries, while foreign retailers such as Carrefour are having an increasing presence in Britain**

- **new technology has facilitated developments such as teleshopping.**

Figure 9.65 is evidence of the decline of independent grocery stores and the growing dominance of the large supermarkets. For example, in 1950 Sainsbury's had 244 stores in Britain each with an average of just 2000 sq ft of floorspace, stocking 550 different products. In 1996, Sainsbury's had 355 supermarkets with the average store now covering 30 000 sq ft. This retailing giant stocks 19 000 different products. As stores have increased in size they have extended their product range, thus affecting more and more shops in traditional centres. In recent years the large retailers have also moved into services, particularly in the financial sector.

The most startling growth has been in superstores which are frequently sited in accessible out-of-town locations. The big chains are keen to open more but they have faced increasing opposition in recent years because:

- **they take customers away from rural shops and traditional shopping areas in towns and cities**

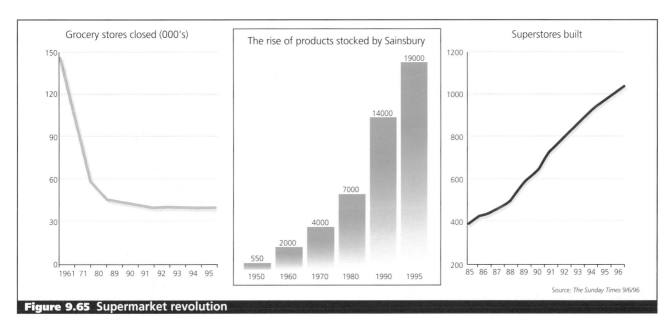

Grocery stores closed (000's)

The rise of products stocked by Sainsbury

Superstores built

Source: *The Sunday Times* 9/6/96

Figure 9.65 Supermarket revolution

■ they are often built on greenfield sites which local people would like to keep in its original land use

■ they have huge parking areas and generate large volumes of traffic.

In terms of location, decentralisation has been the main trend. This has resulted in the partial decline of many central business districts. New planned shopping centres and retail parks have sprung up in both suburban and out-of-town locations. Accessibility has been the main factor in site selection. The planned shopping centres of the modern era originated in North America in the 1950s before arriving in Europe a decade later. Initially most planned shopping centres in Britain were of limited size; located in town centres and allied to new pedestrian precincts. It was not until the 1970s that larger schemes, often in New Towns, were constructed. At this time only Brent Cross, near the start of the M1 in north London, could be considered as an out-of-town centre serving a regional market. However, it was not long before other centres of a similar size appeared elsewhere.

The next significant advance in size was the construction of MetroCentre at Gateshead, following the 'multi-purpose mega-mall' concept from North America. For a while it was the largest shopping centre in Western Europe, occupying 160 000 m^2. There is free parking for 10 000 cars and new bus and rail stations were built for non-motorists. There are 100 buses an hour and 69 trains daily. Other shopping centres of this magnitude soon followed, including Lakeside at Thurrock and Meadowhall in Sheffield. The overall objective of this new size of shopping centre was to create an extended shopping experience that would encourage families to think of a visit as a day out. Not only are a full range of goods and services on offer but there are numerous eating places and many leisure opportunities.

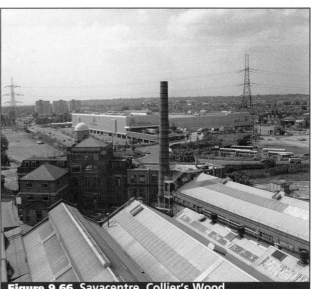

Figure 9.66 Savacentre, Collier's Wood

Figure 9.67 Country Oak retail park, Crawley

The popularity of supermarkets and superstores is evident from the profits made by the retail giants. However, not all sections of society have benefited from the trends in retailing in recent decades. The government is concerned that the lack of shops selling reasonably priced goods on many urban estates forces people on low incomes to travel long distances and often pay inflated prices. Under the government's scheme to tackle social exclusion, it is trying to encourage the large retailers to open more shops in deprived urban areas which have been identified as 'food deserts'.

Increasingly, the large electrical, computer, furniture, carpet and DIY stores have grouped together in retail parks which are frequently found in the outer areas of towns and cities. Usually sited in highly accessible locations, retail parks provide a large number of free parking spaces.

Questions

1 **(a)** Describe the UK's balance of trade in services between 1987 and 1997 (Figure 9.61).
 (b) Identify some of the fastest growing sectors of the UK's service economy.
 (c) Suggest reasons for the success of these sectors.
 (d) Attempt to explain the geographical pattern of the UK's trade in services.

2 **(a)** How significant is the City as a global financial centre?
 (b) What are the reasons for the City's strength in the financial services?
 (c) Discuss the strategies adopted by the City to maintain its position.
 (d) Outline the most recent threat to the City's position (Figure 9.62).

index